ENGINEERING PRINCIPLES FOR FOOD PROCESSING TECHNOLOGY AND PRODUCT REALIZATION

ENGINEERING PRINCIPLES FOR FOOD PROCESSING TECHNOLOGY AND PRODUCT REALIZATION

Edited by

Leonardo Sepúlveda Torre, PhD

Porteen Kannan, PhD

A. K. Haghi, PhD

AAP | APPLE ACADEMIC PRESS

First edition published 2025

Apple Academic Press Inc.
1265 Goldenrod Circle, NE,
Palm Bay, FL 32905 USA

760 Laurentian Drive, Unit 19,
Burlington, ON L7N 0A4, CANADA

CRC Press
2385 NW Executive Center Drive,
Suite 320, Boca Raton FL 33431

4 Park Square, Milton Park,
Abingdon, Oxon, OX14 4RN UK

Apple Academic Press exclusively co-publishes with CRC Press, an imprint of Taylor & Francis Group, LLC

Library and Archives Canada Cataloguing in Publication

CIP data on file with Canada Library and Archives

Library of Congress Cataloging-in-Publication Data

CIP data on file with US Library of Congress

ISBN: 978-1-77491-575-2 (hbk)
ISBN: 978-1-77491-576-9 (pbk)
ISBN: 978-1-03271-392-2 (ebk)

About the Editors

Leonardo Sepúlveda Torre, PhD
Professor, School of Chemistry at the Universidad Autónoma de Coahuila, Saltillo, Coahuila, México

Leonardo Sepúlveda Torre, PhD, is a full Professor at the School of Chemistry at the Universidad Autónoma de Coahuila, Saltillo, Coahuila, México, where he is also a leader of the Bioprocesses and Microbial Biochemistry Group. His postgraduate studies were related to topics on food biotechnology in the Food Research Department of the UAdeC. In 2011, he made a research stay at the Institute of Biotechnology and Bioengineering at the University of Minho (Uminho), Braga, Portugal. He also worked as a collaborator at the Center of Biological Engineering in Uminho, Braga, Portugal, with the link project "Biotechnologies for Regional Food Biodiversity in Latin America." He has participated as author and co-author in 35 scientific articles in indexed journals, several books, and 24 book chapters, and he holds two national patents.

Porteen Kannan, PhD
Assistant Professor, Department of Veterinary Public Health, Madras Veterinary College, Tamil Nadu Veterinary and Animal Sciences University, India

Porteen Kannan, PhD, is an Assistant Professor in the Department of Veterinary Public Health at Madras Veterinary College, Tamil Nadu Veterinary and Animal Sciences University, India. The research activities of Dr. Kannan include food safety and antimicrobial resistance. He did his postdoctoral studies at the US Department of Agriculture, Maryland, USA, with a specialization in foodborne pathogens. He has published his work in both national and international journals. He is actively involved in mentoring both MVSc and PhD students.

A. K. Haghi, PhD

Research Associate, Department of Chemistry, University of Coimbra, Portugal

A. K. Haghi, PhD, is a retired professor and has written, co-written, edited or co-edited more than 1000 publications, including books, book chapters, and papers in refereed journals with over 3800 citations and h-index of 32, according to the Google Scholar database. He is currently a research associate at the University of Coimbra, Portugal. Professor Haghi has received several grants, consulted for several major corporations, and is a frequent speaker to national and international audiences. He is the founder and former Editor-in-Chief of the *International Journal of Chemoinformatics and Chemical Engineering and Polymers Research Journal.* Professor Haghi has worked as an editorial board member of many international journals. He has served as a member of the Canadian Research and Development Center of Sciences and Cultures. He has supervised several PhD and MSc theses at the University of Guilan (UG) and cosupervised international doctoral projects. Professor Haghi holds a BSc in urban and environmental engineering from the University of North Carolina (USA) and holds two MSc degrees, one in mechanical engineering from North Carolina State University (USA) and another one in applied mechanics, acoustics, and materials from the Université de Technologie de Compiègne (France). He was awarded a PhD in engineering sciences at the Université de *Franche-Comté* (France). He is a persistent reviewer of leading international journals.

Contents

Contributors

Y. K. Reyes Acosta
Facultad de Ciencias Químicas de la Universidad Autónoma de Coahuila, Saltillo, Coahuila, México

Cristóbal N. Aguilar
Bioprocesses and Bioproducts Group, Food Research Department, School of Chemistry, Universidad Autónoma de Coahuila, Saltillo, Coahuila, México

C. N. Aguilar Gonzales
Facultad de Ciencias Químicas de la Universidad Autónoma de Coahuila, Saltillo, Coahuila, México
Bioprocesses & Bioproducts Group, Food Research Department, School of Chemistry, Universidad Autónoma de Coahuila, Saltillo, Coahuila, México

J. A. Ascacio-Valdés
Bioprocesses and Bioproducts Group, Food Research Department, School of Chemistry, Universidad Autónoma de Coahuila, Saltillo, Coahuila, México

Rituparna Banerjee
ICAR-National Research Centre on Meat, Chengicherla, Hyderabad, Telangana, India

Sonia Yesenia Silva Belmares
Department of Food Research, School of Chemical Sciences, Autonomous University of Coahuila, Saltillo, Coahuila, Mexico

Juan Buenrostro-Figueroa
Research Center in Food and Development, Cd. Delicias, Chihuahua, México

Juan C. Bustillos-Rodríguez
Tecnológico Nacional de México, Instituto Tecnológico de Cd. Cuauhtémoc, Cd. Cuauhtémoc, Chihuahua, México

Teresa Darlen Carrillo-Castillo
School of Biological Science, Universidad Autonoma de Coahuila, Carretara Torreón-Matamoros, Torreón, Coahuila, Mexico

N.D. Cerda-Cejudo
Bioprocesses and Bioproducts Group, Food Research Department, School of Chemistry, Universidad Autónoma de Coahuila, Saltillo, Coahuila, México

R. I. Narro Cespedes
Facultad de Ciencias Químicas de la Universidad Autónoma de Coahuila, Saltillo, Coahuila, México

A. R. Chandni
Division of Veterinary Public Health, ICAR-IVRI, Izatnagar, Uttar Pradesh, India

Rebeca Villanueva Chávez
Department of food research, School of Chemical Sciences, Autonomous University of Coahuila, Saltillo, Coahuila, Mexico

M. L. Chávez-González
Bioprocesses and Bioproducts Group, Food Research Department, School of Chemistry, Universidad Autónoma de Coahuila, Saltillo, Coahuila, México

Celia Chávez-Mendoza
Centro de Investigación en Alimentación y Desarrollo A C, Coordinación Tecnología de Productos Hortofrutícolas y Lácteos, Delicias, Chihuahua, México

Luis Enrique Cobos-Puc
Department of Food Research, School of Chemical Sciences, Autonomous University of Coahuila, Saltillo, Coahuila, Mexico

J. C. Contreras-Esquivel
Bioprocesses & Bioproducts Group, Food Research Department, School of Chemistry, Universidad Autónoma de Coahuila, Saltillo, Coahuila, México
Facultad de Ciencias Químicas de la Universidad Autónoma de Coahuila, Saltillo, Coahuila, México

Samir Das
ICAR Research Complex for NEH Region, Umiam, Meghalaya, India

Himani Dhanze
Division of Veterinary Public Health, ICAR-IVRI, Izatnagar, Uttar Pradesh, India

L.E. Estrada-Gil
Bioprocesses & Bioproducts Group, Food Research Department, School of Chemistry, Universidad Autónoma de Coahuila, Saltillo, Coahuila, México

A. C. Flores-Gallegos
Bioprocesses & Bioproducts Group, Food Research Department, School of Chemistry, Universidad Autónoma de Coahuila, Saltillo, Coahuila, México

Sandeep Ghatak
ICAR Research Complex for NEH Region, Umiam, Meghalaya, India

M. Govea-Salas
Laboratory of Nanobiociences, School of Chemistry, Universidad Autónoma de Coahuila, Ing. J. Cárdenas Valdéz S/N, República, Saltillo, México

Alexandro Guevara-Aguilar
Centro de Investigación en Alimentación y Desarrollo A C, Coordinación Tecnología de Productos Hortofrutícolas y Lácteos, Delicias, Chihuahua, México

Ayerim Y. Hernández-Almanza
School of Biological Science, Universidad Autonoma de Coahuila, Carretara Torreón-Matamoros, Torreón, Coahuila, Mexico

Francisco Hernández-Centeno
Universidad Autónoma Agraria Antonio Narro, Department of Food Science and Technology, Saltillo, Coahuila, México

Anna Iliná
Department of food research, School of Chemical Sciences, Autonomous University of Coahuila, Saltillo, Coahuila, Mexico

Ana L. Izábal-Carvajal
Bioprocesses and Bioproducts Group, Food Research Department, School of Chemistry, Universidad Autónoma de Coahuila, Saltillo, Coahuila, México

M. Suman Kumar
Division of Veterinary Public Health, ICAR-IVRI, Izatnagar, Uttar Pradesh, India

Cristian Torres León
School of Biological Science, Universidad Autonoma de Coahuila, Carretara Torreón-Matamoros, Torreón, Coahuila, Mexico

Erika Flores Loyola
Facultad de Ciencias Biológicas, Universidad Autónoma de Coahuila, Coahuila, México

Naveena Basappa Maheswarappa
ICAR-National Research Centre on Meat, Chengicherla, Hyderabad, Telangana, India

Diana B. Muñiz Márquez
Universidad Autónoma de San Luis Potosí, Facultad de Estudios Profesionales de la Zona Huasteca, San Luis Potosí, México

Jolanta E. Marszalek
Facultad de Ciencias Biológicas, Universidad Autónoma de Coahuila, Coahuila, México

W. E. Cruz Martinez
Facultad de Ciencias Químicas de la Universidad Autónoma de Coahuila, Saltillo, Coahuila, México

R. Reyna Martinez
Colegio de Estudios Científicos y Tecnológicos del Estado de Coahuila Plantel San Antonio de las Alazanas, Arteaga, Coahuila, México

Alaín Martínez-Pérez
Tecnológico Nacional de México/Instituto Tecnológico de Durango, Durango, México

A. Arun Prince Milton
ICAR Research Complex for NEH Region, Umiam, Meghalaya, India

Kiran Mohan
Department of Livestock Products Technology, Veterinary College, KVAFSU, Bidar, Karnataka, India

Emilio Ochoa-Reyes
Research Center in Food and Development, Cd. Delicias, Chihuahua, México

G. Bhuvana Priya
ICAR Research Complex for NEH Region, Umiam, Meghalaya, India

Agustina Ramírez-Moreno
Facultad de Ciencias Biológicas, Universidad Autónoma de Coahuila, Coahuila, México

Lucio Rodríguez-Sifuentes
Facultad de Ciencias Biológicas, Universidad Autónoma de Coahuila, Coahuila, México

S. Wilfred Ruban
Department of Livestock Products Technology, Veterinary College Bengaluru, KVAFSU, Bidar, Karnataka, India
Division of Veterinary Public Health, ICAR-IVRI, Izatnagar, Uttar Pradesh, India
Veterinary College, KVA&FSU, Hebbal, Bengaluru, Karnataka, India

Eliseo Sanchez-Loredo
Bioprocesses and Bioproducts Research Group, Food Research Department, School of Chemistry, Universidad Autónoma de Coahuila, Saltillo, Coahuila, Mexico

Daniela M. Sánchez-Pérez
Instituto Tecnológico de Torreón (ITT)-DEPI, Carretera Torreón-San Pedro, Torreón, Coahuila, México

Elda Patricia Segura-Ceniceros
Department of food research, School of Chemical Sciences, Autonomous University of Coahuila, Saltillo, Coahuila, Mexico

Leonardo Sepúlveda
Bioprocesses and Bioproducts Group, Food Research Department, School of Chemistry, Universidad Autónoma de Coahuila, Saltillo, Coahuila, México

K. Srinivas
ICAR Research Complex for NEH Region, Umiam, Meghalaya, India

Julio C. Tafolla-Arellano
Universidad Autónoma Agraria Antonio Narro, Department of Food Science and Technology, Saltillo, Coahuila, México

Juan M. Tirado-Gallegos
Tecnología de Alimentos de Origen Animal, Facultad de Zootecnia y Ecología, Universidad Autónoma de Chihuahua, Chihuahua, Chihuahua, México

L. Sepúlveda Torre
Facultad de Ciencias Químicas de la Universidad Autónoma de Coahuila, Saltillo, Coahuila, México

Cristian Torres-León
Research Center and Ethno biological Garden, Universidad Autónoma de Coahuila, Unidad Torreón, Viesca, Coahuila, México

V. A. Cepeda Tovar
Facultad de Ciencias Químicas de la Universidad Autónoma de Coahuila, Saltillo, Coahuila, México

José Eduardo Urquizo-Contreras
School of Biological Science, Universidad Autonoma de Coahuila, Carretara Torreón-Matamoros, Torreón, Coahuila, Mexico

S. S. Reddy Vakamalla
Division of Veterinary Public Health, ICAR-IVRI, Izatnagar, Uttar Pradesh, India

Jorge E. Wong-Paz
Universidad Autónoma de San Luis Potosí, Facultad de Estudios Profesionales de la Zona Huasteca, San Luis Potosí, México

Abbreviations

ABF	alkali-treated bamboo fibers
AC	azodicarbonamide
AF	agave fibers
AFB1	aflatoxin-sensitive aptamer
AFM	atomic force microscopy
AFW	agricultural food wastes
BF	bamboo fibers
BHA	butylated hydroxyanisole
BHT	butylated hydroxytoluene
BW	beeswax
CA	controlled atmosphere
CCA	conventional CA
CDC	Centers for Disease Control and Prevention
ChNC	chitin nanocrystals
ChNCs	chitin nanocomposites
CM	compression molded
CMC	carboxy methyl cellulose
CNC	cellulose nanocrystals
CNT	carbon nanotubes
CPC	centrifugal partition chromatography
CRV	carvacrol
DAEC	diffuse adhering *Escherichia coli*
DCS	dynamic control system
DE	diatomaceous earth
DMA	dynamic thermogravimetric study
DMTA	dynamic mechanical thermal analysis
DSC	differential scanning calorimetry
ECSO	epoxidized cottonseed oil
EIEC	enteroinvasive
ELO	epoxidized linseed oil
EPEC	enteropathogenic
ETEC	enterotoxigenic
EVA	ethylene vinyl acetate
EVOH	ethanol vinyl alcohol
FAO	Food and Agriculture Organization
FCCA	fast conventional CA
FCNs	flax cellulose nanocrystals

FDM	fused deposition modeling
FEN	fennel
FE-SEM	field-emission scanning electron microscopy
FG	fish gelatin
GAE	gallic acid equivalents
GMA	glycidyl methacrylate
GRAS	generally recognized as safe
GSLCs	glioma stem-like cells
GT	genotypes
HDPE	high-density polyethylene
HEV	hepatitis E virus
HHDP	hexahydroxydiphenic acid
HLCNCs	high lignin containing cellulose nanocrystals
HPC	hydroxypropyl cellulose
HPLC	high-performance liquid chromatography
HPMC	hydroxypropyl methylcellulose
HSCC	high-speed countercurrent chromatography
HSPs	heat shock proteins
HUS	hemolytic uremic syndrome
HUVEC	human umbilical vein endothelial cells
IPG	immobilized pH gradient
LA	lactic acid
LC	liquid chromatography
LCST	lowest critical solution temperature
LECA	low-ethylene CA
LEO	lavender essential oil
LO	linseed oil
LT	*Larrea tridentata*
MAE	microwave-assisted extraction
MC	methylcellulose
MD	machine direction
MFLC	microfibrillated lignocellulose
MFW	microfibrillated wood
MIC	minimal inhibitory concentration
MLPC	medium pressure liquid chromatography
MMT	montmorillonite
MRSA	methicillin-resistant *Staphylococcus aureus*
MS	mass spectrometry
NCC	nanocrystalline cellulose
NDGA	nordihydroguaiaretic acid
NFS	nanocomposite-forming solutions
NG	nanographene
OEO	orange essential oil

OHF	olive husk flour
OMMT	organomontmorillonite
PAAc	poly(acrylic acid)
PAAm	polyacrylamide
PAMPS	poly(2-acrylamide-2-methylpropanesulfonic
PBH	polyhydroxybutyrate
PET	polyethylene terephthalate
PLA	Polylactic acid
PO	*Posidonia oceanica*
PP	pomegranate peel
PPE	pomegranate peel extract
PS	polystyrene
PUF	polyurethane foam
PUFA	polyunsaturated fatty acids
R-BC	rotomolded biocomposites
REO	rosemary essential oil
RH	relative humidity
RM	rotomolded
RTKs	receptor tyrosine kinase
RVA	Rapid Visco Analyser
SAP	superabsorbent polymer
SBO	soybean oil
SCB	sugar bagasse pulp
SE	solvent extraction
SEM	scanning electron microscopy
SFA	saturated fatty acids
SFE	supercritical fluid extraction
SLS	static light scattering
SPH	superporous hydrogels
SSF	solid-state fermentation
STEC	Shiga-toxin-producing *Escherichia coli*
STEC	Shigatoxigenic *Escherichia coli*
SWC	subcritical water extraction
TBHQ	tertiary butylated hydroquinone
TC	crystallization temperature
TD	transverse direction
TEM	transmission electron microscopy
Tg	glass transition temperatures
TGA	thermogravimetric analysis
Tm	melting temperature
TP	tea polyphenols
TPS	thermoplastic starch
TVB-N	total volatile compounds

UAE	ultrasound-assisted extraction
UBF	untreated bamboo fibers
ULO	ultra-low oxygen
ULOCA	ultra-low oxygen CA
VLF	venlafaxine
WF	wood flour
WHC	water holding capacity
WPC	wood-structured composites
WVP	water vapor permeability

Preface

This title is a comprehensive research-oriented book designed to cover essential elements of food process engineering, and it discusses the most critical achievements in food science along with innovations that are changing the food industry. It links the key concepts of food engineering and science. All chapters have been updated to provide information about the modern methods for engineering processing along with food safety control and their properties. It combines engineering and product quality/safety concepts with emphasis on practical usefulness of preservation processes as well as process control.

This title focuses on the chemical features of food products, food microbiology, packaging, processing, distribution of quality foods, and preservation. It also reviews the most practical approaches into food product design with practical discussions.

This book is divided into three sections in different areas of food process engineering:

> Section 1 covers a wide range of food packaging and storage engineering processes.
>
> Section 2 describes food biotechnology with an emphasis on novel food processes.
>
> Section 3 contains case studies to illustrate the engineering application of technologies that are discussed.

It is an indispensable reference book for food industry professionals and is a valuable research-oriented volume for postgraduate students in food science and technology.

SECTION 1

FOOD PACKAGING AND STORAGE ENGINEERING

CHAPTER 1

Advances in the Development of Hydrogels as an Intelligent Food Packaging for Loading and Release of Natural Compounds

TERESA DARLEN CARRILLO-CASTILLO,
JOSÉ EDUARDO URQUIZO-CONTRERAS, CRISTIAN TORRES LEÓN, and
AYERIM Y. HERNÁNDEZ-ALMANZA

School of Biological Science, Universidad Autonoma de Coahuila, Carretara Torreón-Matamoros, Torreón, Coahuila, Mexico

ABSTRACT

Hydrogels are networks formed by polymeric chains, cross-linked chemically or by physical interaction. This three-dimensional structure allows them to absorb large amounts of water or other fluids giving them a stable structure and a moist consistency. These functional properties of hydrogels make them suitable for countless practical applications. From tissue engineering to food science, the ability of hydrogels to respond to stimuli has been harnessed, which can be designed and modified. Similarly, its interconnected networks allow extracts, oils, and other molecules to be encapsulated and subsequently released at specific sites that improve the shelf life of foods. When targeting the food industry, the main concern is to use naturally derived polymer-based hydrogels for food safety and to reduce the environmental impact of single-use packaging. Therefore, this chapter focuses on materials of natural origin that have proven to be more appropriate to form polymeric hydrogels with application in the food industry. The application is oriented toward the loading and release of active compounds that provide benefits to food

Engineering Principles for Food Processing Technology and Product Realization. Leonardo Sepúlveda Torre, Porteen Kannan, A. K. Haghi, PhD (Eds.)

improving its organoleptic characteristics, shelf life, and slowing down the growth of pathogenic microorganisms.

1.1 INTRODUCTION

Intelligent food packaging is a new type of material to packaging that includes protective barrier for food, detection, and recording functions (Yang et al., 2021). Currently, attention to this packaging is increasing due to the advantages it represents for the food industry, as well as its friendly characteristics with the environment. They can communicate the conditions of the packaged product on real time, but they do not interact with the food. However, the real-time monitoring requires the use of biosensors that allow the information to be transmitted to the consumer (Yang et al., 2021; Müller and Schmid, 2019). In food industry, hydrogels are a great alternative to be used in intelligent packaging system or like carriers of natural compounds (Batista et al., 2019). Although currently most hydrogels are produced from synthetic polymers that guarantee greater physical-chemical stability than natural ones, hydrogels based on natural compounds are making their way into the industry because they are renewable, have biological properties, present versatile structures, and in addition to that they can be produced on a large scale due to the abundance of the raw material (Batista et al., 2019). This chapter describes the physicochemical characteristics of hydrogels as well as their potential use in food packaging systems. On the other hand, the advantages that these intelligent systems have shown in various applications for food preservation are discussed.

1.1.1 HYDROGELS CHARACTERISTICS AND THEIR APPLICATION AREAS

Hydrogels are cross-linked, hydrophilic polymer networks with a wide range of applications due to their high biocompatibility, hydrophilicity, adaptability, and responsiveness (Heimbuck et al., 2018). They are classified depending on their parent materials, their physical properties, the nature of swelling, their method of polymerization, the ionic charges on the network, the rate of biodegradation, and the nature of crosslinking (Ullah et al., 2015). The three-dimensional (swollen) hydrogels structure of hydrogels comprises a solid phase, based on a polymer or mixtures of polymers, and a liquid phase, usually water or biological fluids (Mahinroosta et al., 2018). In the swollen state exposed to certain pressures, its three-dimensional structure allows the absorption of large quantities of the target fluids without changes in its structure (Hu et al., 2019). They are also known to be very versatile materials as they can self-assemble into a variety of shapes including micelles, nanogels, microgels, beads, films, membranes, fibers, and sponges (Crini et al., 2019).

1.1.1.1 CHEMICAL AND PHYSICAL CROSSLINKING

When hydrogels are formed by chemical crosslinking, the polymer chains interact with each other to form permanent bonds through the action of crosslinking agents (Hu et al., 2019). To initiate chemical crosslinking, it is necessary to introduce a low molecular weight crosslinking agent together with a polymer into the reaction mixture. In the absence of crossover points, hydrophilic linear polymer chains dissolve in water, due to the thermodynamic compatibility of the polymer chain and water (Ullah et al., 2015). Thanks to these cross-linked structures, hydrogels do not disintegrate during swelling. The covalent bonds formed in chemical crosslinking are irreversible and because they are interconnected, they are very strong; therefore, they cannot be dissolved again and are thermally stable (Crini et al., 2019).

The main methods used to achieve chemical crosslinking are radical polymerization, reaction of complementary groups, grafting, and enzymatic reactions (Nezhad-Mokhtari et al., 2019). The hydrophilicity of the network is due to the presence of hydrophilic groups such as $-NH_2$, $-COOH$, $-OH$, $-CONH_2$, $-CONH-$ and $-SO_3H$, capillary effect, and osmotic pressure. Crosslinking agents are generally small multifunctional molecules that react with polymers, favoring interaction with their previously activated functional groups. Chemical cross-linking results in permanent three-dimensional structures, giving the hydrogel thermal stability and mechanical strength (Peppas and Hoffman, 2020). These hydrogels are not degradable and cannot be easily recycled, and since after polymerization they need to be purified to remove residues of crosslinking agents and other chemical compounds; they have not been considered for application in the food industry, due to a possible toxicity.

Hydrogels obtained by physical methods are called hydrogels complexed with polyelectrolytes since they result from ionic interactions between polymers (Barroso et al., 2019). That is, crosslinks are formed by physical domain bonds: hydrogen bonds, hydrophobic interaction, and/or ionic complexation. For example, polymer chains containing amino groups ensure the cationic nature required to interact with the anionic group of other polyelectrolytes and form mixtures with a wide range of macromolecules, especially polysaccharides and proteins (Mujtaba et al., 2019). These types of physical interactions are very suitable because they allow the melting of solvents, the modification of the volume after the process, they are easy to manufacture, remodel, biodegrade and, since they do not use chemical cross-linking agents, they do not present toxicity, which makes them biocompatible (Liu et al., May 2019). The reversibility of physical bonds, that is, their ability to intertwine and then flow again, is caused by some specific stimulus, to which the polymer chains respond. This behavior is what makes them ideal for use in dynamic systems such as food packaging. However, achieving this physical crosslinking crucially depends on two conditions: (1) the interactions between the polymer chains must be strong enough to favor the existence of binding sites during the formation of the three-dimensional network; and (2) the assembled three-dimensional network must allow the absorption of a considerable amount of water in the polymeric network (Mahinroosta et al., 2018). Without these two conditions, the hydrogel will not form and would not be useful for use, despite being nontoxic and environmentally friendly.

The choice of the type of crosslinking to be used (chemical or physical) is based on the intended application of the hydrogel obtained, since, in addition to its reversibility, modifying the degree of crosslinking modulates the mechanical resistance, its absorption capacity, and its electrical charge. Hydrogels can also acquire a neutral, ionic, or amphiphilic charge, which will influence their capacity to retain absorbed water (Batista et al., 2019). That is why in hydrogel science, the general aspects to take into account are: the base polymers, the fabrication of the structure, the response of the hydrogel, and the applications of the hydrogel (Li et al., 2021).

1.1.1.2 SUPERPOROUS AND SUPERABSORBENT HYDROGELS

Superporous and superabsorbent hydrogels represent an innovative next-generation category developed for hydrogels, but they still represent disadvantages to be applied in the food area, since their conformation requires high polymer density and chemical agents such as crosslinkers (Batista et al., 2019). Mention is made of them, since their properties can be very useful, once a methodology that does not include toxic agents is developed in the future.

Superporous hydrogels (SPH) were initially developed as a drug delivery system to be applied in the gastric environment and for the drug to remain inside the pores without being completely released. Its mechanism is to swell instantly by rapidly absorbing water and maintaining its integrity by the osmotic pressure generated by its porous structure, while slowly releasing the active ingredient with which it was loaded (Kumari et al., 2020).

Superabsorbent hydrogel, also named superabsorbent polymer (SAP), is a type of hydrogel that can absorb water up to 500–1500 g/g, while the absorption capacity of ordinary hydrogels does not exceed 1000 g/g, and retains it even under mechanical pressure (Zhang et al., 2011). This occurs because when the hydrophilic groups of the polymer chains come into contact with water, they become ionized, which promotes electrostatic repulsion leading to a higher degree of swelling (Ma and Wen, 2020). The crosslinked network prevents water loss and thus maintains the expansion equilibrium, but the swelling process is limited by the mechanical strength of the matrix because the interactions between the polymer chains become weaker. In some cases, the increase in temperature improves the degree of swelling because it allows a separation of the chains.

1.1.1.3 APPLICATIONS AREAS

Hydrogels are currently the subject of considerable scientific research due to their potential in high-tech applications in the biomedical, pharmaceutical, biotechnological, bioseparation, biosensor, agricultural, oil recovery, and cosmetic fields (Ullah et al., 2015). During the evolution of their applications, the traditional use of hydrogels in the food area has been active packaging for moisture control generated by fresh fruits, vegetables, and meat products (Bodbodak and Rafiee, 2016). But the development of new materials and their ability to encapsulate compounds in their network has opened up more applications in packaging systems such as the release of bioactive agents and sensors, among others (Figure 1.1).

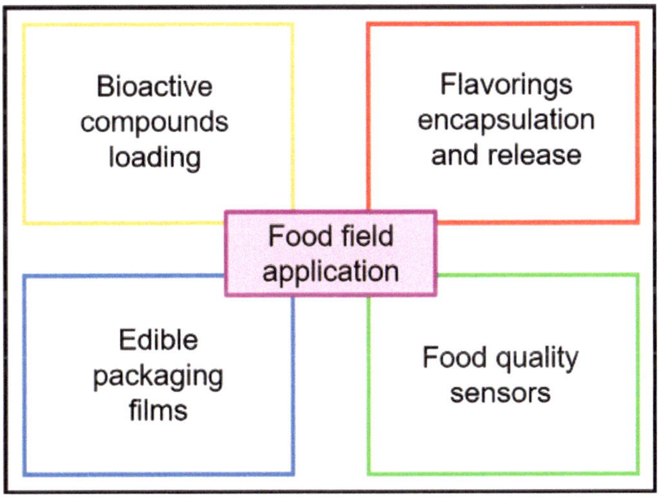

FIGURE 1.1 Application of hydrogels in the food field.

The method of making films for the application of hydrogels in food packaging involves melting the polymer, generally without the addition of chemical crosslinking agents, but during the process, temperature is the key physical agent that must be controlled (Kalia, 2015).

The inclusion of active biocompounds and nanoparticles in the films has been for the purpose of enhancing food preservation by taking advantage of the antimicrobial and antioxidant capacity that these compounds and nanoparticles possess. Studies have shown that its inclusion in the polymeric structure of the hydrogel has also had an effect on improving the mechanical strength of the film and its permeability to water of the hydrogel (Batista et al., 2019).

The use of hydrogels as a carrier for flavorings or as part of an intelligent packaging system with direct application in food matrices has been developed in the work of some researchers:

Food freshness indicator (Baek et al., 2017).

- Information generated about the freshness of food.
- Based on the production of metabolites in food.
- Detection of changes in pH, chemical degradation, or microbial growth.
- Stability and retention of volatile substances (Kwan and Davidov-Pardo, 2017).
- Encapsulation of flavors (nanoemulsions in hydrogels).
- Controlled release occurs by pH modification (e.g., in contact with saliva).

Improvement of the bioavailability of lipophilic compounds (Park et al., 2017).

- Incorporation of lipophilic bioactive compounds (e.g., β-carotene) in the food matrix, improving its bioavailability.

1.2 HYDROGELS BASED ON BIOPOLYMERS TO REDUCE ENVIRONMENTAL IMPACT

Due to their versatility and low cost, synthetic polymers have been widely used in all areas of industry; nevertheless, this and the poor waste management of these materials have resulted in a wide variety of environmental problems (Geyer, 2020). Consequently, many alternatives have emerged to mitigate these negative effects. One of these alternatives consists of using polymers of biological origin instead of those produced synthetically, since they are widely distributed in nature and, due to their characteristics, are environmentally friendly (Shen et al., 2020). These types of polymers, also called biopolymers, are macromolecules made from monomers synthesized in biological systems such as plants, animals, and microorganisms. Therefore, biopolymers represent a biodegradable, biocompatible, and easily renewable material (Varghese et al., 2019). The use of natural polymers as a basis for the creation of hydrogels is a field with enormous potential in areas such as medicine, the food industry, and cosmetics. Mainly because in these areas the use of synthetic polymers has been limited due to their poor biodegradability and potential toxicity (Bao et al., 2019). Natural polymers can be classified into six groups: polysaccharides, proteins, polynucleotides, polyisoprenes, polyesters, and lignins (Varghese et al., 2019). Some of these natural polymers used in the elaboration of hydrogels are discussed in the following section.

1.2.1 PROTEINS

Proteins are biomolecules composed of a chain of amino acids linked by peptide bonds; these are essential macromolecules found in all living organisms and can represent up to fifty percent of the dry weight of their cells. In the food industry, proteins have been widely used due to their functional properties, biodegradability, nontoxicity, and nutritional value. One of these functional properties is the ability to form gels at different scales ranging from micro to macro and has applications in preserving or increasing the bioavailability of functional and bioactive compounds in food (Bourbon et al., 2019). The main proteins used in the elaboration of hydrogels are described below.

1.2.1.1 ANIMAL-DERIVED PROTEINS

Animals and their products represent a well-known source of protein. Table 1.1 describes the main types of animal proteins to produce hydrogels.

1.2.1.2 PLANT-DERIVED PROTEINS

In the food industry, plant-based proteins represent a cheap, abundant, biodegradable, and nontoxic source of natural polymers with fewer environmental implications than animal-based proteins. This is due to the large amounts of water and feed needed to produce the proteins in animals (Lopes et al., 2022).

TABLE 1.1 Biopolymers Derived from Animal Proteins to Form Hydrogels.

Biopolymer	Characteristics	Application
Collagen	Protein present in the extracellular matrix and the most abundant in the animal kingdom. It is obtained from tissues from marine, bovine, and porcine species (Sarrigiannidis et al., 2021).	Colorimetric sensor. Deliver bioactive compounds (Shi et al., 2018).
Gelatin	Protein derived through the denaturation of collagen either by chemical or by physical ways (Irastorza et al., 2021).	Encapsulation of essential oils with antimicrobial activity (Goudoulas et al., 2022).
Caseins	Protein present in milk. They are negatively charged phosphorylated proteins, rich in the amino acid proline and resistant to heat denaturation (Panahi and Baghban-Salehi, 2014).	Encapsulation of active compounds (Sadiq et al., 1965).
Whey protein	Protein derived from milk. This fraction of milk proteins is composed of α-lactalbumin (α-Lg), β-lactoglobulin (β-Lg), and bovine serum albumin (Sadiq et al., 1965).	Have been reported to have the ability to form hydrogels that can be used in the food industry for encapsulation and delivery of bioactive compounds (Abaee et al., 2017).

1.2.1.2.1 Soybean Proteins

The main component of soybeans is protein (35–40%). This fraction is composed of glycinin (11s globulins) and β-conglycinin (7s globulins). Soy protein is used in industry and research as soy protein isolate, which is a purified and concentrated version of the protein (Sadiq et al., 1965). This biomolecule is widely used due to its low cost, availability, and nutritional properties. However, these gels exhibit weak mechanical characteristics and rapid degradation. In response to this effect, the use of chemical and enzymatic crosslinkers represents a viable solution to improve the characteristics of these materials (Maltais et al., 2009).

1.2.1.2.2 Pea Proteins

The seeds of this legume contain about 30% protein consisting of legumin 11s and vicilin 7s globulins. Like soy protein, pea protein isolate is the derived product used in the food industry (Djoullah and Saurel, 2021). This protein has been used in the development of hydrogels with food applications mainly because of its sustainable production, non-allergenic status, and tunable properties (Meng and Cloutier, 2014).

1.2.2 POLYSACCHARIDES

These polymers are carbohydrates and composed by the union of monosaccharides through glycosidic bonds (Liu et al., 2021). Polysaccharides as base biomaterials in the elaboration of hydrogels represent a nontoxic, biodegradable, biocompatible, and stable alternative. Furthermore, they have good mechanical properties (Varghese et al., 2019). Consequently, animals, plants, algae, and microorganisms have attracted attention in recent years because it is possible to isolate a wide variety of polysaccharides from them (Liu et al., 2021).

1.2.2.1 PLANT-DERIVED POLYSACCHARIDES

The use of plants as a source of polysaccharides is probably one of the most characteristic approaches in biotechnology. Plant-derived polysaccharides are considered a viable alternative for obtaining raw materials that can be used in a wide variety of biotechnological processes such as the production of gels. This is due to their potential to be produced sustainably, their high availability, and their low cost (Bahú et al., 2022).

1.2.2.1.1 Cellulose

Cellulose is the most abundant polysaccharide in nature. In the food industry, this polymer is characterized by its versatility, which is a consequence of the wide variety of cellulose derivatives used in this field, such as methylcellulose (MC), carboxy methyl cellulose (CMC), hydroxypropyl cellulose (HPC), or hydroxypropyl methylcellulose (HPMC) (Ghosh and Katiyar, 2019). Hydrogels made from cellulose and its derivatives are mainly used for packaging, although their ability as delivery carriers of bioactive compounds and their capacity to respond to stimuli have also been studied (Yan et al., 2020). However, these materials still represent a series of significant challenges to be solved. The main one is finding a suitable solvent for this material that is insoluble in water (Ghosh and Katiyar, 2019). Despite this drawback, hydrogels have been developed for the production of smart packaging, such as the work carried out by Shaghaleh et al., who obtained a hydrogel with pH/thermo-responsive behavior and antimicrobial properties, made from TEMPO oxidized nano fibrillated cellulose and food-grade cationic-modified poly(N-isopropyl acrylamide-co-acrylamide). The hydrogel was able to release preservative agents in response to stimuli generated by a climacteric fruit (Shaghaleh et al., 2021).

1.2.2.1.2 Starch

This biopolymer is found in granules form consisting of amylose and amylopectin (Zhu, 2017). Starch is characterized by its ability to form gels in the presence of water and heat. For this reason, this polysaccharide is widely used as a food additive or as a packaging material. However, processes such as starch retrogradation and the loss of mechanical properties in hydrogels caused by low temperatures represent a series of drawbacks that need to be resolved to be used (Yang et al., 2021). Despite this, the use of hydrogels based on starch obtained from different sources is studied in the elaboration of systems to encapsulate bioactive compounds which are also present in hydrogels as copolymers together with other polysaccharides or proteins (Ghosh and Katiyar, 2019).

1.2.2.1.3 Algae Polysaccharides

In recent years, algae have been the source for obtaining a great variety of important biopolymers such as agarose, alginate, or carrageenans, to mention a few. The main advantage of these polysaccharides over those obtained from other sources is their abundance, since even due to this fact these biomolecules are considered a base resource in the next generation of energy and bioproduct generation (Lin and Jiao, 2012).

1.2.2.1.4 Alginate

Alginate is a polysaccharide synthesized mainly by brown seaweeds, although it is known that some microorganisms are also capable of producing it. This heteropolysaccharide is formed by α-D-mannuronic acid and β-L-guluronic acid. Furthermore, one of the most significant properties of this biopolymer is its ability to produce hydrogels in the presence of divalent cations (generally Ca^{2+}), a process known as ionotropic gelation (Dallabona et al., 2020). Hydrogels obtained from alginate are probably the best known and most widely implemented due to the facilities that their creation process involves. In this context, alginate hydrogels have been widely used at nano-, micro-, and macro- scales in the encapsulation of different bioactive compounds and the creation of food packaging (films and coatings) (Osojnik Črnivec and Poklar Ulrih, 2019). An example of its application in the development of smart packaging was the work of Sutthasupa et al.; they elaborated a hydrogel with a mixture of alginate and methylcellulose cross-linked with Ca^{2+}. This packaging functioned as a spoilage indicator for ground pork. This effect was achieved by incorporating bromothymol blue dye, which is sensitive to pH, into the hydrogel matrix. The result was a hydrogel with the ability to react with a color change in the presence of total volatile compounds (TVB-N) when the pH becomes basic, working in this way as a viable alternative in the field of active packaging and with application in meat products (Sutthasupa et al., 2021).

1.2.2.1.5 Carrageenans

Carrageenans are another representative polysaccharide of this group and are produced by red seaweeds. This polysaccharide is characterized by being sulfated and having six different forms depending on its composition and origin. These forms are kappa, iota, lambda, mu, un, and theta (Lin and Jiao, 2012). However, only the kappa, lambda, and iota variants are studied for the elaboration of hydrogels due to their gelling properties (Berton et al., 2020). The gelation of these polymers generally occurs through cationic crosslinkers such as metal ions or other polymers. This process, as with alginate hydrogels, is a relatively simple process (Zhang et al., 2016).

1.2.2.2 ANIMAL-DERIVED POLYSACCHARIDES

Chitosan is the main example of biopolymers of animal origin which have properties to form hydrogels. Chitin is an amino polysaccharide that can be obtained from the exoskeleton of insects, crustaceans, and mollusks and in the cell wall of fungi. Chemically, chitin is made up of N-acetyl glucosamine units; however, chitosan is formed by the deacetylation of this polymer, which is a linear polymer made up of N-acetyl-D-glucosamine and D-glucosamine units (Maddaloni and Vassalini, 2020). This polysaccharide is the most important derivative of chitin and is widely used in the food industry as a material to produce films, coatings, and hydrogels. These hydrogels have great potential for applications due to their safe and innocuous nature, their response to pH, and their ability to encapsulate and release natural compounds in their matrix (Qu and Luo, 2020).

1.2.2.3 MICROORGANISM-DERIVED POLYSACCHARIDES

Microbial polysaccharides, due to their diversity and properties, have positioned themselves commercially as a resource of great importance in different processes; xanthan gum is probably the most representative polymer of the group. It is an extracellular polysaccharide produced by the Gram-negative bacterium Xanthomonas campestris by fermentation of sugars (Osojnik Črnivec and Poklar Ulrih, 2019). It has the ability to form hydrogels by physical and chemical crosslinking. However, due to their stability, chemically cross-linked hydrogels are the most studied, mainly in the encapsulation and controlled release of compounds (Tao et al., 2016).

1.2.3 SYNERGIES BETWEEN BIOPOLYMERS TO ENHANCE HYDROGEL PROPERTIES

Throughout this section, the importance of natural polymers in the development of hydrogels was reviewed, highlighting their great variety, safety, biodegradability, and high availability. However, the main drawbacks in the use of these materials were also raised, where the mechanical properties, instability to external factors, and water retention capacity are the main limitations in the use and manufacture of these gels. Consequently, in response to these problems, the combination of two or more natural polymers has emerged as a viable alternative for the improvement of these materials (Liu et al., 2021). Therefore, on this basis, it is possible to create hydrogels using protein–protein, protein–polysaccharide, polysaccharide–polysaccharide mixtures, and even systems containing three of these biopolymers in addition to hydrogels that embed fibers in their structure, which can be food grade for further application in this field (Khalesi et al., 2021). These associations between polymers are possible thanks to the bonds that can form between them and their ability to form complexes through noncovalent and covalent interactions. The examples of these systems include soy protein–whey protein, gelatin–whey protein, cellulose–starch, and casein–carrageenans, among many other hydrogels (Liu et al., 2021).

1.2.4 *RELEVANCE OF THE RESIDUAL ORIGIN OF BIOPOLYMERS*

It is a reality that the use of biopolymers for the elaboration of hydrogels is a promising and environmentally friendly approach. Furthermore, due to the characteristics shown, these materials have significant advantages over synthetic polymers. In this context, the use of natural polymers in the production of hydrogels can also have another positive effect on environmental awareness if we consider the fact that these biopolymers can be obtained from waste produced by the food industry itself. In fact, according to FAO data, around 1.3 billion tons of food are discarded each year, representing a potential wasted source of biomolecules such as proteins and carbohydrates (Ranganathan et al., 2020). Therefore, the use of biopolymers in the production of hydrogels not only represents an alternative to environmental care, providing materials that in the long term do not present the same problems as synthetic polymers but also making greater use of raw materials obtained from the food industry waste that compromise the environment. Thus, biopolymers represent a diverse, safe, nonpolluting source of raw materials to produce food-grade hydrogels with the potential to be produced in a sustainable manner.

1.3 SMART HYDROGELS, THEIR DESIGN TO RESPOND TO STIMULI

Smart hydrogels are those that show significant physicochemical change in response to small changes in the environment. These stimuli include physical environmental conditions, such as temperature, electric and magnetic fields, light, pressure, and sounds, and the chemical conditions of the system, such as pH, ionic strength, solvent composition, and molecular species (Dehshahri et al., 2021). These are stimuli that change the degree of swelling or deswelling of the hydrogel in the form of volume collapse or phase transition (Figure 1.2). Such changes are reversible; therefore, hydrogels are capable of returning to their initial state as soon as the stimulus is removed. That is why they can be "reusable." These responses of hydrogels to the environment provide a variety of applications that can be exploited in the field of food packaging.

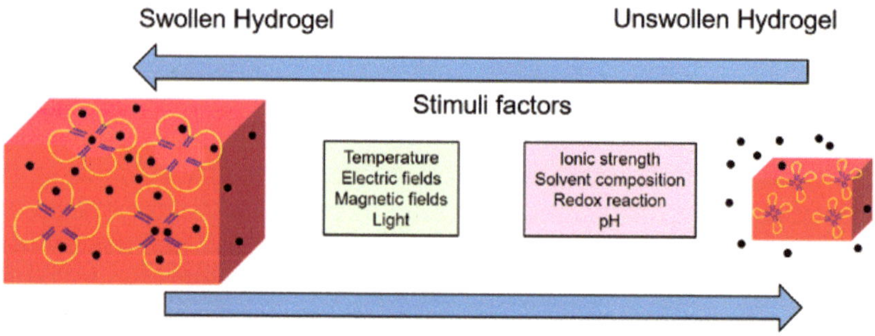

FIGURE 1.2 Response of smart hydrogels to stimuli.

Physical stimuli change the molecular interactions at the critical starting points of the hydrogel, thus undergoing changes in its growth actions, network structure, mechanical strength, and permeability. For their part, chemical stimuli change the interactions between polymer chains and between polymer chains and solvents at the molecular level.

1.3.1 DESIGN OF SMART HYDROGELS

Smart hydrogels have been designed to swell or shrink in response to physical, chemical, or biological cues from the environment or even a combination of these. The development of these smart hydrogels is a meticulous process that takes into account polymer type, polymer molecular weight, crosslinker molecule, and crosslink density to suit a specific application and provide a combination of the properties listed above. The stimuli disturb the intermolecular interactions of the hydrogels formed in an aqueous environment, resulting in changes in the conformation of the polymeric chains, which causes macroscopic swelling or contractions. To make these chains respond to stimuli, they are designed using functional groups, adapting the blocks of different types of polymers, type and quantity of the crosslinking molecule, and the integration of nanoparticles to broaden their functionalities and the scope of their application (Ding et al., 2020).

1.3.1.1 THERMOSENSITIVE HYDROGELS

Its conformational change is based on the phase transition behavior of thermosensitive polymers. Hydrogels with inverse thermosensitivity are insoluble in water at room temperature, that is, above their lowest critical solution temperature (LCST). This causes the aggregation of unsolvated polymeric chains (Taylor et al., 2017), as happens with poly (N-isopropylacrylamide) (PNIPAAm), when the temperature is above its LCST (32°C), its chains they shrink due to their decreased solubility in water (Heskins and Guillet, 1968). The example of a polymer with a higher critical solution temperature (UCST) response is polyacrylamide (PAAm), when the ambient temperature is above its UCST; the strength of the water–polymer interactions caused by the bonds of hydrogen becomes stronger than the polymer–polymer interactions, causing an increase in the solubility of the hydrogel (Seuring and Agarwal, 2012). If it is chosen to combine polymers of the two types, a hydrogel is achieved that has a dual response to temperatures, above the UCST (15°C) and below the LCST (41°C) (Chang et al., 2009).

1.3.1.2 PH-SENSITIVE HYDROGELS

These types of hydrogels are developed with polyelectrolytes containing weak acidic or basic functional groups. These functional groups only allow the passage of ions with opposite charges, creating an osmotic pressure difference in the hydrogel network, inducing its deformation (Firestone and Siegel, 1991). pH-sensitive hydrogels are classified as acidic or basic depending on the pendant group they have on their chains. Acids possess carboxylic groups, phosphonic acid groups, or boronic acid groups (Wu et al., 2010). For their part, basic hydrogels have basic groups such as amines or amide groups (Mukhopadhyay et al., 2014).

When the pH value surrounding the acid hydrogel is greater than the dissociation constant (Ka), the acid pendant groups release protons, obtaining a negative charge, causing an increase in the hydrophilic nature of the hydrogels, and allowing the network to swell. When the pH decreases, the effect is reversible. In the case of the basic hydrogel, its pendant groups accept protons at a pH below the dissociation constant (Kb), remaining positively charged. This increases the hydrophilic nature causing an electrostatic repulsion between the polymer chains, causing the hydrogel to expand (Buenger et al., 2012). Amphoteric hydrogels have also been created, with groups of both acidic and basic monomers, which allows the hydrogel to only remain swollen in a specific pH range, and contract at any other value (Abou Taleb, 2013).

1.3.1.3 SALT-SENSITIVE HYDROGELS

These hydrogels change when the salinity in the solution varies; the polyelectrolyte networks expand in pure water and contract in saline solutions. Its response mechanism is similar to that of pH-sensitive hydrogels. When the hydrogel is in low salinity solutions, its ionizable groups dissociate causing a high charge density. This leads to an intermolecular repulsion between the chains of the hydrogel, causing a volumetric expansion. When the salt concentration increases, counterions are being added, which attenuate the intermolecular repulsion, causing the collapse of the polymeric network (Xu et al., 2008).

1.3.1.4 ELECTROSENSITIVE HYDROGELS

They are a type of electroactive polymers that, when receiving a voltage, result in a nonuniform distribution of ions on the hydrogel layer, which causes an osmotic pressure difference, resulting in changes in curvature of the hydrogel layer. This electrical responsiveness of hydrogels can be engineered using ionizable functional groups. Some known electroactive polymers are poly(acrylic acid) (PAAc) (Moschou et al., 2004), poly(4-hydroxybutyl acrylate) (poly(4-HBA)) (Kwon et al., 2010), and poly(2-acrylamide-2-methylpropanesulfonic) (PAMPS) (Yang et al., 2017). Ionizable functional groups are located both in the backbone of the chain and in the side chains, which provide abundant fixed charges that will react rapidly to electric fields, resulting in a short response time (approx. 0.1 s) (Mirfakhrai et al., 2007). If, in addition, ionic compounds are added to the hydrogel matrix, the concentration of mobile ions increases, inducing an even faster response of the hydrogel to electric fields (Ding et al., 2020).

1.3.1.5 PHOTOSENSITIVE HYDROGELS

This type is designed by incorporating photoactive fractions within the hydrogel matrix (Li et al., 2019), which leads the hydrogel to have a photothermal excitation or a reversible crosslinking reaction. For the reversible crosslinking mechanism, photoisomerization (ring opening/closing reactions) and photocleavage (induced by photolabile protecting groups) are two predominant approaches that generate sequential steps of crosslinking, resulting in reversible contraction–expansion of polymeric chains (Kloxin et al., 1979).

Dehydration–hydration processes are also a physical response in the hydrogel that can be achieved by selective lighting to cause localized heating. Similarly, if photosensitive nanomaterials are incorporated into the hydrogel matrix, dehydration processes can be accelerated and intensified. These photothermal nanomaterials convert light irradiation into heat dissipation, inducing hydrogel deformation. Some known nanomaterials, which have already been used for their photothermal capacity, are gold (Aida et al., 2018), carbon-based materials (Yang et al., 2019) and organic compounds such as cyanines (Zhang et al., 2016).

1.4 RESPONSE OF HYDROGELS TO CHANGES IN FOOD

When a smart hydrogel is used as part of a smart packaging system, its main purpose is to provide information about the freshness of the contained food products or as part of a simple detection method to determine the presence of contaminants. Aflatoxin, for example, is a mycotoxin that can develop in many agricultural products. Zhao et al designed a smart hydrogel by cross-linking the polymer chains with an aflatoxin-sensitive aptamer (AFB1) and by loading the hydrogel with urease. When the system is exposed to concentrations of AFB1, the hydrogel collapses releasing urease into solution, causing changes in pH. This pH change can be easily measured, and indirectly determines the concentration of AFB1 present (Zhao et al., 2018).

1.4.1 FOOD CHARACTERISTICS THAT ARE IMPROVED WITH THE APPLICATION OF SMART HYDROGELS

Hydrogels are ideal for improving the characteristics of foods because of their softness, elasticity, absorbent nature, flexibility, and hygroscopic nature (Manzoor et al., 2022). Hydrogels can be used in food preservation, pharmaceuticals, agriculture, and food packaging. Intelligent hydrogels alter their structure and change phase to outside stimuli and can be used for innovative food applications. Food characteristics that hydrogels primarily influence are antimicrobial, gas barrier, nutritional, functional, flavor, and color properties.

1.4.1.1 ANTIMICROBIAL PROPERTIES

Food preservation's primary objective has been controlling microbial populations, specifically pathogenic microorganisms (Torres-León et al., 2022). Food safety is a big problem for humans, especially bacteria that cause foodborne diseases (Torres-León et al., 2022). One way to inhibit the growth of microorganisms in food is by implementing antimicrobial food packaging systems. Hydrogels formulated with biopolymers and nanomaterials have been used in food packaging to improve microbial stability and prolong the shelf life (Beigmohammadi et al., 2016). The carrageenan-based hydrogel films with metallic nanoparticles exhibited antibacterial activity against *Escherichia coli*, and *Listeria monocytogenes* (Oun and Rhim, 2017).

1.4.1.2 GAS BARRIER PROPERTIES

Nowadays, the short shelf life of fruit and vegetables is one of the biggest trading problems (Torres-León et al., 2018). The edible coatings have been broadly used to extend the shelf life of fruits, vegetables, cheese, mushrooms, rainbow trout fillets, chicken fillets, and processed potatoes (Chacon et al., 2022). Hydrogels can control moisture loss and transpiration produced by the food (Leyva-Jiménez et al., 2023). Multicomponent hydrogel films (composed of agar, κ-carrageenan, konjac glucomannan powder, and nanoclay) applied to fresh spinach packaging reveal a high potential for the use as an antifogging film for packaging highly respiring agricultural produce (Rhim and Wang, 2013).

1.4.1.3 NUTRITIONAL PROPERTIES

Hydrogels may also be used to encapsulate and control the release of nutrients (Yang et al., 2022). Oil in water-hydrogelled emulsions and ethylcellulose organogels replaced beef fat in emulsion-type meat batters. Reformulated products improved the fatty acid profile (decrease in saturated fatty acids (SFA) from 11.8% to ≈2% and an increase in polyunsaturated fatty acids (PUFA) from 0.3% to ≈5%) (Alejandre et al., 2019). Dos Santos et al. (2020) demonstrated that emulsion gels based on pork skin, inulin, α-cyclodextrin, polydextrose, and bamboo fiber are alternative ingredients to replace pork back fat in manufacturing emulsified meat products. The fat content in sugar-snap cookies was reduced by replacing a commercial shortening with a hydrogel formed by mixing a canola oil/candelilla oleogel (5 g/100 g) and a gelatinized corn starch hydrogel (5 g/100 g) in a 1:1 ratio (Barragán-Martínez et al., 2022).

1.4.1.4 FUNCTIONAL PROPERTIES

The hydrogels in food are a vehicle system for the antimicrobials, antioxidants, flavorings, colorants, and bioactive components. The hydrogel loaded with 5% pomegranate peel extract showed good antioxidant (81.13%) and antimicrobial activities (Guo et al., 2019). In recent years, polyphenols (as naturally derived antioxidants) have gained growing interest in the food industry. Phenolic compounds are secondary metabolites produced by plants and exhibit a wide range of functional properties (Torres-león et al., 2017). Trifković et al. (2014) designed thyme polyphenols loaded chitosan hydrogels to prolong the release of polyphenols in simulated gastrointestinal conditions. The release of encapsulated polyphenols in simulated gastrointestinal fluids was extended to 3 h.

1.4.1.5 FLAVOR PROPERTIES

The hydrogels can also be used to encapsulate and control the release of flavor compounds in food. Soluble-filled hydrogels can encapsulate a flavor and release the flavor in the presence of artificial saliva. pH-sensitive hydrogels can dissociate in the presence of saliva; this leads to the release and perception of encapsulated flavors (Amiryousefi et al., 2016). Filled hydrogels were determined to encapsulate and release flavor (Kwan and Davidov-Pardo, December 2017).

1.4.1.6 COLOR PROPERTIES

Food additives are used for color enrichment in the food industry. The use of a hydrogel formed by mixing a canola oil/candelilla oleogel (5 g/100 g) and a gelatinized corn starch hydrogel (5 g/100 g) in the cookie preparation reduced the effective content of lipids, reducing in this way the extent of redness and yellowness reactions, which was reflected in the color parameters (Barragán-Martínez et al., 2022). Hydrogels are also used as a biosensor in innovative packaging. Intelligent hydrogel packaging may contain a colorimetric indicator that changes color in response to a change in pH in the food, which is related to quality attributes. Lu et al. (2020) formulated sugarcane bagasse nanocellulose hydrogel as a colourimetric freshness indicator for intelligent food packaging.

1.5 THE HYDROGELS' ABILITY TO BE LOADED WITH NATURAL COMPOUNDS

As has been described so far, hydrogels are a versatile material, which is why they also have the ability to be loaded with natural compounds, to later release them in a controlled manner, thus expanding their applications. Due to the amphiphilic characteristic of their chains and the polarity of their networks, hydrogels can be loaded with both hydrophilic and hydrophobic compounds. Essential oils can be incorporated by means of emulsifying agents, which stabilize them in the hydrogel matrix. Natural aqueous-based extracts have a natural affinity with the hydrophilic part of the hydrogel, so their incorporation can be done by simply incorporating it into the aqueous medium that will serve to swell the hydrogel. This is in order to develop intelligent systems for the delivery of bioactive compounds in food when a change occurs in the microenvironment of the packaging or outside of it.

1.5.1 LOADED WITH ESSENTIAL OILS

In addition to their utility as flavorings and food additives, essential oils can exhibit excellent antibacterial and antimicrobial activity. Its incorporation in hydrogels has been for the purpose of creating systems that can help food packaging to prolong the storage life of food by inhibiting bacterial growth and avoiding oxidative processes.

From the research carried out, using hydrogels loaded with oils, which had antibacterial properties, Goudoulas et al. developed gelatin hydrogels containing microdroplets of rosemary essential oil (REO) and orange essential oil (OEO). They stabilized the droplets using the surfactant Tween®80. The gels/REO inhibited the growth of *Bacillus cereus* and *Escherichia coli* (contaminating meat products bacteria); meanwhile, the gels/OEO inhibited the growth of *B. cereus*, although to a minor degree than REO. These hydrogels loaded with microdispersed essential oils are a product that has the potential to be used as a coating for meat products and prolong their storage life (Goudoulas et al., 2022).

Lavender essential oil (LEO) also has antibacterial properties, but its application is limited due to its high volatility. Therefore, Deng et al. investigated the possibility of using hydrogel particles to encapsulate and control the release of LEO. The oil was encapsulated in oil-in-water emulsions, which were then loaded onto hydrogel particles to finally cross-link them with Ca^{2+}. The results were that the hydrogel improved the stability of LEO during storage and controlled its release by changes in pH. The authors suggested that this system can be incorporated into food packaging as a new aroma delivery system with controlled release of essential oils (Deng et al., 2020).

On the other hand, the essential oils of garlic (GEO) and holy basil (HBEO) are commonly used in food products for their antimicrobial activity, but there are studies in which the synergistic actions of two oils improve potency and spectrum of activity antimicrobial. A common problem with these oils is their chemical deterioration due to light, temperature, and exposure to oxygen. To overcome these drawbacks, Torpol et al. encapsulated a mixture of GEO and HBEO in pectin beads, which were coated with chitosan and finally cross-linked with $CaCl_2$ using the ionic gelation method. The beads achieved a mean encapsulation efficiency of 70% of the oil mix and cumulative release after storage at 5°C for 15 days. The inhibitory effect was tested by depositing the pearls (4.4 g) in cellulose bags and applying it to *Bacillus cereus, Clostridium perfringens, Escherichia coli, Pseudomonas florescens*, *Listeria monocytogenes*, and *Staphylococcus aureus*, demonstrating an inhibitory effect on all microorganisms. In view of these results, they suggest that pearls can be included in storage bags or food packaging to provide microbiological safety (Torpol et al., 2019).

1.5.2 LOADED WITH NATURAL EXTRACTS

Due to their biocompatible properties, hydrogels are ideal for the delivery of phenolic compounds extracted from plant material since they can maintain the stability of the compounds that are incorporated into them, improving the nutritional value, color, and aroma of the food product. For example, curcumin is a polyphenol with antimicrobial, anticancer, and anti-inflammatory activity, but with low solubility; therefore, encapsulation in vehicles such as hydrogels can increase its solubility, keep it stable, and allow controlled and targeted delivery to the food of interest (Corkovi et al., 2021). Thus, many of the phenolic and volatile compounds of interest that have an antioxidant and antimicrobial effect can be used to prolong the shelf life of foods, as long as their properties are maintained. This is how the strategy of preserving them within hydrogel matrices seems to be a good strategy.

Malaysian herbal extract is high in phenols and exhibits antioxidant activities. Wan Yajana et al. extracted these biocomposites and added them to an edible film developed with semirefined carrageenan and glycerol. Once the film was obtained, meat burgers were wrapped with it and stored in the refrigerator (4 ± 2°C) for 14 days. Lipid deterioration was significantly less than the product that was not film-wrapped with the herbal extract. Due to this evidence and that of other tests they carried out, they concluded that this film can be used as a packaging material to improve the quality characteristics of meat (Wan Yahaya et al., 2019).

Pomegranate extract is another biocompound that has been studied and successfully incorporated into hydrogel matrices. As in the work of Maroufi et al., who created a polymeric hydrogel pad based on chitosan, guar gum, and modified dialdehyde, the pad had an absorption capacity of up to 12,000% of its dry weight and loading with 5% pomegranate peel extract (PPE) showed good antioxidant activity (81.13%) and antimicrobial efficacy against gram-negative bacteria of *Escherichia coli* and gram-positive of *Staphylococcus aureus*. Taking these results into consideration, they concluded that the pomegranate extract-loaded hydrogel was a potentially suitable material for absorbing liquid at the bottom of fresh meat, poultry, and fish package trays to increase shelf life (Maroufi et al., 2021).

Tea polyphenols (TP) are the main active ingredient in tea, recognized for their antioxidant, antiviral, and antibacterial biological activities. These extracts have been widely used as additives in the pharmaceutical and food industries because they have stable chemical properties and good reactivity with biological macromolecules (Matsui, 2015). That is why it has been studied as an active ingredient in polysaccharide-based films. Gao et al. developed a film for the use in the food industry using pectin, chitosan, and tea polyphenols, cross-linked with calcium ions. They determined its antioxidant activity in vitro and its antiseptic and color-protective capacity was evaluated in freshly cut beef cubes. The film exhibited ability to scavenge free radicals, inhibitory ability against bacteria and to prevent deterioration in meat color. These results led to the conclusion that the film may have the value to be used as a packaging film and slow release material, with the application in the food and pharmaceutical industries (Gao et al., 2019).

1.5.3 EXTRACTS THAT FUNCTION AS INDICATORS

Some extracts have the property of responding colorimetrically to changes in the chemical environment of the food. That is why, in addition to their anti-oxidant and antimicrobial effects, they can be used as indicators in real time by incorporating them into the hydrogels that are part of a food container. Taking advantage of this peculiarity, Pereira et al. used a mixture of chitosan polymers and polyvinyl alcohol to create a film for food packaging. This film was added with anthocyanins extracted from red cabbage (*Brassica oleracea*) to develop a time-temperature indicator. They carried out color variation tests on the film in relation to the change in pH. They applied the indicator to a sample of pasteurized milk and the film showed evident changes in its color when the storage temperature increased, indicating changes in the chemical composition of the milk as evidence of its deterioration (Pereira et al., 2015).

Another novel colorimetric indicator used the ability of the *Arnebia euchroma* root extract to change color in the presence of volatile basic nitrogen (TVB-N), which manifests itself as the fish meat deteriorates (increase in pH), being an indicator of its freshness in real time. The dye was incorporated into a red algae agar matrix to form a film that was applied directly to Wuchang goldfish (*Megalobrama amblycephala*). The test was carried out at refrigeration temperatures (4°C) and at room temperature (25°C). The results showed that the indicator film exhibited a noticeable color change consistent with the deterioration threshold of TVB-N content. This system appears to be a simple, nondestructive, and visual method to estimate the freshness of fish during storage and transportation (Huang et al., 2019).

Finally, some articles mention that in the food industry most of the hydrogels used are based on proteins and carbohydrates, since the use of synthetic polymers has been related to possible cytotoxicity. In a following section, the types of biopolymers that have been used to create hydrogels with application in smart packaging for food packaging will be detailed.

1.6 FUTURE PERSPECTIVES FOR HYDROGELS IN THE FOOD INDUSTRY

Despite the multiple works carried out to develop hydrogels suitable for application in the food industry and all the natural materials that have been used, it is still an area rich in opportunities for the creation of new types of packaging, innovative for the entire range of food and at the same time, friendly to the environment.

1.6.1 OTHER WAYS TO CREATE HYDROGEL PLATFORMS

The synthesis of hydrogels-type nanogels and microgels has been extensively studied for controlled release of drugs, but could be adapted to food preservation to release active compounds.

Packaging labels can be printed using injectable hydrogels, which can be created from natural materials such as pectin. The color can be added to the hydrogel and these colorants can be completely of natural origin, so that their contact with the food does not have toxic effects. These natural dyes could naturally degrade over time along with the shelf life of the food so that the consumer is informed in real time that they should consume it before the packaging acquires a certain intensity of color.

1.6.2 OTHER PARTICLES THAT CAN BE INCLUDED IN HYDROGELS

Although this chapter focused on materials of natural origin and biocomposites, graphene and graphene nanomaterials are another very good alternative because they are used in biomedicine for being a non-cytotoxic and biocompatible material. Therefore, it could be included in smart food packaging since it is classified as an active and smart material.

In the same way, silver nanoparticles have wide applications in water treatment, agriculture, medicine, drug administration, among others. The activity that has attracted the most attention of these nanoparticles is their antimicrobial function, which is why polysaccharide hydrogels can be developed together with silver nanoparticles, for the production of new antibacterial films with biocompatibility and stability of silver nanoparticles.

1.6.3 OTHER STIMULI TO WHICH HYDROGELS CAN RESPOND

Hydrogels that respond to the presence of bacterial metabolites (sensitive to enzymes) are beginning to be explored. Its sensitivity is specific, so it can be a qualitative and quantitative method. The current disadvantage is that it takes a relatively long time, so the area of interest is to develop systems that do it in a short time, since bacterial contamination in food represents a great threat to human health. The sensitivity of hydrogels can also focus on the separation of proteins and lipids. Its hydrophobic character can selectively adsorb a hydrophobic target under certain conditions and thereby make the separation. This answer can be applied to the benefit of the food industry.

If the field outside of food packaging is opened up a bit, thermally stable hydrogels may find applications in food processing. These hydrogels can be used for the separation of oil and water and other waste produced during food processing. They also help masking of unpleasant tastes and the prevention of scaling and boilers, pipes and heat exchange surfaces. The combination of thermoresistant hydrogels has very unusual rheological and thermal properties and therefore resists much higher temperatures without melting.

Magnetic response is another attribute of smart hydrogels, but much research remains to be done for practical application in the food area.

1.7 CONCLUSIONS

The large amount of work that has been carried out, using smart hydrogels, has obtained promising results to be a safe and sustainable alternative to plastic materials that are currently still used in excess, such as food packaging. The long-term goal must be to achieve 100 percent biodegradable packaging. Their elimination does not cause burden to the environment, they can be created with materials obtained from agro-industrial waste and they can be reincorporated into the environment in a natural way.

Something that is important to consider is that, despite all the advantages that the use of biomaterials for food packaging can represent and the additional benefits provided by the inclusion of bioactive components, it cannot be ignored how it can affect the basic functionality of the packaging. A package must have mechanical resistance, permeability to water and oxygen, protection against light, and preferably transparency to be attractive to the consumer. The alteration of the packaging can affect the brightness, appearance, flavor, and even the crispness of the food, something that can be negative in consumer preference.

Another disadvantage that must be overcome is that, although stimuli-responsive hydrogels are already widely known and elaborated, as well as the way to load them with bioactive compounds, it is necessary to develop suitable manufacturing methods to produce these hydrogels commercially, a scale that is economically profitable for the industry.

KEYWORDS

- **hydrogel**
- **food packaging**
- **polymers**
- **active compounds**
- **microorganisms**
- **food safety**

REFERENCES

Abaee, A.; Mohammadian, M.; Jafari, S. M. Whey and Soy Protein-Based Hydrogels and Nano-Hydrogels as Bioactive Delivery Systems. *Trends Food Sci. Technol.* **2017,** *70*, 69–81. https://doi.org/10.1016/j.tifs.2017.10.011

Abou Taleb, M. F. Radiation Synthesis of Multifunctional Polymeric Hydrogels for Oral Delivery of Insulin. *Int. J. Biol. Macromol.* **2013,** *62*, 341–347. https://doi.org/10.1016/j.ijbiomac.2013.09.004

Aida, T.; Sun, Z.; Yamauchi, Y.; Araoka, F.; Kim, Y. S.; Bergueiro, J.; Ishida, Y.; Ebina, Y.; Sasaki, T.; Hikima, T. An Anisotropic Hydrogel Actuator Enabling Earthworm-like Directed Peristaltic Crawling. *Angew. Chem.* **2018,** *130* (48), 15772–15776. https://doi.org/10.1002/ange.201810052

Alejandre, M.; Astiasarán, I.; Ansorena, D.; Barbut, S. Using Canola Oil Hydrogels and Organogels to Reduce Saturated Animal Fat in Meat Batters. *Food Res. Int.* **2019,** *122*, 129–136. https://doi.org/10.1016/J.FOODRES.2019.03.056

Amiryousefi, M. R.; Mohebbi, M.; Golmohammadzadeh, S.; Koocheki, A. Encapsulation of Caffeine in Hydrogel Colloidosome: Optimization of Fabrication, Characterization and Release Kinetics Evaluation. *Flavour Fragr. J.* **2016,** *31* (2), 163–172. https://doi.org/10.1002/FFJ.3297

Baek, S.; Kim, D.; Jeon, S. L.; Seo, J. Preparation and Characterization of PH-Responsive Poly(N,N-Dimethyl Acrylamide-Co-Methacryloyl Sulfadimethoxine) Hydrogels for Application as Food Freshness Indicators. *React. Funct. Polym.* **2017,** *120* (September), 57–65. https://doi.org/10.1016/j.reactfunctpolym.2017.09.003

Bahú, J. O.; de Andrade, L. R. M.; de Melo Barbosa, R.; Crivellin, S.; da Silva, A. P.; Souza, S. D. A.; Cárdenas Concha, V. O.; Severino, P.; Souto, E. B. Plant Polysaccharides in Engineered Pharmaceutical Gels. *Bioengineering* **2022,** *9* (8), 376. https://doi.org/10.3390/bioengineering9080376

Bao, Z.; Xian, C.; Yuan, Q.; Liu, G.; Wu, J. Natural Polymer-Based Hydrogels with Enhanced Mechanical Performances: Preparation, Structure, and Property. *Adv. Healthc. Mater.* **2019,** *8* (17), 1–11. https://doi.org/10.1002/adhm.201900670

Barragán-Martínez, L. P.; Román-Guerrero, A.; Vernon-Carter, E. J.; Alvarez-Ramirez, J. Impact of Fat Replacement by a Hybrid Gel (Canola Oil/Candelilla Wax Oleogel and

Gelatinized Corn Starch Hydrogel) on Dough Viscoelasticity, Color, Texture, Structure, and Starch Digestibility of Sugar-Snap Cookies. *Int. J. Gastron. Food Sci.* **2022,** *29*, 100563. https://doi.org/10.1016/J.IJGFS.2022.100563

Barroso, N.; Guaresti, O.; Pérez-álvarez, L.; Ruiz-rubio, L. Self-Healable Hyaluronic Acid/ Chitosan Polyelectrolyte Complex Hydrogels and Multilayers. *Eur. Polym. J.* **2019,** *120* (September), 109268. https://doi.org/10.1016/j.eurpolymj.2019.109268

Batista, R. A.; Espitia, P. J. P.; de Quintans, J. S. S.; Freitas, M. M.; Cerqueira, M. Â.; Teixeira, J. A.; Cardoso, J. C. Hydrogel as an Alternative Structure for Food Packaging Systems. *Carbohydr. Polym.* **2019,** *205*, 106–116. https://doi.org/https://doi.org/10.1016/j.carbpol.2018.10.006

Batista, R. A.; Perez Espitia, P. J.; Quintans, J. D. S. S.; Machado, M.; Ângelo, M.; António, J.; Cordeiro, J. Hydrogel as an Alternative Structure for Food Packaging Systems. *Carbohydr. Polym.* **2019,** *205*, 106–116.

Beigmohammadi, F.; Peighambardoust, S. H.; Hesari, J.; Azadmard-Damirchi, S.; Peighambardoust, S. J.; Khosrowshahi, N. K. Antibacterial Properties of LDPE Nanocomposite Films in Packaging of UF Cheese. *LWT - Food Sci. Technol.* **2016,** *65*, 106–111. https://doi.org/10.1016/J.LWT.2015.07.059

Berton, S. B. R.; de Jesus, G. A. M.; Sabino, R. M.; Monteiro, J. P.; Venter, S. A. S.; Bruschi, M. L.; Popat, K. C.; Matsushita, M.; Martins, A. F.; Bonafé, E. G. Properties of a Commercial κ-Carrageenan Food Ingredient and Its Durable Superabsorbent Hydrogels. *Carbohydr Res* **2020,** *487*, 107883. https://doi.org/10.1016/j.carres.2019.107883.

Bodbodak, S.; Rafiee, Z. *Recent Trends in Active Packaging in Fruits and Vegetables*; Elsevier Inc., 2016. https://doi.org/10.1016/B978-0-12-804313-4.00003-7

Bourbon, A. I.; Pereira, R. N.; Pastrana, L. M.; Vicente, A. A.; Cerqueira, M. A. Protein-based Nanostructures for Food Applications. *Gels* **2019,** *5* (1), 1–18. https://doi.org/10.3390/gels5010009

Buenger, D.; Topuz, F.; Groll, J. Hydrogels in Sensing Applications. *Prog. Polym. Sci.* **2012,** *37*, 1678–1719. https://doi.org/10.1016/j.progpolymsci.2012.09.001

Chacon, W.; Paz-arteaga, S.; Torres-león, C.; Ayala Valencia, G. Gums-Based Coatings Applied to Extend the Shelf Life of Foods: A Review. *J. Polym. Environ.* **2022,** *2020*.

Chang, Y.; Chen, W.; Yandi, W.; Shih, Y.; Chu, W. Dual-Thermoresponsive Phase Behavior of Blood Compatible Zwitterionic Copolymers Containing Nonionic Poly (N-Isopropyl Acrylamide). *Biomacromolecules* **2009,** *10* (8), 2092–2100.

Corkovi, I.; Pichler, A.; Šimunovi, J.; Kopjar, M. Hydrogels: Characteristics and Application as Delivery Systems of Phenolic and Aroma Compounds. *Foods* **2021,** *10* (6), 1252.

Crini, G.; Torri, G.; Lichtfouse, E.; Kyzas, G. Z.; Wilson, L. D.; Morin-Crini, N. Dye Removal by Biosorption Using Cross-Linked Chitosan-Based Hydrogels. *Environ. Chem. Lett.* **2019,** *17* (4), 1645–1666. https://doi.org/10.1007/s10311-019-00903-y

Dalponte Dallabona, I.; de Lima, G. G.; Cestaro, B. I.; Tasso, I. D. S.; Paiva, T. S.; Laureanti, E. J. G.; Jorge, L. M. D. M.; da Silva, B. J. G.; Helm, C. V.; Mathias, A. L.; Jorge, R. M. M. Development of Alginate Beads with Encapsulated Jabuticaba Peel and Propolis Extracts to Achieve a New Natural Colorant Antioxidant Additive. *Int. J. Biol. Macromol.* **2020,** *163*, 1421–1432. https://doi.org/10.1016/j.ijbiomac.2020.07.256

Dehshahri, A.; Kumar, A.; Madamsetty, V. S.; Uzieliene, I.; Tavakol, S.; Azedi, F.; Fekri, H. S.; Zarrabi, A.; Mohammadinejad, R.; Thakur, V. K. New Horizons in Hydrogels for Methotrexate Delivery. *Gels* **2021,** *7* (1), 1–20. https://doi.org/10.3390/gels7010002

Deng, X.; Chen, J.; Chen, W. Hydrogel Particles as a Controlled Release Delivery System for Lavender Essential Oil Using PH Triggers. *Colloids Surf. A* **2020,** *603*, 125134. https://doi.org/10.1016/j.colsurfa.2020.125134

Ding, M.; Jing, L.; Yang, H.; Machnicki, C. E.; Fu, X.; Li, K.; Wong, I. Y.; Chen, P. Multifunctional Soft Machines Based on Stimuli-Responsive Hydrogels: From Freestanding Hydrogels to Smart Integrated Systems. *Mater. Today Adv.* **2020,** *8*, 1–19. https://doi.org/10.1016/j.mtadv.2020.100088

Djoullah, A.; Saurel, R. Controlled Release of Riboflavin Encapsulated in Pea Protein Microparticles Prepared by Emulsion-Enzymatic Gelation Process. *J. Food Eng.* **2021,** *292*. https://doi.org/10.1016/j.jfoodeng.2020.110276

Dos Santos, M.; Ozaki, M. M.; Ribeiro, W. O.; Paglarini, C. D. S.; Vidal, V. A. S.; Campagnol, P. C. B.; Pollonio, M. A. R. Emulsion Gels Based on Pork Skin and Dietary Fibers as Animal Fat Replacers in Meat Emulsions: An Adding Value Strategy to Byproducts. *LWT* **2020,** *120*, 108895. https://doi.org/10.1016/J.LWT.2019.108895

Firestone, B. A.; Siegel, R. A. Kinetics and Mechanisms of Water Sorption in Hydrophobic, Ionizable Copolymer Gels. *J. Appl. Polym. Sci.* **1991,** *43*, 901–914.

Gao, H. X.; He, Z.; Sun, Q.; He, Q.; Zeng, W. C. A Functional Polysaccharide Film Forming by Pectin, Chitosan, and Tea Polyphenols. *Carbohydr. Polym.* **2019,** *215*, 1–7. https://doi.org/10.1016/j.carbpol.2019.03.029

Geyer, R. *Production, Use, and Fate of Synthetic Polymers*; Elsevier Inc., 2020. https://doi.org/10.1016/b978-0-12-817880-5.00002-5

Ghosh, T.; Katiyar, V. *Cellulose-Based Hydrogel Films for Food Packaging*; 2019. https://doi.org/10.1007/978-3-319-77830-3_35

Goudoulas, T. B.; Vanderhaeghen, S.; Germann, N. Micro-Dispersed Essential Oils Loaded Gelatin Hydrogels with Antibacterial Activity. *LWT - Food Sci. Technol.* **2022,** *154*, 1–10. https://doi.org/10.1016/j.lwt.2021.112797

Guo, J. S.; Fang, F.; Yan, P.; Chen, Y. P. Sludge Reduction Based on Microbial Metabolism for Sustainable Wastewater Treatment. *Bioresour. Technol.* **2020,** *297* (September 2019), 122506. https://doi.org/10.1016/j.biortech.2019.122506

Heimbuck, A. M.; Priddy-Arrington, T. R.; Sawyer, B. J.; Caldorera-Moore, M. E. Effects of Post-Processing Methods on Chitosan-Genipin Hydrogel Properties. *Mater. Sci. Eng. C* **2019,** *98* (April 2018), 612–618. https://doi.org/10.1016/j.msec.2018.12.119

Heskins, M.; Guillet, J. E. Solution Properties of Poly (N-Isopropylacrylamide). *J. Macromol. Sci.: Part A- Chem.* **1968,** *2*, 1441–1455.

Hu, W.; Wang, Z.; Xiao, Y.; Zhang, S.; Wang, J. Advances in Crosslinking Strategies of Biomedical Hydrogels. *Biomater. Sci.* **2019,** *7* (3), 843–855. https://doi.org/10.1039/c8bm01246f

Huang, S.; Xiong, Y.; Zou, Y.; Dong, Q.; Ding, F.; Liu, X. A Novel Colorimetric Indicator Based on Agar Incorporated with Arnebia Euchroma Root Extracts for Monitoring Fish Freshness. *Food Hydrocoll.* **2019,** *90*, 198–205. https://doi.org/10.1016/j.foodhyd.2018.12.009

Irastorza, A.; Zarandona, I.; Andonegi, M.; Guerrero, P.; de la Caba, K. The Versatility of Collagen and Chitosan: From Food to Biomedical Applications. *Food Hydrocoll.* **2021,** *116*, 1–11. https://doi.org/10.1016/j.foodhyd.2021.106633

Kalia, S. *Polymeric Hydrogels as Smart Biomaterials (Springer Series on Polymer and Composite Materials)*; Springer International Publishing, 2015.

Khalesi, H.; Lu, W.; Nishinari, K.; Fang, Y. Fundamentals of Composites Containing Fibrous Materials and Hydrogels: A Review on Design and Development for Food Applications. *Food Chem.* **2021,** *364* (June), 130329. https://doi.org/10.1016/j.foodchem.2021.130329

Kloxin, A. M.; Kasko, A. M.; Salinas, C. N.; Anseth, K. S. Photodegradable Hydrogels for Dynamic Tuning of Physical and Chemical Properties. *Science (1979)* **2013,** *324* (2009), 59–63. https://doi.org/10.1126/science.1169494

Kumari, P. V. K.; Sharmila, M.; Rao, Y. S. Super Porous Hydrogels: A Review. *J. Pharm. Res. Int.* **2020,** 153–165. https://doi.org/10.9734/jpri/2020/v32i1330595

Kwan, A.; Davidov-Pardo, G. Controlled Release of Flavor Oil Nanoemulsions Encapsulated in Filled Soluble Hydrogels. *Food Chem.* **2018,** *250* (December 2017), 46–53. https://doi.org/10.1016/j.foodchem.2017.12.089

Kwon, G. H.; Choi, Y.; Park, Y.; Woo, H.; Lee, B.; Kim, H.; Lee, S. Electrically-Driven Hydrogel Actuators in Microfluidic Channels: Fabrication, Characterization, and Biological Application. *Lab-on-a-Chip* **2010,** *10*, 1604–1610. https://doi.org/10.1039/b926443d

Leyva-Jiménez, F. J.; Oliver-Simancas, R.; Castangia, I.; Rodríguez-García, A. M.; Alañón, M. E. Comprehensive Review of Natural Based Hydrogels as an Upcoming Trend for Food Packing. *Food Hydrocoll.* **2023,** *135*, 108124. https://doi.org/10.1016/J.FOODHYD.2022.108124

Li, J.; Jia, X.; Yin, L. Hydrogel: Diversity of Structures and Applications in Food Science Hydrogel: Diversity of Structures and Applications in Food Science. *Food Rev. Int.* **2021,** *37* (3), 313–372. https://doi.org/10.1080/87559129.2020.1858313.

Li, L.; Scheiger, J. M.; Levkin, P. A. Design and Applications of Photoresponsive Hydrogels. *Adv. Mat. Res.* **2019,** *31*, 1–17. https://doi.org/10.1002/adma.201807333

Lin, J.; Jiao, G.; Kermanshahi-pour, A. Algal Polysaccharides-Based Hydrogels: Extraction, Synthesis, Characterization, and Applications. *Mar. Drugs* **2012,** 253–264. https://doi.org/10.1201/b13868-22

Liu, K.; Chen, Y. Y.; Zha, X. Q.; Li, Q. M.; Pan, L. H.; Luo, J. P. Research Progress on Polysaccharide/Protein Hydrogels: Preparation Method, Functional Property and Application as Delivery Systems for Bioactive Ingredients. *Food Res. Int.* **2021,** *147*, 1–16. https://doi.org/10.1016/j.foodres.2021.110542

Liu, Z.; Liu, C.; Sun, X.; Zhang, S.; Yuan, Y.; Wang, D.; Xu, Y. Fabrication and Characterization of Cold-Gelation Whey Protein-Chitosan Complex Hydrogels for the Controlled Release of Curcumin. *Food Hydrocoll.* **2020,** *103* (May 2019), 105619. https://doi.org/10.1016/j.foodhyd.2019.105619

Lopes, P. M. P.; Moldovan, D.; Moldovan, M.; Carpa, R.; Saroşi, C.; Păşcuţă, P.; Moldovan, A. M.; Fechete, R.; Popescu, V. New Composite Hydrogel Based on Whey and Gelatin Crosslinked with Copper Sulphate. *Materials* **2022,** *15* (7), 1–21. https://doi.org/10.3390/ma15072611

Lu, P.; Yang, Y.; Liu, R.; Liu, X.; Ma, J.; Wu, M.; Wang, S. Preparation of Sugarcane Bagasse Nanocellulose Hydrogel as a Colourimetric Freshness Indicator for Intelligent Food Packaging. *Carbohydr. Polym.* **2020,** *249* (August), 116831. https://doi.org/10.1016/j.carbpol.2020.116831

Maddaloni, M.; Vassalini, I.; Alessandri, I. Green Routes for the Development of Chitin/Chitosan Sustainable Hydrogels. *Sustain. Chem.* **2020,** *1* (3), 325–344. https://doi.org/10.3390/suschem1030022

Mahinroosta, M.; Jomeh, Z.; Allahverdi, A.; Shakoori, Z. Hydrogels as Intelligent Materials: A Brief Review of Synthesis, Properties and Applications. *Mater. Today Chem.* **2018,** *8*, 42–55. https://doi.org/10.1016/j.mtchem.2018.02.004

Maltais, A.; Remondetto, G. E.; Subirade, M. Soy Protein Cold-Set Hydrogels as Controlled Delivery Devices for Nutraceutical Compounds. *Food Hydrocoll.* **2009,** *23* (7), 1647–1653. https://doi.org/10.1016/j.foodhyd.2008.12.006

Manzoor, A.; Dar, A. H.; Pandey, V. K.; Shams, R.; Khan, S.; Panesar, P. S.; Kennedy, J. F.; Fayaz, U.; Khan, S. A. Recent Insights into Polysaccharide-Based Hydrogels and Their Potential Applications in Food Sector: A Review. *Int. J. Biol. Macromol.* **2022,** *213*, 987–1006. https://doi.org/10.1016/J.IJBIOMAC.2022.06.044

Maroufi, L. Y.; Tabibiazar, M.; Ghorbani, M.; Jahanban-Esfahlan, A. Fabrication and Characterization of Novel Antibacterial Chitosan/Dialdehyde Guar Gum Hydrogels Containing Pomegranate Peel Extract for Active Food Packaging Application. *Int. J. Biol. Macromol.* **2021,** *187* (July), 179–188. https://doi.org/10.1016/j.ijbiomac.2021.07.126

Matsui, T. Condensed Catechins and Their Potential Health-Bene Fi Ts. *Eur. J. Pharmacol.* **2015,** *765*, 495–502. https://doi.org/10.1016/j.ejphar.2015.09.017

Meng, Y.; Cloutier, S. *Gelatin and Other Proteins for Microencapsulation*; Elsevier Inc., 2014. https://doi.org/10.1016/b978-0-12-404568-2.00020-0

Mirfakhrai, T.; Madden, J. D. W.; Baughman, R. H. Polymer Artificial. *Materials Today,* **2007,** *10* (4), 30–38

Moschou, E. A.; Peteu, S. F.; Bachas, L. G.; Madou, M. J.; Daunert, S. Artificial Muscle Material with Fast Electroactuation under Neutral PH Conditions. *Chem. Mater.* **2004,** *16*, 2499–2502.

Mujtaba, M.; Morsi, R. E.; Kerch, G.; Elsabee, M. Z.; Kaya, M.; Labidi, J.; Khawar, K. M. Current Advancements in Chitosan-Based Film Production for Food Technology; A Review. *Int. J. Biol. Macromol.* **2019,** *121*, 889–904. https://doi.org/10.1016/j.ijbiomac.2018.10.109

Mukhopadhyay, P.; Sarkar, K.; Bhattacharya, S.; Bhattacharyya, A.; Mishra, R.; Kundu, P. P. PH Sensitive N-Succinyl Chitosan Grafted Polyacrylamide Hydrogel for Oral Insulin Delivery. *Carbohydr. Polym.* **2014,** *112*, 627–637. https://doi.org/10.1016/j.carbpol.2014.06.045

Müller, P.; Schmid, M. Intelligent. Packaging in the Food Sector: A Brief Overview. *Foods* **2019,** *8* (1), 16.

Nezhad-Mokhtari, P.; Ghorbani, M.; Roshangar, L.; Soleimani Rad, J. Chemical Gelling of Hydrogels-Based Biological Macromolecules for Tissue Engineering: Photo- and Enzymatic-Crosslinking Methods. *Int. J. Biol. Macromol.* **2019,** *139*, 760–772. https://doi.org/10.1016/j.ijbiomac.2019.08.047

Osojnik Črnivec, I. G.; Poklar Ulrih, N. *Nano-Hydrogels of Alginate for Encapsulation of Food Ingredients*; 2019. https://doi.org/10.1016/B978-0-12-815663-6.00013-6

Oun, A. A.; Rhim, J. W. Carrageenan-Based Hydrogels and Films: Effect of ZnO and CuO Nanoparticles on the Physical, Mechanical, and Antimicrobial Properties. *Food Hydrocoll.* **2017,** *67*, 45–53. https://doi.org/10.1016/J.FOODHYD.2016.12.040

Panahi, R.; Baghban-Salehi, M. *Protein-Based Hydrogels*; 2014. https://doi.org/10.4032/9789814463218

Park, S.; Mun, S.; Kim, Y. R. Effect of Xanthan Gum on Lipid Digestion and Bioaccessibility of β-Carotene-Loaded Rice Starch-Based Filled Hydrogels. *Food Res. Int.* **2018,** *105* (August 2017), 440–445. https://doi.org/10.1016/j.foodres.2017.11.039

Peppas, N. A.; Hoffman, A. S. Hydrogels. In *Biomaterials Science*, 4th ed.; 2020, No. 1; pp 153–166. https://doi.org/10.1016/B978-0-12-816137-1.00014-3.

Pereira, V. A.; de Arruda, I. N. Q.; Stefani, R. Active Chitosan/PVA Films with Anthocyanins from Brassica Oleraceae (Red Cabbage) as Time–Temperature Indicators for Application in Intelligent Food Packaging. *Food Hydrocoll.* **2015,** *43*, 180 188. https://doi.org/10.1016/j.foodhyd.2014.05.014

Qu, B.; Luo, Y. Chitosan-Based Hydrogel Beads: Preparations, Modifications and Applications in Food and Agriculture Sectors—A Review. *Int. J. Biol. Macromol.* **2020,** *152*, 437–448. https://doi.org/10.1016/j.ijbiomac.2020.02.240

Ranganathan, S.; Dutta, S.; Moses, J. A.; Anandharamakrishnan, C. Utilization of Food Waste Streams for the Production of Biopolymers. *Heliyon* **2020,** *6* (9), 1–13. https://doi.org/10.1016/j.heliyon.2020.e04891

Rhim, J. W.; Wang, L. F. Mechanical and Water Barrier Properties of Agar/κ-Carrageenan/Konjac Glucomannan Ternary Blend Biohydrogel Films. *Carbohydr. Polym.* **2013,** *96* (1), 71–81. https://doi.org/10.1016/J.CARBPOL.2013.03.083

Sadiq, U.; Gill, H.; Chandrapala, J. Casein Micelles as an Emerging Delivery System for Bioactive Food Components. *Foods* **2021,** *10* (8), 1965. https://doi.org/10.3390/foods10081965

Sarrigiannidis, S. O.; Rey, J. M.; Dobre, O.; González-García, C.; Dalby, M. J.; Salmeron-Sanchez, M. A Tough Act to Follow: Collagen Hydrogel Modifications to Improve Mechanical and Growth Factor Loading Capabilities. *Mater. Today Bio.* **2021,** *10*, 1–22. https://doi.org/10.1016/j.mtbio.2021.100098

Seuring, J.; Agarwal, S. First Example of a Universal and Cost-E Ff Ective Approach: Polymers with Tunable Upper Critical Solution Temperature in Water and Electrolyte Solution. **2012**.

Shaghaleh, H.; Hamoud, Y. A.; Xu, X.; Liu, H.; Wang, S.; Sheteiwy, M.; Dong, F.; Guo, L.; Qian, Y.; Li, P.; Zhang, S. Thermo-/PH-Responsive Preservative Delivery Based on TEMPO Cellulose Nanofiber/Cationic Copolymer Hydrogel Film in Fruit Packaging. *Int. J. Biol. Macromol.* **2021,** *183*, 1911–1924. https://doi.org/10.1016/j.ijbiomac.2021.05.208

Shen, M.; Song, B.; Zeng, G.; Zhang, Y.; Huang, W.; Wen, X.; Tang, W. Are Biodegradable Plastics a Promising Solution to Solve the Global Plastic Pollution? *Environ. Pollut.* **2020,** *263*, 114469. https://doi.org/10.1016/j.envpol.2020.114469.

Shi, M.; Shi, Y. L.; Li, X. M.; Yang, R.; Cai, Z. Y.; Li, Q. S.; Ma, S. C.; Ye, J. H.; Lu, J. L.; Liang, Y. R.; Zheng, X. Q. Food-Grade Encapsulation Systems for (-)-Epigallocatechin Gallate. *Molecules* **2018,** *23* (2), 1–17. https://doi.org/10.3390/molecules23020445

Sutthasupa, S.; Padungkit, C.; Suriyong, S. Colorimetric Ammonia (NH3) Sensor Based on an Alginate-Methylcellulose Blend Hydrogel and the Potential Opportunity for the Development of a Minced Pork Spoilage Indicator. *Food Chem.* **2021,** *362*, 130151. https://doi.org/10.1016/j.foodchem.2021.130151

Tao, Y.; Zhang, R.; Xu, W.; Bai, Z.; Zhou, Y.; Zhao, S.; Xu, Y.; Yu, D. Rheological Behavior and Microstructure of Release-Controlled Hydrogels Based on Xanthan Gum Crosslinked with Sodium Trimetaphosphate. *Food Hydrocoll.* **2016,** *52*, 923–933. https://doi.org/10.1016/j.foodhyd.2015.09.006

Taylor, M. J.; Tomlins, P.; Sahota, T. S. Thermoresponsive Gels. *Gels* **2017,** *3* (4), 1–31. https://doi.org/10.3390/gels3010004

Torpol, K.; Sriwattana, S.; Sangsuwan, J.; Wiriyacharee, P.; Prinyawiwatkul, W. Optimising Chitosan—Pectin Hydrogel Beads Containing Combined Garlic and Holy Basil Essential Oils and Their Application as Antimicrobial Inhibitor. *Food Sci. Technol.* **2019,** *54* (6), 2064–2074. https://doi.org/10.1111/ijfs.14107

Torres-León, C.; Aguilar, C. Food Preservation. In *Quantitative Methods and Analytical Techniques in Food Microbiology*; Sepulveda, L., N. Aguilar, C., Porteen, K., Haghi, A., Eds.; Apple Academic Press Inc.: Abingdon, Oxon, OX14 4RN UK, 2022; pp 39–57.

Torres-León, C.; Sepulveda, L.; Aguilar, C. N. Food and Diseases: What to Know in the Fight to Ensure Food Safety. In *Quantitative Methods and Analytical Techniques in Food Microbiology*; Leonardo, S. T., Cristóbal, N. A., Porteen Kannan, A. K. H., Ed.; Apple Academic Press Inc.: Abingdon, Oxon, OX14 4RN UK, 2022; pp 57–75.

Torres-león, C.; Ventura-Sobrevilla, J.; Serna-Cock, L.; Ascacio-Valdés, J. A. J. A.; Contreras-Esquivel, J.; Aguilar, C. N. Pentagalloylglucose (PGG): A Valuable Phenolic Compound with Functional Properties. *J. Funct. Foods* **2017,** *37*, 176–189. https://doi.org/10.1016/j.jff.2017.07.045

Torres-León, C.; Vicente, A. A.; Flores-López, M. L.; Rojas, R.; Serna-Cock, L.; Alvarez-Pérez, O. B.; Aguilar, C. N. Edible Films and Coatings Based on Mango (Var. Ataulfo) by-Products to Improve Gas Transfer Rate of Peach. *LWT - Food Sci. Technol.* **2018,** *97*, 624–631. https://doi.org/10.1016/j.lwt.2018.07.057

Trifković, K. T.; Milašinović, N. Z.; Djordjević, V. B.; Krušić, M. T. K.; Knežević-Jugović, Z. D.; Nedović, V. A.; Bugarski, B. M. Chitosan Microbeads for Encapsulation of Thyme (Thymus Serpyllum L.) Polyphenols. *Carbohydr. Polym.* **2014,** *111*, 901–907. https://doi.org/10.1016/J.CARBPOL.2014.05.053

Ullah, F.; Othman, M. B. H.; Javed, F.; Ahmad, Z.; Akil, H. M. Classification, Processing and Application of Hydrogels: A Review. *Mater. Sci. Eng. C* **2015,** *57*, 414–433. https://doi.org/10.1016/j.msec.2015.07.053

Varghese, S. A.; Rangappa, S. M.; Siengchin, S.; Parameswaranpillai, J. *Natural Polymers and the Hydrogels Prepared from Them*; Elsevier Inc., 2019. https://doi.org/10.1016/B978-0-12-816421-1.00002-1

Wan Yahaya, W. A.; Yazid, N. A.; Mohod Azman, N. A.; Almajano, M. P. Antioxidant Activities and Total Phenolic Content of Malaysian Herbs as Components of Active Packaging Film in Beef Patties. *Antioxidants* **2019,** *8* (7), 1–15.

Wu, W.; Mitra, N.; Yan, E. C. Y.; Zhou, S. Multifunctional Hybrid Nanogel for and Self-Regulated Insulin Release at Physiological PH. *ACS Nano* **2010,** *4* (8), 4831–4839.

Xu, Y.; Bolisetty, S.; Drechsler, M.; Fang, B.; Yuan, J.; Ballauff, M.; Mu, A. H. E. PH and Salt Responsive Poly (N, N-Dimethylaminoethyl Methacrylate) Cylindrical Brushes and Their Quaternized Derivatives. *Polymer (Guildf)* **2008,** *49*, 3957–3964. https://doi.org/10.1016/j.polymer.2008.06.051

Yan, L.; Wang, L.; Gao, S.; Liu, C.; Zhang, Z.; Ma, A.; Zheng, L. Celery Cellulose Hydrogel as Carriers for Controlled Release of Short-Chain Fatty Acid by Ultrasound. *Food Chem.* **2020,** *309*, 125717. https://doi.org/10.1016/j.foodchem.2019.125717

Yang, C.; Liu, Z.; Chen, C.; Shi, K.; Zhang, L.; Ju, X.-J.; Wang, W.; Xie, R.; Chu, L.-Y. Reduced Graphene Oxide-Containing Smart Hydrogels with Excellent Electro-Response and Mechanical Property for Soft Actuators. *ACS Appl. Mater. Interfaces,* **2017,** *9*, 15758–15767.

Yang, J.; Shen, M.; Luo, Y.; Wu, T.; Chen, X.; Wang, Y.; Xie, J. Advanced Applications of Chitosan-Based Hydrogels: From Biosensors to Intelligent Food Packaging System. *Trends Food Sci. Technol.* **2021,** *110*, 822–832. https://doi.org/https://doi.org/10.1016/j.tifs.2021.02.032

Yang, M.; Yuan, Z.; Liu, J.; Fang, Z.; Fang, L.; Yu, D.; Li, Q. Photoresponsive Actuators Built from Carbon-Based Soft Materials. *Adv. Opt. Mater.* **2019,** *7* (16), 1–30. https://doi.org/10.1002/adom.201900069

Yang, Z.; Chen, L.; McClements, D. J.; Qiu, C.; Li, C.; Zhang, Z.; Miao, M.; Tian, Y.; Zhu, K.; Jin, Z. Stimulus-Responsive Hydrogels in Food Science: A Review. *Food Hydrocoll.* **2022,** *124*, 107218. https://doi.org/10.1016/j.foodhyd.2021.107218

Yang, Z.; Chen, X.; Xu, Z.; Ji, N.; Xiong, L.; Sun, Q. Anti-Freezing Starch Hydrogels with Superior Mechanical Properties and Water Retention Ability for 3D Printing. *Int. J. Biol. Macromol.* **2021,** *190*, 382–389. https://doi.org/10.1016/j.ijbiomac.2021.08.235

Zhang, J.; Liu, S.; Hui, X.; Xie, Z.; Jing, X. Cyanine-Curcumin Assembling Nanoparticles for Near-Infrared Imaging and Photothermal Therapy. *ACS Biomater. Sci. Eng.* **2016,** *2* (11), 1942–1950. https://doi.org/10.1021/acsbiomaterials.6b00315

Zhang, Z.; Zhang, R.; Chen, L.; McClements, D. J. Encapsulation of Lactase (β-Galactosidase) into κ-Carrageenan-Based Hydrogel Beads: Impact of Environmental Conditions on Enzyme Activity. *Food Chem.* **2016,** *200*, 69–75. https://doi.org/10.1016/j.foodchem.2016.01.014

Zhao, M.; Wang, P.; Guo, Y.; Wang, L.; Luo, F.; Qiu, B.; Guo, L.; Su, X.; Lin, Z.; Chen, G. Detection of a Fl Atoxin B 1 in Food Samples Based on Target-Responsive Aptamer-Cross-Linked Hydrogel Using a Handheld PH Meter as Readout. *Talanta* **2018,** *176*, 34–39. https://doi.org/10.1016/j.talanta.2017.08.006

Zhu, F. Encapsulation and Delivery of Food Ingredients Using Starch Based Systems. *Food Chem* **2017,** *229*, 542–552. https://doi.org/10.1016/j.foodchem.2017.02.101

CHAPTER 2

Bionanocomposites Based on Thermoplastic Starch for the Food Industry

JUAN M. TIRADO-GALLEGOS[1], EMILIO OCHOA-REYES[2], JUAN C. BUSTILLOS-RODRÍGUEZ[3], JULIO C. TAFOLLA-ARELLANO[4], FRANCISCO HERNÁNDEZ-CENTENO[4], and JUAN BUENROSTRO-FIGUEROA[2]

[1]*Tecnología de Alimentos de Origen Animal, Facultad de Zootecnia y Ecología, Universidad Autónoma de Chihuahua, Chihuahua, Chihuahua, México*

[2]*Research Center in Food and Development, Cd. Delicias, Chihuahua, México*

[3]*Tecnológico Nacional de México, Instituto Tecnológico de Cd. Cuauhtémoc, Cd. Cuauhtémoc, Chihuahua, México*

[4]*Universidad Autónoma Agraria Antonio Narro, Department of Food Science and Technology, Saltillo, Coahuila, México*

ABSTRACT

Food packaging plays an important role in the determination of shelf life of food. The principal functions are to contain and protect the food during their storage and distribution. Also, the packaging informs the consumer about the food properties and conditions of use. For this purpose, most of the materials used as food packaging are the synthetic packaging materials. However, these materials are nonbiodegradable, and their excessive use promotes ecological problems. For this reason, there is an increasing interest

Engineering Principles for Food Processing Technology and Product Realization. Leonardo Sepúlveda Torre, Porteen Kannan, A. K. Haghi, PhD (Eds.)

in the development of biodegradable packaging from starch alone or as part of a composite, which offers great potential to replace the conventional food packaging due to its versatility, availability, and low cost. Nevertheless, starch thermoplastic materials present poor mechanical and barrier properties compared with synthetic polymers. Research groups have tried to overcome these limitations adding additives. In this context, the nanotechnology represents a great potential to design new packaging materials with improved physicochemical properties. In this chapter, we analyze and summarize the recent advances in the use of nanomaterials like nanocrystals, nanoparticles, nanoclays, among others on the design of bionanocomposites obtained from thermoplastic starch and their application as food packaging.

2.1 INTRODUCTION

Nanotechnology is an important tool to develop and improve advanced materials. Nanotechnology implies to work on a scale of about 1–100 nm to the characterization, preparation, and influence of material structures, devices, and systems with new properties according to their nanostructures (Azeredo, 2009; Ashfaq et al., 2022; Sharma et al., 2020). As a result, nanocomposited materials have a large surface-to-volume ratio and surface activity, which allows adding to compatible polymers matrix with different loadings, and this linkage will improve the mechanical, thermal stability, and electrical conductivity. For that, nanomaterials are favorable to use in food packaging materials to produce smart and active packaging (Youssef and El-Sayed, 2018).

The environmental problems derived from petroleum-based polymers, biopolymers (naturally occurring polymers obtained from renewable resources) have emerged as renewable packaging materials such as a diverse set of lipids (wax and free fatty acids), proteins (casein, gluten, and whey), polysaccharides (alginates, chitosan, cellulose, and starch derivates), polyphenols, and polymers produced by bacteria, fungi, plants, and animals (Kaplan, 1998). Within the biopolymers produced from polysaccharides, the use of starch and its derivates has been especially interesting.

Starch is a biopolymer used to obtain biocomposite materials with several advantages: is renewable source, completely biodegradable, nontoxicity, easily available, and low cost (Mendes et al., 2016; Fahmy et al., 2020; Ahari et al., 2021; Gamage et al., 2022). Thermoplastic starch (TPS) is obtained after disruption and plasticization of native starch using plasticizers such as glycerol, water, and sorbitol. TPS present poor mechanical properties and a high-water sensibility; however, these disadvantages can be overcome using reinforcing agents in the starch matrix, such as cellulosic fibers, whiskers, and nanofibers (Babaee et al., 2015; Nasseri and Mohammadi, 2014). In this chapter, recent advances in research and development of bionanocomposites obtained from thermoplastic starch with nanomaterials are analyzed and summarized, as well as their application in food packaging.

2.2 PREPARATION AND CHARACTERIZATION OF NANOCRYSTALS AND NANOPARTICLES

2.2.1 *SYNTHESIS OF NANOCRYSTALS*

Usually, the term "nanocrystals" involved particles in a crystalline state, in most of the reported cases, in the strict sense, an amorphous nanoparticle should not be referred to as nanocrystal (Malamatari et al., 2018). The changing physical is the most important properties of materials in a nano-size regime, being this a major challenge in the nanomaterials science directly related with processing method to modify the most important characteristics of nanocrystal systems such as size and shape (Byrappa et al., 2008; Mohanraj and Chen, 2006). Nanoparticles derived from natural resources, specifical polysaccharides, such as starch, cellulose, and chitin, are sources of nanocrystals for food and nonfood applications. These polysaccharides are obtained at a low cost from biomass available in large quantities, renewable, biodegradable, some properties like their thermal behavior, processability, and ability to assemble into ordered structures provide the potential for compatibilizing nanocrystals and directing their assembly, being these advantages of natural polysaccharide nanocrystals over inorganic nanoparticles (Skaff and Emrick, 2004; Labet et al., 2007; Lin et al., 2012; Boufi et al., 2018). In the case of nanocrystals produced from starches, the remnants nanocrystals differ in size and shape in function of the amylose/amylopectin ratio, the chain-length distribution of amylopectin, and the crystalline arrangement (X-ray diffraction pattern) (de la Concha et al., 2018). The polymer nanocrystals can be ranging in crystal size from about tens to hundred nanometers, to obtain crystals uniform is crucial the standardization of the nanocrystallization processes to further control the size, distribution and the shape (Oikawa et al., 2004). Besides, the synthesis, development, and material characterization of novel material to conform nanocrystals have attracted the research interest of diverse engineering areas as physic and chemistry (Tjong, 2013). The nanocrystals from polysaccharides are obtained mainly through chemical modifications, for example, the acid hydrolysis by hydrochloric or sulfuric acid as reagents. First, the biomass is generally submitted to a bleaching treatment with NaOH to purify the cellulose, starch, or chitin by removing other constituents as the amorphous region, the bleached material is then disintegrated in water, and the resulting suspension is submitted to a hydrolysis treatment which produces nanocrystals which have high crystallinity and nanoscale platelet morphology. By another side, the conditions of hydrolysis during the extraction process (such as the type of acid, acid concentration, temperature, and time) affect the size and yield of nanocrystals, including chemical modification, triggered changes in solubility or polarity, as well as surface and thermal properties. Despite this, acid hydrolysis is mostly utilized in the preparation of polysaccharide nanocrystals (Lin et al., 2012; Lin et al., 2011; Dufresne, 2008). On the other hand, during acid hydrolysis, abundant hydroxyl groups are generated on the surface of nanocrystals which contribute to their positive surface chemical properties and are therefore an essential route to altering the surface structure, regulating surface properties, and developing functional materials (Huang et al., 2015). In this sense, the hierarchical organization of polysaccharides such as cellulose, starch, and chitin in natural materials along with their semi-crystalline structure has been a key factor in the preparation of nanocrystals through controlled acid hydrolysis (Haaj et al., 2016). Finally, nanocrystals originated from biopolymers have drawn considerable attention because they may be utilized as novel and biofunctional materials in diverse industries (Kim et al., 2013).

2.2.2 SYNTHESIS OF NANOPARTICLES

The nanotechnology has gained high interest for its versatile applications in various fields, because many polysaccharide nanoparticles have recently been intensively studied (Huh et al., 2017; Korbekandi and Iravani, 2013). Polysaccharides are natural biopolymers which are widely used for applications in food, medical, agricultural, and environmental industries, which have been recognized because of their outstanding biocompatible, biodegradable, and renewable properties; on the other hand, these present partly crystalline structures conferring some attributes to conform nanoparticles (Le Corre et al., 2010; Wang et al., 2017). Likewise, the use of natural polymers is promising for solving resource shortages and minimizing the environmental problems compared with conventional plastics (Jiang et al., 2016). Diverse polymeric materials for preparing nanoparticles have been used such as chitosan, starch, alginate, and pectin, among others, through mechanical, physical, and chemical methods. Among them are acid hydrolysis, ultrasonic treatment, supercritical fluid extraction, solvent evaporation, precipitation, and emulsion crosslinking. Basically, these processes involve the conversion of solid to solid or liquid to solid or gas to solid with the aim of making polymeric particles in specific size ranges. Nowadays, identification of new sources of biopolymers and new types of preparation method has become relevant to fabricate biopolymer particles with novel or improved functional attributes. For example, the new methods are combining physical treatments with chemical reaction (Byrappa et al., 2008; Boufi et al., 2018; de la Concha et al., 2018; El-Sheikh, 2017; Liu et al., 2008; Joye and McClements, 2014). Acid hydrolysis has been widely used for the preparation of nanoparticles from polysaccharides, mainly due to simplicity and easy to control, typically using hydrochloric or sulfuric acid. However, this technique is time-consuming with nanoparticles yield relatively low (Kim et al., 2015). Ultrasonic treatment has been employed to physical polymers modification of microparticles into nanoparticles in a range from 30 to 100 nm by vibration fracturing; this method is effective and environmentally friendly avoiding any chemical additive. Furthermore, this treatment can be used in combination with acid hydrolysis to prevent the formation of aggregates during the hydrolysis; on the other hand, the use of high-intensity ultrasound for the generation and application of nano-sized materials has become increasingly widespread and widely adopted for particle size reduction in emulsification, dispersion, and deagglomeration as well as milling (Boufi et al., 2018; Chang et al., 2017; Haaj et al., 2013). Some methods such as solvent extraction-evaporation, solvent diffusion, and organic phase separation require the use of organic solvents which are hazardous to the environment and physiological systems. Another important tool of materials processing in the last two decades has been the supercritical fluid technology, usually defined as a solvent at a temperature above its critical temperature, where the fluid remains a single phase regardless of pressure (Byrappa et al., 2008; Mohanraj and Chen, 2006). The precipitation method is simplicity and high reproducibility, involves a successive addition of a dilute polysaccharide solution into a nonsolvent, or inversely, however, there are some disadvantages such as the generation of wastes highly diluted polysaccharide solutions and larger volume of non-solvents affecting the efficiency negatively and increasing the cost (Chang et al., 2017; Chin et al., 2011).

2.2.3 CHARACTERIZATION OF NANOCRYSTALS AND NANOPARTICLES

Recently, the design of particles with specific properties has recently been driven by the application of nanotechnology principles to foods; the dimensions of a biopolymer particle alter its functional performance in foods, for example, particle size influences the bulk physicochemical properties of foods (optics, rheology, and stability), and encapsulation characteristics (e.g., loading, retention, and release); however, some characteristics remain unknown or not well understood (Joye and McClements, 2014). Many techniques have been explored to elucidate the structure and spectroscopic characters of nanocrystals and nanoparticles. However, as the basic parameter, size and morphology are first considered in many studies and are measured by either light scattering or microscopy methods (Malamatari et al., 2018; Joye and McClements, 2014). The characterization of shape and size distribution of polysaccharide nanocrystals and nanoparticles can be investigated using static light scattering (SLS) that relies on the measurement of the intensity of scattered light waves as a function of scattering angle, and then, utilization of an appropriate mathematical model to convert the scattering pattern into a particle size distribution. The models used assume that the scattering particles are spherical, homogeneous, and noninteracting with a well-defined refractive index. However, in practice, many types of biopolymer particles do not conform to these assumptions. Besides, sample preparation often requires dilution and stirring, which may alter the integrity or aggregation state of biopolymer particles. The microscopy methods have been convenient to characterize nanocrystal and nanoparticles due to the variety of morphology and geometrical dimensions as length, diameter, thickness, and aspect ratio of polysaccharide due to the different sources and extracting methods. Among them are transmission electron microscopy (TEM), atomic force microscopy (AFM), and field-emission scanning electron microscopy (FE-SEM). Besides, the use of scanning electron microscopy (SEM) provides a more extensive morphological inspection of polysaccharide nanocrystal reinforced which leads to conclusions about the homogeneity of the composite, the presence of voids, the dispersion level of the nanoparticles within the continuous matrix, the presence of aggregates, sedimentation, and the possible orientation of particles (Lin et al., 2012; Dufresne, 2008; Joye and McClements, 2014).

TABLE 2.1 Obtention and Application of Nanocrystals and Nanoparticles of Polysaccharides.

Source of extraction of nanocrystals and nanoparticles	Application	Author
Waxy corn starch Cotton linter pulp	As reinforces in the waterborne polyurethanes matrix.	Wang et al. (2010)
Sago starch	The synthesis method of precipitation is fast and easy to perform, the mean sizes of starch nanoparticles could be modulated by the synthesis parameters such as the use of surfactants during the synthesis process.	Chin et al. (2011)
Cellulose Pea starch	Drug release.	Lin et al. (2011)
Standard corn starch Waxy corn starch	The use of a physical method as ultrasonication in nanoparticles formation leads to a high yield in nanoparticles and does not require any chemical treatment.	Haaj et al. (2013)
Standard corn starch Waxy corn starch	The combination of acid hydrolysis at low temperature and ultrasonic treatment is an effective process for the mass production of crystalline starch nanoparticles	Kim et al. (2013)
Cellulose	Ingredient to gels formation	Hu et al. (2014)
Standard Corn starch Taro starch	As improvers in the mechanical properties and thermal stability of films.	Dai et al. (2015)
Potato starch	As a filler in composite films. Water vapor transition rate was reduced.	Gujral et al., (2021)
Amadumbe starch	As improver in thermal stability, tensile strength and decreased water vapor permeability of films.	Mukurumbira et al., (2017)
Waxy corn starch Amaranth starch	Acid hydrolysis was used for nanocrystal production, the arrangement of starch components in the crystalline lamella is related to starch granule size, producing nanocrystals with different structures.	de la Concha et al. (2018)

2.3 TECHNIQUES FOR THE DEVELOPMENT OF NANOCOMPOSITES

2.3.1 CASTING

Several techniques have been developed to nanocomposites elaboration of natural polymers such as starch and cellulose. Starch processing by casting and extrusion are the most used at the laboratory and industrial scales.

The casting is the main technique at the laboratory level of production of starch-based films in an aqueous medium due to simplicity, ensuring the proper mixing of the nanofillers with a starch matrix. Basically, it consists in pouring on small plates or molds an aqueous suspension of gelatinized starch by heating, plasticizer, and other additives, followed by a drying process to eliminate the water to obtain a solid nanocomposite film. The resulting average size of films is associated with both masses of suspension poured it and mold or plate (García et al., 2015).

Several studies have reported particular parameters both gelatinization and drying varying temperatures and times, which influenced and showed detrimental effects on film formation. Thus, the mixing absorption and time are significant to ensure homogeneous dispersion of the fillers in the starch matrix (Madhumitha et al., 2018). For example, biodegradable starch/clay nanocomposite films for food packaging applications were made by casting since film blowing presented various issues such as low melt toughness and films became brittle and sticky straight after production due to the high-water content (Avella et al., 2005). In this sense, thermoplastic starch-waxy maize starch nanocrystals nanocomposites were processed by casting, mixing native waxy maize starch granules, glycerol, and distilled water. Then, the mixture was heated at 150°C for 10 min in a reactor, under pressure and mechanical stirring. The suspension of waxy maize starch nanocrystals was added at 40°C. Finally, the mixture was evaporated in a nonventilated oven at 40°C for 24 h (Angellier et al., 2006).

Besides, starch-based nanocomposites reinforced with flax cellulose nanocrystals (FCNs) were produced by casting solution process which consisted in gelatinization at 100°C for 30 min, after cooling down to about 70°C; the resulting paste was degassed under vacuum to remove the remaining air and cast in a polystyrene square Petri dish, followed by drying in a vented oven at 50°C. The results showed that the FCNs played an important role in improving the mechanical properties and water resistance of the starch-based materials (Cao et al., 2008). Also, the casting process was employed to sago starch biocomposite filled with nanocrystalline cellulose (NCC) from rattan biomass, where starch, glycerol, and acetic acid were mixed and heated at 70°C for 30 min to gelatinize. The water was evaporated at room temperature overnight after that solution poured into an acrylic mold. The film was formed uniformly and removed smoothly (Nasution et al., 2018).

Bio-nanocomposite films based on cassava starch or chitosan with a constant thickness, reinforced with montmorillonite or bamboo nanofibers, were prepared by the casting technique. The solution was drying at 32 ± 2°C for approximately 24 h, in an oven with forced air circulation, after all films were stored at a controlled relative humidity of 75% for not less than 48 h at 25°C. The nanofibers increased the tensile strength in 50% of starch films. Also, cassava starch films showed a better response to nanostructure process through the mechanical properties as well as good interaction between the polymer matrix and bamboo nanofibers (Llanos and Tadini, 2018).

Although there is the extensive utilization of this technique, the main limiting factor is the scalability from laboratory to industrial levels due to control on the sample film thickness and uniformity. The polymer needs to be soluble in the same solvent used for dispersion and swelling of the filler, also it is slow and expensive. For large-scale manufacture, an extrusion process is usually used.

2.3.2 EXTRUSION

Extrusion compounding is the most interesting method for industrial processing due to easy scale-up and the possibility of further molding of the materials; it is also an energy-efficient system achieving breakdown of the starch granule structure through a combination of high shear, temperature, and pressure to successfully melt starch. Typical single- or doublé-screw extruders in general form consist of a hopper, barrel, feed screw, thermocouples, and dies (García et al., 2015).

The extrusion can be classified into two types: reactive extrusion and extrusion-cooking. Reactive extrusion is a concurrent reaction in extrusion processing of polymers (Kim et al., 2015). This method is an efficient, versatile, cost-effective, and environmentally-friendly method to produce new materials with enhanced performance properties such as the interface interactions between used components (Formela et al., 2018). For example, a novel extrusion method for the preparation of TPS and TPS/CNF considered as green nanocomposites was reported by Ghanbari et al. (2018), consisting of a two-step process. First, the main components of the matrix were manually premixed and extruded by a co-rotating twin-screw extruder, with a screw speed of 80 rpm. The barrel temperatures of the extruder were controlled at different temperatures while the temperature of the die was set at 125°C. The extruded strand was passed through a water bath, granulated, and dried at 105°C for 24 h. In the second step, the resulting granules were subsequently compression-molded under following conditions: press temperature 190°C, pressure during heating 3.5 MPa, heating time 10 min, and cooling time 5 min.

The results strongly showed the advantages of reinforcing TPS with CNFs, a natural, cheap, and abundant green nanocomposite using extrusion process.

Moreover, biodegradable polylactide and thermoplastic starch blends were prepared with drug release purposes in which the granulates form was blown using a single-screw extruder with screw diameter D = 20 mm (L:D = 30:1) connected to blown film unit with a ring-shaped die to prepare biodegradable films. The blown film extrusion process was carried out at a temperature range from 155°C to 175°C and a screw speed of 44 rpm. Films in the range 50–90 mm of thickness, which could be used for drug delivery systems (Karolina et al., 2018) were obtained. On another side, the extrusion-cooking commonly is related to food industry and used in direct extrusion snacks, cereal flakes, and diverse breakfast foods produced from cereal, among others (García et al., 2015). However, it has been employed in foamed starch-based materials under various conditions: temperature, humidity, type of the die, screw rotational speed, various raw materials, and additives blends. For example, the mixtures based on potato starch and with the addition the foaming agent Plastron foam PDE and poly(vinyl) alcohol PVA presented the best properties and processing costs under this method (Combrzyński et al., 2018). Thus, with these methods we can improve nanocomposites elaboration of natural polymers and its properties.

2.4 INFLUENCE OF INCORPORATION OF NANOMATERIALS IN THE PROPERTIES OF STARCH NANOCOMPOSITES

The thermoplastic starch packaging designed at the laboratory presents poor properties compared with polymers derived from petroleum. For this reason, most of the time some additives must be added to the formulation to enhancing their functional properties (Nasseri and Mohammadi, 2014). The additives mots used in the formulation of bioplastics are lipids, surfactants, antioxidants, antimicrobials, and recently, nanomaterials. There are many materials with nanomaterials incorporated (nanocomposites) food packages commercialized in the world, while others are in development. The addition of nanomaterials to polymers can improve the mechanical, barrier, and thermal properties, among others (Babaee et al., 2015; Ahmed et al., 2012). However, according to Anglès and Dufresne (2000) it is not easy to model the properties of starch nanocomposites because they are formed by different components: starch, nanomaterial, plasticizer, and water. These four components can be present in crystalline, amorphous, or liquid phase, which promotes different competitive interactions between them. These factors should be considered in the study of rheological, mechanical, barrier, and thermal properties

2.4.1 RHEOLOGICAL AND PASTING PROPERTIES

Rheological properties offer information to understand the relation between the nanomaterials and thermoplastic starch (TPS) matrix in solution. Moreover, for the potential scaling of films production, these properties play a key factor in the fluid transport and therefore in the pumps and impellers design. On the other hand, the pasting properties of starch dispersions are commonly evaluated by rapid visco analyzer (RVA); this technique involves subjecting the starch dispersion to heating–cooling controlled cycles under constant agitation.

Most of the studies of bionanocomposites have been conducted to final product properties. Therefore, there are limited studies that measured the evolving dynamic rheological properties of nanocomposite forming solutions (NFS) during and after gelatinization. Chiou et al. (2005) studied the dynamic rheological properties of high solids starch–nanoclay formulations containing different types of starches (wheat, potato, corn, and waxy corn) and hydrophilic (Cloisite Na^+) and hydrophobic (Cloisite 30B, 10A, and 15A) nanoclays. The nanocomposite with Cloisite Na^+ presented the more gel-like behavior at higher clay concentrations than the other hydrophobic samples. The authors suggested that during the gelatinization process, the lixiviated amylose established strong interactions with the Cloisite Na^+. Montmorillonite (MMT) is the most used nanoclay for the design of films with food applications. In a recent study, Romero-Bastida et al. (2018) evaluated the effect of MMT on the rheological properties of nanocomposite-forming solutions (NFS) based on corn starch with different amylose contents (30 and 70%). The flow curves evidenced a non-Newtonian behavior with flow index lower than one ($n < 1$, pseudoplastic fluid) with a shear thinning behavior, characterized by a reduction of apparent viscosity when increasing the shear rate. When the amylose content increases, the flow index (n) decreased as a consequence of interactions between starch components. Regarding dynamic properties, the storage module (G') was higher than loss modulus (G'') suggesting a gel-like behavior. The addition of MMT increased the storage modulus, which indicated reinforcing of starch matrix.

On the other hand, there are reports about how the inclusion of nanomaterials in NFS affect their pasting properties. In this sense, Chang et al. (2010)

reported that the pasting peak viscosity of nanocomposite-forming solutions (NFS) decreased with the addition of chitin-NP. The pasting viscosity increased as the chitin-NP content was higher. The authors suggest that it was the result of higher interactions between chitin-NP and starch. Similar results were observed in plasticized starch bionanocomposites reinforced with chitosan nanoparticles (Chang et al., 2010).

2.4.2 MECHANICAL PROPERTIES

The nanocomposites consist of two fractions, one in mayor percent designed as "matrix," and the other minor percent known as "charge." The presence of nanomaterials promotes nanocomposites with a great interfacial area, which affects their mechanical properties. The main mechanical properties of materials are tensile strength (TE, the maximum force required to rupture the material), Young's modulus (MY, proportionality coefficient between applied stress and strain), and elongation at the break (EB, is the ratio between changed length and initial length after breakage).

There are many reports about the influence of incorporation of nanomaterials of different sources on the mechanical properties of thermoplastic starch films. For instance, the addition of cellulose–starch nanocomposites promoted increments up to 1052 (MPa) and 2909 (GPa) in TS and YM, respectively. In contrast, the EB decreased around 79% (Grande et al., 2009). Lu et al. (2006) observed increments in YM and TS from 56 to 480 MPa and from 2.8 to 6.9 MPa when the content of ramie crystallites increased from 0 to 20% wt in the wheat TPS matrix. This behavior was associated with the great number of three-dimensional networks of intermolecular hydrogen bonding interactions between the same cellulose crystals and cellulose crystals with the TPS matrix. Similar results were reported by Ma et al. (2009) in the TPS pea matrix reinforced with ZnO nanoparticles, as the nano-ZnO content varied from 0 to 4 wt%, the TS and YM increased from 3.94 to 10.80 MPa and from 49.80 to 137.00 MPa, respectively.

Other authors reported starch nanoparticles and nanocrystals for obtaining nanocomposites, as the same form that the above studies mentioned; the inclusion of these nanomaterials promotes increments in the TS and YM (Angellier et al., 2006; Viguié et al., 2007). According to the authors, the chemical similarity between starch nanocrystals and TPS matrix allows strong interactions between starch nanocrystals/matrix and nanocrystals/nanocrystals. In studies with chitin nanoparticles (Chitin-NP) used for obtaining starch nanocomposites, there was an increment in the TS from 2.84 to 7.79 MPa when the concentration of Chitin-NP varied from 0 to 5%. However, the EB decreased from 59.3% to 19.3% (Chang et al., 2010). As the same form that when starch nanoparticles were used as filler, the Chitin-NP establishes strong interfacial interactions with the TPS matrix as a consequence of the similar structure of chitin and starch. The decrement in the EB is a consequence of the decreased flexibility of the starch molecules.

Another material used as filler in the TPS matrix is the nanoclays (Layered silicates), which are economical, environmentally friendly, and naturally abundant (Li et al., 2006). This material can form competitive interactions in the starch matrix. The plasticizer can be absorbed on the clay surface, affecting the mechanical properties of the nanocomposite (Chivrac et al., 2008). Chung et al. (2010) observed that the addition of montmorillonite or laponite to the TPS matrix enhances the YM and TS until 65 and 30%, respectively, compared with the unfilled starch material. This was attributed to the good dispersion of clay in the starch matrix.

2.4.3 BARRIER PROPERTIES

The barrier properties of starch films depend on their microstructure and chemical nature of components. Compared with synthetic polymers, thermoplastic starch matrix presents high water vapor permeability (WVP) values. This disadvantage is the result of the great number of hydroxyl groups present in the matrix components (starch and glycerol), which interact with the migrating water molecules. For food applications, the moisture exchange with the environment must be minimal.

García et al. (2009) observed that the addition of waxy starch nanocrystals (2.5%wt) in nanocomposites of cassava starch reduced the WVP by 40%. This behavior was associated with the fact that the nanocrystals were distributed into the TPS matrix, generating a tortuous path for the water vapor diffusion. Then, the same research group (García et al., 2011) prepared bionanocomposites made from thermoplastic waxy maize starch reinforced with waxy maize starch nanocrystals at the same quantity previously evaluated (2.5%wt). The results were opposite to that observed using a cassava starch matrix, the addition of 2.5%wt waxy maize starch nanocrystals increased the WVP by 79% (from 3.8 to 6.8 × 10–10 g s^{-1} m^{-1} Pa^{-1}). This phenomenon was a consequence of the formation of preferential paths for water diffusion with high concentrations of OH where the nanocrystals are glue–glycerol bonded, which increased the WVP. The addition and concentration of montmorillonite in corn starch films plasticized with glycerol reduced the WVP (Slavutsky et al., 2012; Wang et al., 2009). The author mentioned that it was the result of the strong interactions of montmorillonite with the starch and the formation of a tortuous path. The addition of chitosan nanoparticles (chitosan-NP) from 0 to 6%wt into TPS films from potato starch reduced the WVP from 5.8 to 3.5 × 10^{-10} g m^{-1} s^{-1} Pa^{-1}. However, when the chitosan-NP concentration increased up to 8%wt, the WVP slightly increased (Chang et al., 2010). The addition of chitosan-NP at low levels allows good dispersion of nanoparticles in the TPS matrix, which promotes a tortuous path for the water molecules. These observations were similar to those reported by the same group research (Chang et al., 2010) in TPS nanocomposites with chitin nanocrystals (chitin-NC) from 0 to 5%wt. The WVP decreased with the chitin-NC increment from 1 to 4%wt, but when the concentration of nanocrystals increased until 5%wt, the WVP increased. Low levels of chitin-NC dispersed well in the TPS matrix; however, at high concentrations, the chitin-NC form aggregates that facilitie water vapor permeation. On the other hand, the addition of cellulose nanofibers isolated from kenaf bast fibers (*Hibiscus cannabinus*) improved the WVP by the production of denser and less porous materials. Moreover, compared with the neat TPS, the interactions between cellulose nanofibers and TPS matrix restrict the moisture diffusion rate (Babaee et al., 2015).

The permeability to gases such as oxygen plays a key role in food packaging since this gas causes deterioration of food by oxidation. With the incorporation of the nanofiller, starch-based materials generally are improved. For instance, the addition of starch nanocrystals (2.5 and 5 wt%) reduced the oxygen permeance in the TPS matrix (González et al., 2015). This behavior has also been reported in bionanocomposites reinforced with natural and an organically modified montmorillonite (Zeppa et al., 2009). In another hand, compared with the neat matrix, the incorporation of chitin-NC (up to 20 wt%) decreased by about 30% the oxygen permeability (Salaberria et al., 2015). According to the authors, the increase in the intermolecular interactions in the matrix due to the addition of nanofillers leads to a high oxygen barrier.

2.4.4 THERMAL PROPERTIES

The study of thermal properties allows understanding the processing temperature range and use of materials. The glass transition temperature (T_g)

is the most important thermal property of polymers. Materials under their T_g behave as brittle. Thermal glass transition can be measured by dynamic mechanical thermal analysis (DMTA) or differential scanning calorimetry (DSC). Thermoplastic starch plasticized by polyols like glycerol results in a complex and heterogeneous system with a rich phase of glycerol dispersed in another phase rich in starch (continuous phase), where each of them has its own T_g value (Lu et al., 2006; Dufresne and Castaño, 2017). These values are affected by the type and concentration of nanomaterial used as filler. Cao et al. (2008) studied nanocomposites from pea starch reinforced with cellulose nanocrystals of flax fiber (5, 10, 15, 20, 25, and 30 wt%). The authors reported that increments in the loading of cellulose nanocrystal promoted variations from 43.3 to 48.8 in T_{g2} (starch-rich phase), whereas the T_{g1} (glycerol-rich phase) remained without variations. This behavior was similar to that previously reported by Lu et al. (2006); they observed increments in T_{g2} from 26.8 to 55.7°C as a consequence of the incorporation of ramie nanocrystallites (0–40%wt) into the thermoplastic wheat starch matrix. In general, the addition of cellulose nanocrystals increases the glass transition temperature. This phenomenon is a consequence of the strong molecular interactions by hydrogen bonding between starch and nanocrystals, which results in low mobility of starch chains (Chang et al., 2010; Lu et al., 2006; Dufresne and Castaño, 2017; Cao et al., 2008).

Nanoclays are nanomaterials widely used for food packaging design; among them, the most used is the montmorillonite (Chung et al., 2010; Sinha Ray and Okamoto, 2003). González et al. (2015) prepared thermoplastic starch bionanocomposites reinforced with starch nanocrystals. The authors reported that the thermal stability of bionanocomposites decreased with starch nanocrystals added. It was explained by the high affinity between glycerol and waxy starch nanocrystals. Moreover, after acid hydrolysis for nanocrystal production, some acidic sulfate ester groups should be remained attached to waxy starch nanocrystals surface, which seems to act as catalyzers of the thermal decomposition, decreasing the degradation temperature for nanocomposites with the filled matrix.

Another thermal property of materials is their thermal stability, which is evaluated by thermogravimetric analysis (TGA) (Chang et al., 2010). Chang et al. (2010) added nanoparticles of chitosan (chitosan-NP) to the TPS matrix at concentrations from 0 to 4 wt%. They reported that the addition of chitosan-NP reduced the mass loss at the onset temperature. The T_{max} increased with chitosan-NP increments, obtaining 315, 321, and 328°C, for 0, 2, and 4 wt.% CN, respectively. The authors attributed the higher thermal stability in the TPS matrix with chitosan-NP to the strong interfacial interactions established between them. González et al. (2015) prepared thermoplastic starch bionanocomposites reinforced with starch nanocrystals. The thermal stability of bionanocomposites decreased with starch nanocrystals added. This was explained by the high affinity between glycerol and waxy starch nanocrystals. Moreover, after acid hydrolysis for nanocrystal production, some acidic sulfate ester groups should be remained attached to waxy starch nanocrystals surface, which seems to act as catalyzers of the thermal decomposition, decreasing the degradation temperature for nanocomposites with filled matrix (García et al., 2011; González et al., 2015).

2.5 APPLICATION OF STARCH NANOCOMPOSITES AS FOOD PACKAGING

In recent decades, the demand for "ready-to-eat," "ready-to-cook," or "ready-to-use" products has grown considerably without ever stopping the aspect of food security implicit in its production, commercialization, and consumption (He and Hwang, 2016; Carbone et al., 2016). In the specific case of food packaging and preservation, applications have been envisioned ranging from a simple cover, as a mechanical barrier to moisture or damping to prevent or reduce physical damage, to active and intelligent packaging, which interact with the food to extend its shelf life. For example, combining materials with antioxidant or antimicrobial compounds. On another hand, it is used to communicate information to the consumer through sensory messages, such as color change, which may indicate changes in the integrity of the matrix food or chemical composition, whether favorable or unfavorable for the consumption of the product. Thus, nanotechnology promises to revolutionize the food packaging (Chellaram et al., 2014).

2.5.1 COATINGS

The formulation and use of starch-based covers combined with other nanometer-scale materials that could be used to preserve the sensory and nutritional quality of foods, especially those marketed fresh, or minimally processed. Falguera et al. (2011) reported that these covers have to meet certain characteristics that make them practical for their use, such as a good moisture barrier with certain permeability to gases and above all to microorganisms that corrupt foods (Valdés et al., 2015). They should also comply with the damping function to reduce physical damage due to friction between the product units and the packaging, and even during their handling and commercialization in the logistics chain (Han, 2014). Also, additional characteristics are required such as be edible or easily removed for subsequent safe consumption and not causing or with reducing the negative impact on the environment (Cazón et al., 2017; De Ancos et al., 2015).

The covers of starch composites have been used for the conservation of fresh or minimally processed fruits and vegetables (Valdés et al., 2015; Quintero et al., 2010); they can act as an alternative to modified atmospheres, regulating the respiration of vegetables depending on the properties of the cover (Park et al., 2014). Also, the adhesive properties have been explored in some food packaging, such as that used in hamburger wrappers and other products (Mbey et al., 2012; Hamad et al., 2018). The success of this type of materials is markedly conditioned by the handling of the products; an inadequate handling can seriously compromise the covers protection to the food by damaging or removing them prematurely from the surface of the product, even, due to the nature of the materials, for example the starch does not favor the functionality of the coatings in fresh foods, resulting in an opposite effect on the expected (Quintero et al., 2010).

2.5.2 RIGID PACKAGING

The current consumer is very familiar with rigid packaging, and packaging, have been extensively used, for example, the transparent plastic boxes, to preserved and reduce physical damage due to rough handling. However, most containers and packaging of this type are commonly made from petroleum-based plastics and are generally used once, promoting environmental problems (Cavero and Arroyo, 2017). To face these issues, thermoplastic starch (TPS) nanocomposites have been developed from TPS in combination with nanometric structures, such as silica nanoparticles (Arayaphan et al., 2020; de Azêvedo et al., 2021), nanoclays (Aouada et al., 2013; Müller et al., 2012; de Jara et al., 2020), organic nanofibers (Babaee et al., 2015), among others, in order to achieve rigid packaging (Salaberria et al., 2014). The result is promising, according to important properties such as the optical, mechanical, and biodegradability (Mohammadi Nafchi et al., 2013; Pelissari et al., 2017). An interesting food industry area is the packaging of fruits and vegetables for fresh consumption, as has been reported by Campos-Requena et al. (2017). However, it cannot be ruled out all other uses of packaging of rigid material made with conventional plastics especially about dry products (for example, cookies, nuts, dried meat, cheeses, etc.) (Ghoshal, 2018) and ready-to-eat products (De Ancos et al., 2015).

Another type of rigid materials based on starch nanocomposites are plastic foams (Ahmadzadeh et al., 2017). These materials can be applied in the design of individual containers and buffers for foods such as fresh whole fruits, to prevent and reduce the mechanical damages caused by the handling of the same from the harvest to the consumption, reducing some negative effects in their maturation process to allow proper transpiration.

2.5.3 FLEXIBLE PACKAGING

At present day, there are no containers that are used more than those that are flexible, which, together with the rigid containers, make up most of the universe of packaging used in the food production industry (González and Alvarez Igarzabal, 2015; Podshivalov et al., 2017). Plastic films are the main example of this, and because of the problems posed by its wide use and origin (petroleum derivatives) (Mittal et al., 2016), they have been developing alternatives based on starch nanocomposites with other materials, such as organic nanofibers (González et al., 2015), other polysaccharides such as chitosan (Chang et al., 2010), talc nanoparticles (Castillo et al., 2019). All these materials with encouraging results in terms of their chemical and mechanical properties and with high advantages in terms of their biodegradability (Kim et al., 2015) make these types of materials an economical and promising option for the substitution of the use of flexible plastics derived from conventional nonrenewable sources (Ayana et al., 2014).

The applications of films starch-based nanocomposites go from individual packages (Nathalie et al., 2017) to packaging with greater capacity for the transport of food, such as traditional plastic bags (López et al., 2015). They could be used in the vacuum packaging of fresh meat and dairy products, such as cheeses, once the problem of hydrophilicity and low mechanical resistance has been solved, characteristics of starch films (Mendes et al., 2016). Due to their gases and water vapor barrier properties achieved due to the combination of materials, they can also be used in packaging of minimally processed vegetables (De Ancos et al., 2015).

2.5.4 ACTIVE AND INTELLIGENT PACKAGING

Active and intelligent packagings have been developed and evaluated in recent decades as one of the complete options aimed at extending the shelf life of products and all their intrinsic properties (flavor, color, nutrients, texture, etc.) (Barzegar et al., 2014; Goudarzi et al., 2017), and at the same time interact with the consumer, providing information in real time about the state of conservation of the packaged food product (Mihindukulasuriya and Lim, 2014).

For the formulation of materials for active starch-based packaging, including nanocomposites, antimicrobial agents, antioxidants, or double action have been included, mainly compounds of plant origin, such as essential oils of different species (Atarés and Chiralt, 2016; Ribeiro-Santos et al., 2017). In this sense, there are reposts with rosemary (Estevez-Areco et al., 2019), oregano (Aguilar-Sánchez et al., 2019), peptides, such as nisin (Meira et al., 2017), as well as inorganic compounds, such as ZnO nanoparticles (Arifin et al., 2022), TiO_2 (Rong et al., 2021), silver (Srikhao et al., 2021), and gold (Rovina et al., 2021), although traditional conservatives, such as potassium sorbate, have not been neglected (Barzegar et al., 2014).

Its applications are aimed especially at packaging food susceptible to contamination with pathogenic microorganisms, such as fresh or minimally processed meat and fish (Wu et al., 2010), soft cheeses (Meira et al., 2016), foods of lower aqueous activity, susceptible to attack by molds, such as fresh bread (Barzegar et al., 2014). The applications have also been made in the trapping of harmful compounds in foods, such as cholesterol, for which tests have been carried out on the packaging of milk (González and Alvarez Igarzabal, 2015). Another application that has been explored is that of starch films with antioxidants used in butter preservation (Hernández et al., 2017).

Some intelligent packaging has been designed to reduce certain compounds, such as blue methylene or methyl orange (Mallakpour and Rashidimoghadam, 2017) and some natural pigments, such as anthocyanins (Choi et al., 2017), which usually indicate an alteration in the pH of the food, due to exposure to an inadequate temperature or microbial development (Goudarzi et al., 2017), indicating the freshness of the packaged food (Imran et al., 2010), thus warning the consumer about the convenience or not of acquiring or ingesting it. The applications of this type of material could be in a very broad sense; the materials described above can be functionalized to be used as active and/or intelligent packaging (Ranjan et al., 2014; Usman et al., 2016). They have been successfully tested, for example, in the packaging of pork (Choi et al., 2017), fresh pasteurized milk (Liu et al., 2017), and fresh fish (Zhai et al., 2017).

The advantages that this would imply, from the prevention of food waste, the increase of its shelf-life, the prevention of foodborne diseases, and the reduction of environmental contamination (Othman, 2014) by having materials of accelerated biodegradability and that it is possible to produce at low cost by being based on materials of an organic nature (Carbone et al., 2016). However, even with all the benefits reported in the various studies, there are others that warn about the inappropriate use of this type of material in food packaging, since the active ingredients used could also affect the health of the consumers if they are used in inappropriate concentrations, especially those that include heavy metals in their composition (Ranjan et al., 2014; Dimitrijevic et al., 2015), although the base materials are not toxic for humans or animals (Woehl et al., 2010), so it is necessary to study exhaustively the impact of these materials on the health of consumers and their regulation before adopting them in a massive way.

2.6 CONCLUSIONS

Throughout this chapter, it has become evident that the nanotechnology has gained high interest for its versatile applications for preparing starch food packaging. Also, it has been indicated that the starch can be considered a suitable natural polymer for preparing biodegradable matrices of thermoplastic starch reinforced with nanoparticles (nanocomposites). Several techniques have been developed for nanocomposites elaboration of natural polymers, such as casting and extrusion are the most used at the laboratory and industrial scales with the purpose of improving the production of natural polymer nanocomposites. There are few studies regarding the rheological characterization of nanocomposite forming solutions because it is not easy to model the properties of nanocomposites, as it consists of different components that come in different forms. However, there are starch-based nanocomposite films combined with various materials that do not present problems of hydrophobicity and little mechanical resistance and can be used in the food industry.

KEYWORDS

- **thermal properties**
- **starch nanoparticles**
- **mechanical properties**
- **cellulose nanocrystals**

REFERENCES

Aguilar-Sánchez, R.; Munguía-Pérez, R.; Reyes-Jurado, F.; Navarro-Cruz, A. R.; Cid-Pérez, T. S.; Hernández-Carranza, P.; Beristain-Bauza, S. D.; Ochoa-Velasco, C. E.; Avila-Sosa, R. Structural, Physical, and Antifungal Characterization of Starch Edible Films Added with Nanocomposites and Mexican Oregano (*Lippia berlandieri* Schauer) Essential Oil. *Molecules* [Online], **2019**.

Ahari, H.; Anvar, A. A.; Ataee, M.; Naeimabadi, M. Employing Nanosilver, Nanocopper, and Nanoclays in Food Packaging Production: A Systematic Review. *Coatings* [Online], **2021**.

Ahmadzadeh, S.; Nasirpour, A.; Keramat, J.; Desobry, S. Powerful Solution to Mitigate the Temperature Variation Effect: Development of Novel Superinsulating Materials. In *Food Packaging and Preservation, Handbook of Food Bioengineering*, 1st ed.; Grumezescu, A. M.; Holban, A. M., Eds.; Elsevier. International.: 2017; Vol. 9, p 137.

Ahmed, J.; Tiwari, B. K.; Imam, S. H.; Rao, M. A. Starch as Biopolymer and Nanocomposite. In *Starch-based Polymeric Materials and Nanocomposites: Chemistry, Processing, and Applications*; Ahmed, J., Tiwari, B. K., Imam, S. H., Rao, M. A., Eds.; CRC Press: Boca Raton, FL, 2012.

Angellier, H.; Molina-Boisseau, S.; Dole, P.; Dufresne, A. Thermoplastic Starch−Waxy Maize Starch Nanocrystals Nanocomposites. *Biomacromolecules* **2006,** *7* (2), 531–539.

Anglès, M. N.; Dufresne, A. Plasticized Starch/Tunicin Whiskers Nanocomposites. 1. Structural Analysis. *Macromolecules* **2000,** *33* (22), 8344–8353.

Aouada, F. A.; Mattoso, L. H.; Longo, E. A Simple Procedure for the Preparation of Laponite and Thermoplastic Starch Nanocomposites: Structural, Mechanical, and Thermal Characterizations. *J. Thermoplast. Compos. Mater.* **2013,** *26* (1), 109–124.

Arayaphan, J.; Boonsuk, P.; Chantarak, S. Enhancement of Water Barrier Properties of Cassava Starch-based Biodegradable Films Using Silica Particles. *Iran. Polym. J.* **2020,** *29* (9), 749–757.

Arifin, H. R.; Djali, M.; Nurhadi, B.; Azlin-Hasim, S.; Masruchin, N.; Vania, P. A.; Hilmi, A. Corn Starch-Based Bionanocomposite Film Reinforced With ZnO Nanoparticles and Different Types of Plasticizers. *Front. Sustain. Food Syst.* **2022,** *6*.

Ashfaq, A.; Khursheed, N.; Fatima, S.; Anjum, Z.; Younis, K. Application of Nanotechnology in Food Packaging: Pros and Cons. *J. Agric. Food Res.* **2022,** *7,* 100270.

Atarés, L.; Chiralt, A. Essential Oils as Additives in Biodegradable Films and Coatings for Active Food Packaging. *Trends Food Sci. Technol.* **2016,** *48,* 51–62.

Avella, M.; De Vlieger, J. J.; Errico, M. E.; Fischer, S.; Vacca, P.; Volpe, M. G. Biodegradable Starch/Clay Nanocomposite Films for Food Packaging Applications. *Food Chem.* **2005,** *93* (3), 467–474.

Ayana, B.; Suin, S.; Khatua, B. B. Highly Exfoliated Eco-Friendly Thermoplastic Starch (TPS)/poly (lactic acid)(PLA)/Clay Nanocomposites Using Unmodified Nanoclay. *Carbohydr. Polym.* **2014,** *110,* 430–439.

Azeredo, H. M. C. D., Nanocomposites for Food Packaging Applications. *Food Res. Int.* **2009,** *42* (9), 1240–1253.

Babaee, M.; Jonoobi, M.; Hamzeh, Y.; Ashori, A. Biodegradability and Mechanical Properties of Reinforced Starch Nanocomposites Using Cellulose Nanofibers. *Carbohydr. Polym.* **2015,** *132,* 1–8.

Barzegar, H.; Azizi, M. H.; Barzegar, M.; Hamidi-Esfahani, Z. Effect of Potassium Sorbate on Antimicrobial and Physical Properties of Starch–Clay Nanocomposite Films. *Carbohydr. Polym.* **2014,** *110,* 26–31.

Boufi, S.; Haaj, S. B.; Magnin, A.; Pignon, F.; Impéror-Clerc, M.; Mortha, G. Ultrasonic Assisted Production of Starch Nanoparticles: Structural Characterization and Mechanism of Disintegration. *Ultrason. Sonochem.* **2018,** *41,* 327–336.

Byrappa, K.; Ohara, S.; Adschiri, T. Nanoparticles Synthesis Using Supercritical Fluid Technology–Towards Biomedical Applications. *Adv. Drug Deliv. Rev.* **2008,** *60* (3), 299–327.

Campos-Requena, V. H.; Rivas, B. L.; Pérez, M. A.; Figueroa, C. R.; Figueroa, N. E.; Sanfuentes, E. A. Thermoplastic Starch/Clay Nanocomposites Loaded with Essential Oil Constituents as Packaging for Strawberries—In Vivo Antimicrobial Synergy Over Botrytis Cinerea. *Postharvest Biol. Technol.* **2017,** *129,* 29–36.

Cao, X.; Chen, Y.; Chang, P. R.; Muir, A. D.; Falk, G. Starch-based Nanocomposites Reinforced with Flax Cellulose Nanocrystals. *Express Polym. Lett.* **2008,** *2* (7), 502–510.

Cao, X.; Chen, Y.; Chang, P.; Muir, A.; Falk, G. Starch-based Nanocomposites Reinforced with Flax Cellulose Nanocrystals. *Express Polym. Lett.* **2008,** *2* (7), 502–510.

Carbone, M.; Donia, D. T.; Sabbatella, G.; Antiochia, R. Silver Nanoparticles in Polymeric Matrices for Fresh Food Packaging. *J. King Saud Univ. Sci.* **2016,** *28* (4), 273–279.

Castillo, L. A.; López, O. V.; García, M. A.; Barbosa, S. E.; Villar, M. A. Crystalline Morphology of Thermoplastic Starch/Talc Nanocomposites Induced by Thermal Processing. *Heliyon* **2019,** *5* (6), e01877.

Cavero, H. A.; Arroyo, E. Obtencion de Biopolimeros de Papa Como Una Alternativa al Desarrollo de Materiales Inocuos al Medio Ambiente. *Revista Cientifica TECNIA* **2017,** *24* (1), 13–19.

Cazón, P.; Velazquez, G.; Ramírez, J. A.; Vázquez, M. Polysaccharide-Based Films and Coatings for Food Packaging: A REVIEW. *Food Hydrocoll.* **2017,** *68,* 136–148.

Chang, P. R.; Jian, R.; Yu, J.; Ma, X. Fabrication and Characterisation of Chitosan Nanoparticles/Plasticised-Starch Composites. *Food Chem.* **2010,** *120* (3), 736–740.

Chang, P. R.; Jian, R.; Yu, J.; Ma, X. Starch-based Composites Reinforced with novel chitin Nanoparticles. *Carbohydr. Polym.* **2010,** *80* (2), 420–425.

Chang, Y.; Yan, X.; Wang, Q.; Ren, L.; Tong, J.; Zhou, J. High Efficiency and Low Cost Preparation of Size Controlled Starch Nanoparticles Through Ultrasonic Treatment and Precipitation. *Food Chem.* **2017,** *227,* 369–375.

Chellaram, C.; Murugaboopathi, G.; John, A. A.; Sivakumar, R.; Ganesan, S.; Krithika, S.; Priya, G. Significance of Nanotechnology in Food Industry. *APCBEE Procedia* **2014,** *8,* 109–113.

Chin, S. F.; Pang, S. C.; Tay, S. H. Size Controlled Synthesis of Starch Nanoparticles by a Simple Nanoprecipitation Method. *Carbohydr. Polym.* **2011,** *86* (4), 1817–1819.

Chiou, B.-S.; Yee, E.; Glenn, G. M.; Orts, W. J. Rheology of Starch–Clay Nanocomposites. *Carbohydr. Polym.* **2005,** *59* (4), 467–475.

Chivrac, F.; Gueguen, O.; Pollet, E.; Ahzi, S.; Makradi, A.; Averous, L. Micromechanical Modeling and Characterization of the Effective Properties in Starch-based Nano-Biocomposites. *Acta Biomater.* **2008,** *4* (6), 1707–1714.

Choi, I.; Lee, J. Y.; Lacroix, M.; Han, J. Intelligent pH Indicator Film Composed of Agar/Potato Starch and Anthocyanin Extracts from Purple Sweet Potato. *Food Chem.* **2017,** *218,* 122–128.

Chung, Y.-L.; Ansari, S.; Estevez, L.; Hayrapetyan, S.; Giannelis, E. P.; Lai, H.-M. Preparation and Properties of Biodegradable Starch–Clay Nanocomposites. *Carbohydr. Polym.* **2010,** *79* (2), 391–396.

Combrzyński, M.; Mościcki, L.; Mitrus, M.; Kupryaniuk, K.; Oniszczuk, A. Application of Extrusion-cooking Technique for Foamed Starch-based Materials. *BIO Web Conf.* **2018,** *10,* 1–6.

Dai, L.; Qiu, C.; Xiong, L.; Sun, Q. Characterisation of Corn Starch-based Films Reinforced with Taro Starch Nanoparticles. *Food Chem.* **2015,** *174,* 82–88.

De Ancos, B.; González-Peña, D.; Colina-Coca, C.; Sánchez-Moreno, C. Uso de Películas/Recubrimientos Comestibles en los Productos de IV y V Gama. *Revista Iberoamericana de Tecnología Postcosecha* **2015,** *16* (1).

de Azêvedo, L. C.; Rovani, S.; Santos, J. J.; Dias, D. B.; Nascimento, S. S.; Oliveira, F. F.; Silva, L. G. A.; Fungaro, D. A. Study of Renewable Silica Powder Influence in the Preparation of Bioplastics from Corn and Potato Starch. *J. Polym. Environ.* **2021,** *29* (3), 707–720.

de Jara, E. M.; García-Hernández, E.; Quequezana-Bedregal, M.; Arrieta-González, C.; Salgado-Delgado, R.; Lastarria-Tapia, H.; Castañón-Vilca, J. Potato Starch-based Films: Effects of Glycerol and Montmorillonite Nanoclay Concentration. *Rev. Mex. Ing. Quím.* **2020,** *19* (2), 627–637.

de la Concha, B. B. S.; Agama-Acevedo, E.; Nuñez-Santiago, M. C.; Bello-Perez, L. A.; Garcia, H. S.; Alvarez-Ramirez, J. Acid Hydrolysis of Waxy Starches with Different Granule Size for Nanocrystal Production. *J. Cereal Sci.* **2018,** *79,* 193–200.

Dimitrijevic, M.; Karabasil, N.; Boskovic, M.; Teodorovic, V.; Vasilev, D.; Djordjevic, V.; Kilibarda, N.; Cobanovic, N. Safety Aspects of Nanotechnology Applications in Food Packaging. *Proc. Food Sci.* **2015,** *5,* 57–60.

Dufresne, A. Polysaccharide Nano Crystal Reinforced Nanocomposites. *Can. J. Chem.* **2008,** *86* (6), 484–494.

Dufresne, A.; Castaño, J. Polysaccharide Nanomaterial Reinforced Starch Nanocomposites: A Review. *Starch-Stärke* **2017,** *69* (1–2), 1500307.

El-Sheikh, M. A. New Technique in Starch Nanoparticles Synthesis. *Carbohydr. Polym.* **2017,** *176,* 214–219.

Estevez-Areco, S.; Guz, L.; Famá, L.; Candal, R.; Goyanes, S. Bioactive Starch Nanocomposite Films with Antioxidant Activity and Enhanced Mechanical Properties Obtained by Extrusion Followed by Thermo-Compression. *Food Hydrocoll.* **2019,** *96,* 518–528.

Fahmy, H. M.; Salah Eldin, R. E.; Abu Serea, E. S.; Gomaa, N. M.; AboElmagd, G. M.; Salem, S. A.; Elsayed, Z. A.; Edrees, A.; Shams-Eldin, E.; Shalan, A. E. Advances in Nanotechnology and Antibacterial Properties of Biodegradable Food Packaging Materials. *RSC Adv.* **2020,** *10* (35), 20467–20484.

Falguera, V.; Quintero, J. P.; Jiménez, A.; Muñoz, J. A.; Ibarz, A. Edible Films and Coatings: Structures, Active Functions and Trends in their Use. *Trends Food Sci. Technol.* **2011,** *22* (6), 292–303.

Fernández Valdés, D.; Bautista Baños, S.; Fernández Valdés, D.; Ocampo Ramírez, A.; García Pereira, A.; Falcón Rodríguez, A. Películas y Recubrimientos Comestibles: Una Alternativa Favorable en la Conservación Poscosecha de Frutas y Hortalizas. *Rev. Cienc. Téc. Agropecu.* **2015,** *24* (3), 52–57.

Formela, K.; Zedler, Ł.; Hejna, A.; Tercjak, A. Reactive Extrusion of Bio-based Polymer Blends and Composites-Current Trends and Future Developments. *Express Polym. Lett.* **2018,** *12* (1).

Gamage, A.; Thiviya, P.; Mani, S.; Ponnusamy, P. G.; Manamperi, A.; Evon, P.; Merah, O.; Madhujith, T. Environmental Properties and Applications of Biodegradable Starch-based Nanocomposites. *Polymers* [Online], **2022**.

García, N. L.; Famá, L.; D'Accorso, N. B.; Goyanes, S. *Biodegradable Starch Nanocomposites BT—Eco-friendly Polymer Nanocomposites: Processing and Properties*. Thakur, V. K., Thakur, M. K., Eds.; Springer India: New Delhi, 2015; pp 17–77.

García, N. L.; Ribba, L.; Dufresne, A.; Aranguren, M. I.; Goyanes, S. Physico-Mechanical Properties of Biodegradable Starch Nanocomposites. *Macromol. Mater. Eng.* **2009,** *294* (3), 169–177.

García, N. L.; Ribba, L.; Dufresne, A.; Aranguren, M.; Goyanes, S. Effect of Glycerol on the Morphology of Nanocomposites Made from Thermoplastic Starch and Starch Nanocrystals. *Carbohydr. Polym.* **2011,** *84* (1), 203–210.

Ghanbari, A.; Tabarsa, T.; Ashori, A.; Shakeri, A.; Mashkour, M. Preparation and Characterization of Thermoplastic Starch and Cellulose Nanofibers as Green Nanocomposites: Extrusion Processing. *Int. J. Biol. Macromol.* **2018,** *112,* 442–447.

Ghoshal, G. Recent Trends in Active, Smart, and Intelligent Packaging for Food Products A2—Grumezescu, Alexandru Mihai. In *Food Packaging and Preservation*; Holban, A. M. G. a. A. M. Ed.; Academic Press, 2018; pp 343–374.

González, A.; Alvarez Igarzabal, C. I. Nanocrystal-reinforced Soy Protein Films and their Application as Active Packaging. *Food Hydrocoll.* **2015,** *43,* 777–784.

González, K.; Retegi, A.; González, A.; Eceiza, A.; Gabilondo, N. Starch and cellulose nanocrystals together into thermoplastic starch bionanocomposites. *Carbohydr. Polym.* **2015,** *117,* 83–90.

Goudarzi, V.; Shahabi-Ghahfarrokhi, I.; Babaei-Ghazvini, A. Preparation of Ecofriendly UV-Protective Food Packaging Material by Starch/TiO_2 Bio-nanocomposite: Characterization. *Int. J. Biol. Macromol.* **2017,** *95,* 306–313.

Grande, C. J.; Torres, F. G.; Gomez, C. M.; Troncoso, O. P.; Canet-Ferrer, J.; Martínez-Pastor, J. Development of Self-assembled Bacterial Cellulose–Starch Nanocomposites. *Mater. Sci. Eng. C* **2009,** *29* (4), 1098–1104.

Gujral, H.; Sinhmar, A.; Nehra, M.; Nain, V.; Thory, R.; Pathera, A. K.; Chavan, P. Synthesis, Characterization, and Utilization of Potato Starch Nanoparticles as a Filler in Nanocomposite Films. *Int. J. Biol. Macromol.* **2021,** *186,* 155–162.

Haaj, S. B.; Magnin, A.; Pétrier, C.; Boufi, S. Starch Nanoparticles Formation via High Power Ultrasonication. *Carbohydr. Polym.* **2013,** *92* (2), 1625–1632.

Haaj, S. B.; Thielemans, W.; Magnin, A.; Boufi, S. Starch Nanocrystals and Starch Nanoparticles from Waxy Maize as Nanoreinforcement: A Comparative Study. *Carbohydr. Polym.* **2016,** *143,* 310–317.

Hamad, A. F.; Han, J.-H.; Kim, B.-C.; Rather, I. A. The Intertwine of Nanotechnology with the Food Industry. *Saudi J. Biol. Sci.* **2018,** *25* (1), 27–30.

Han, J. H. Chapter 9—Edible Films and Coatings: A Review. In *Innovations in Food Packaging*, 2nd ed.; Han, J. H. Ed.; Academic Press: San Diego, 2014; pp 213–255.

He, X.; Hwang, H.-M. Nanotechnology in Food Science: Functionality, Applicability, and Safety Assessment. *J. Food Drug Anal.* **2016,** *24* (4), 671–681.

Hernández, A. I. M.; González, R. A. O.; Padrón, P.; Araya, Q.; Wexler, G.; Cubero, C. Biodegradable Films with Bioactive Properties. *Rev. Venez. Cienc. Tecnol. Aliment* **2017,** *8* (1), 57–89.

Hu, Z.; Cranston, E. D.; Ng, R.; Pelton, R. Tuning Cellulose Nanocrystal Gelation with Polysaccharides and Surfactants. *Langmuir* **2014,** *30* (10), 2684–2692.

Huang, J.; Chang, P. R.; Dufresne, A. Polysaccharide Nanocrystals: Current Status and Prospects in Material Science. In *Polysaccharide-Based Nanocrystals*; Huang, J., Chang, P. R., Dufresne, A., Eds.; Wiley-VCH: Weinheim, Germany, 2015; Vol. 1, pp 1–14.

Huh, M. S.; Lee, E. J.; Koo, H.; Yhee, J. Y.; Oh, K. S.; Son, S.; Lee, S.; Kim, S. H.; Kwon, I. C.; Kim, K. Polysaccharide-based Nanoparticles for Gene Delivery. *Topics Curr. Chem.* **2017,** *375* (2), 31.

Imran, M.; Revol-Junelles, A.-M.; Martyn, A.; Tehrany, E. A.; Jacquot, M.; Linder, M.; Desobry, S. Active Food Packaging Evolution: Transformation from Micro-to Nanotechnology. *Crit. Rev. Food Sci. Nutr.* **2010,** *50* (9), 799–821.

Jiang, S.; Liu, C.; Wang, X.; Xiong, L.; Sun, Q. Physicochemical Properties of Starch Nanocomposite Films Enhanced by Self-assembled Potato Starch Nanoparticles. *LWT - Food Sci. Technol.* **2016,** *69,* 251–257.

Joye, I. J.; McClements, D. J. Biopolymer-based Nanoparticles and Microparticles: Fabrication, Characterization, and Application. *Curr. Opin. Colloid Interf. Sci.* **2014,** *19* (5), 417–427.

Kaplan, D. L. Introduction to Biopolymers from Renewable Resources. In *Biopolymers from Renewable Resources*; Kaplan, D. L., Ed.; Springer Berlin Heidelberg: Berlin, Heidelberg, 1998; pp 1–29.

Karolina, L.; Anna, T.-H.; Damian, S.; Justyna, O.; Piotr, T.; Jakub, B. Biodegradable Polylactide and Thermoplastic Starch Blends as Drug Release Device—Mass Transfer Study. **2018,** *20,* 75–75.

Kim, H.-Y.; Park, D. J.; Kim, J.-Y.; Lim, S.-T. Preparation of Crystalline Starch Nanoparticles Using Cold Acid Hydrolysis and Ultrasonication. *Carbohydr. Polym.* **2013,** *98* (1), 295–301.

Kim, H.-Y.; Park, S. S.; Lim, S.-T. Preparation, Characterization and Utilization of Starch Nanoparticles PRUEBA Y. *Colloids Surf. B:* **2015,** *126,* 607–620.

Korbekandi, H.; Iravani, S. Biological Synthesis of Nanoparticles Using Algae. In *Green Biosynthesis of Nanoparticles: Mechanisms and Applications*; Rai, M., Posten, C., Eds.; 2013; pp 53–60.

Labet, M.; Thielemans, W.; Dufresne, A. Polymer Grafting Onto Starch Nanocrystals. *Biomacromolecules* **2007,** *8* (9), 2916–2927.

Le Corre, D.; Bras, J.; Dufresne, A. Starch Nanoparticles: A Review. *Biomacromolecules* **2010,** *11* (5), 1139–1153.

Li, T.; Turng, L. S.; Gong, S.; Erlacher, K. Polylactide, Nanoclay, and Core–Shell Rubber Composites. *Polym. Eng. Sci.* **2006,** *46* (10), 1419–1427.

Lin, N.; Huang, J.; Chang, P. R.; Anderson, D. P.; Yu, J. Preparation, Modification, and Application of Starch Nanocrystals in Nanomaterials: A Review. *J. Nanomater.* **2011,** 2011, 1–13.

Lin, N.; Huang, J.; Chang, P. R.; Feng, L.; Yu, J. Effect of Polysaccharide Nanocrystals on Structure, Properties, and Drug Release Kinetics of Alginate-based Microspheres. *Colloids Surf. B: Biointerfaces* **2011,** *85* (2), 270–279.

Lin, N.; Huang, J.; Dufresne, A. Preparation, Properties and Applications of Polysaccharide Nanocrystals in Advanced Functional Nanomaterials: A Review. *Nanoscale* **2012,** *4* (11), 3274–3294.

Liu, B.; Xu, H.; Zhao, H.; Liu, W.; Zhao, L.; Li, Y. Preparation and Characterization of Intelligent Starch/PVA Films for Simultaneous Colorimetric Indication and Antimicrobial Activity for Food Packaging Applications. *Carbohydr. Polym.* **2017,** *157,* 842–849.

Liu, Z.; Jiao, Y.; Wang, Y.; Zhou, C.; Zhang, Z. Polysaccharides-based Nanoparticles as Drug Delivery Systems. *Adv. Drug Deliv. Rev.* **2008,** *60* (15), 1650–1662.

Llanos, J. H. R.; Tadini, C. C. Preparation and Characterization of Bio-nanocomposite Films Based on Cassava Starch or Chitosan, Reinforced with Montmorillonite or Bamboo Nanofibers. *Int. J. Biol. Macromol.* **2018,** *107,* 371–382.

López, O.; Castillo, L. A.; García, M. A.; Villar, M. A.; Barbosa, S. E. Food Packaging Bags based on Thermoplastic Corn Starch Reinforced with Talc Nanoparticles. *Food Hydrocoll.* **2015,** *43,* 18–24.

Lu, Y.; Weng, L.; Cao, X. Morphological, Thermal and Mechanical Properties of Ramie Crystallites—Reinforced Plasticized Starch Biocomposites. *Carbohydr. Polym.* **2006,** *63* (2), 198–204.

Ma, X.; Chang, P. R.; Yang, J.; Yu, J. Preparation and Properties of Glycerol Plasticized-Pea Starch/Zinc Oxide-Starch Bionanocomposites. *Carbohydr. Polym.* **2009,** *75* (3), 472–478.

Madhumitha, G.; Fowsiya, J.; Mohana Roopan, S.; Thakur, V. K. Recent Advances in Starch–Clay Nanocomposites. *Int. J. Polym. Anal. Charact.* **2018,** 1–15.

Malamatari, M.; Taylor, K. M. G.; Malamataris, S.; Douroumis, D.; Kachrimanis, K. Pharmaceutical Nanocrystals: Production by Wet Milling and Applications. *Drug Discov. Today* **2018,** *23* (3), 534–547.

Mallakpour, S.; Rashidimoghadam, S. Starch/MWCNT-vitamin C Nanocomposites: Electrical, Thermal Properties and Their Utilization for Removal of Methyl Orange. *Carbohydr. Polym.* **2017,** *169,* 23–32.

Mbey, J. A.; Hoppe, S.; Thomas, F. Cassava Starch–Kaolinite Composite Film. Effect of Clay Content and Clay Modification on Film Properties. *Carbohydr. Polym.* **2012,** *88* (1), 213–222.

Meira, S. M. M.; Zehetmeyer, G.; Scheibel, J. M.; Werner, J. O.; Brandelli, A. Starch-halloysite Nanocomposites Containing Nisin: Characterization and Inhibition of Listeria Monocytogenes in Soft Cheese. *LWT - Food Sci. Technol.* **2016,** *68,* 226–234.

Meira, S. M. M.; Zehetmeyer, G.; Werner, J. O.; Brandelli, A. A Novel Active Packaging Material based on Starch-halloysite Nanocomposites Incorporating Antimicrobial Peptides. *Food Hydrocoll.* **2017,** *63,* 561–570.

Mendes, J. F.; Paschoalin, R. T.; Carmona, V. B.; Sena Neto, A. R.; Marques, A. C. P.; Marconcini, J. M.; Mattoso, L. H. C.; Medeiros, E. S.; Oliveira, J. E. Biodegradable Polymer Blends based on Corn Starch and Thermoplastic Chitosan Processed by Extrusion. *Carbohydr. Polym.* **2016,** *137,* 452–458.

Mihindukulasuriya, S. D. F.; Lim, L. T. Nanotechnology Development in Food Packaging: A Review. *Trends Food Sci. Technol.* **2014,** *40* (2), 149–167.

Mittal, A.; Garg, S.; Kohli, D.; Maiti, M.; Jana, A. K.; Bajpai, S. Effect of Cross Linking of PVA/Starch and Reinforcement of Modified Barley Husk on the Properties of Composite Films. *Carbohydr. Polym.* **2016,** *151,* 926–938.

Mohammadi Nafchi, A.; Moradpour, M.; Saeidi, M.; Alias, A. K. Thermoplastic Starches: Properties, Challenges, and Prospects. *Starch-Stärke* **2013,** *65* (1-2), 61–72.

Mohanraj, V.; Chen, Y. Nanoparticles-A Review. *Tropi. J. Pharmaceut. Res.* **2006,** *5* (1), 561–573.

Mukurumbira, A. R.; Mellem, J. J.; Amonsou, E. O. Effects of Amadumbe Starch Nanocrystals on the Physicochemical Properties of Starch Biocomposite Films. *Carbohydr. Polym.* **2017,** *165,* 142–148.

Müller, C. M. O.; Laurindo, J. B.; Yamashita, F. Composites of Thermoplastic Starch and Nanoclays Produced by Extrusion and Thermopressing. *Carbohydr. Polym.* **2012,** *89* (2), 504–510.

Nasseri, R.; Mohammadi, N. Starch-based Nanocomposites: A Comparative Performance Study of Cellulose Whiskers and Starch Nanoparticles. *Carbohydr. Polym.* **2014,** *106,* 432–439.

Nasution, A. H.; Harahap, H.; Al Fath, M. T. Physical Properties of Sago Starch Biocomposite Filled with Nanocrystalline Cellulose (NCC) from Rattan Biomass: The Effect of Filler Loading and Co-plasticizer Addition. *IOP Conf. Ser.: Mater. Sci. Eng.* **2018,** *309* (1), 12033–12033.

Nathalie, G.; Stéphane, P.; Jose, M. L.; Yolanda, E.; Carole, G. Nanotechnologies for Active and Intelligent Food Packaging: Opportunities and Risks. In *Nanotechnology in Agriculture and Food Science*; Axelos, M. A.. Voorde, M. H. V. D., Eds.; Wiley-VCH Verlang GmbH & Co. KGaA. 2017; pp 177–196.

Oikawa, H.; Masuhara, A.; Kasai, H.; Mitsui, T.; Sekiguchi, T.; Nakanishi, H. Organic and Polymer Nanocrystals: Their Optical Properties and Function. In *Handai Nanophotonics*; Masuhara, H.; Kawata, S., Eds.; Elsevier: New York, USA, 2004; Vol. 1, pp 205–224.

Othman, S. H. Bio-nanocomposite Materials for Food Packaging Applications: Types of Biopolymer and Nano-sized Filler. *Agric. Agric. Sci. Procedia* **2014,** *2,* 296–303.

Park, H. J.; Byun, Y. J.; Kim, Y. T.; Whiteside, W. S.; Bae, H. J. Processes and Applications for Edible Coating and Film Materials from Agropolymers A2—Han, Jung H. In *Innovations in Food Packaging*, 2nd ed.; Han, J. H., Ed.; Academic Press: San Diego, 2014; pp 257–275.

Pelissari, F. M.; Andrade-Mahecha, M. M.; Sobral, P. J. D. A.; Menegalli, F. C. Nanocomposites based on Banana Starch Reinforced with Cellulose Nanofibers Isolated from Banana Peels. *J. Colloid Interface Sci.* **2017,** *505,* 154–167.

Podshivalov, A.; Zakharova, M.; Glazacheva, E.; Uspenskaya, M. Gelatin/Potato Starch Edible Biocomposite Films: Correlation between Morphology and Physical Properties. *Carbohydr. Polym.* **2017,** *157,* 1162–1172.

Quintero, C.; Falguera, V.; Muñoz, H. Films and Edible Coatings: Importance, and Recent Trends in Fruit and Vegetable Value Chain. *Rev. Tumbaga* **2010,** *5,* 93–118.

Ranjan, S.; Dasgupta, N.; Chakraborty, A. R.; Samuel, S. M.; Ramalingam, C.; Shanker, R.; Kumar, A. Nanoscience and Nanotechnologies in Food Industries: Opportunities and Research Trends. *J. Nanopart. Res.* **2014,** *16* (6), 2464.

Ribeiro-Santos, R.; Andrade, M.; Melo, N. R. D.; Sanches-Silva, A. Use of Essential Oils in Active Food Packaging: Recent Advances and Future Trends. *Trends Food Sci. Technol.* **2017,** *61,* 132–140.

Romero-Bastida, C. A.; Chávez Gutiérrez, M.; Bello-Pérez, L. A.; Abarca-Ramírez, E.; Velazquez, G.; Mendez-Montealvo, G. Rheological Properties of Nanocomposite-Forming Solutions and Film Based on Montmorillonite and Corn Starch with Different Amylose Content. *Carbohydr. Polym.* **2018,** *188,* 121–127.

Rong, L.; Shen, M.; Wen, H.; Ren, Y.; Xiao, W.; Xie, J. Preparation and Characterization of Hyacinth Bean Starch Film Incorporated with TiO_2 Nanoparticles and Mesona Chinensis Benth Polysaccharide. *Int. J. Biol. Macromol.* **2021,** *190,* 151–158.

Rovina, K.; Vonnie, J. M.; Mantihal, S.; Joseph, J.; Halid, N. F. A. Development of Films based on Tapioca Starch/Gold Nanoparticles for the Detection of Organophosphorus Pesticides. *J. Consum. Prot. Food Safety* **2021,** *16* (2), 143–152.

Salaberria, A. M.; Diaz, R. H.; Labidi, J.; Fernandes, S. C. M. Role of Chitin Nanocrystals and nanofibers on Physical, Mechanical and Functional Properties in Thermoplastic Starch Films. *Food Hydrocoll.* **2015,** *46,* 93–102.

Salaberria, A. M.; Labidi, J.; Fernandes, S. C. M. Chitin Nanocrystals and Nanofibers as Nano-sized Fillers into Thermoplastic Starch-based Biocomposites Processed by Melt-mixing. *Chem. Eng. J.* **2014,** *256,* 356–364.

Sharma, R.; Jafari, S. M.; Sharma, S. Antimicrobial Bio-nanocomposites and their Potential Applications in Food Packaging. *Food Cont.* **2020,** *112,* 107086.

Sinha Ray, S.; Okamoto, M. Polymer/Layered Silicate Nanocomposites: A Review from Preparation to Processing. *Prog. Polym. Sci.* **2003,** *28* (11), 1539–1641.

Skaff, H.; Emrick, T. Semiconductor Nanoparticles. In *Nanoparticles: Building Blocks for Nanotechnology*; Rotello, V., Ed.; Kluwer Academic/Plenum Publishers: New York, USA, 2004; pp 27–46.

Slavutsky, A. M.; Bertuzzi, M. A.; Armada, M. Water Barrier Properties of Starch-Clay Nanocomposite Films. *Braz. J. Food Technol.* **2012,** *15* (3), 208–218.

Srikhao, N.; Kasemsiri, P.; Ounkaew, A.; Lorwanishpaisarn, N.; Okhawilai, M.; Pongsa, U.; Hiziroglu, S.; Chindaprasirt, P. Bioactive Nanocomposite Film based on Cassava Starch/ Polyvinyl Alcohol Containing Green Synthesized Silver Nanoparticles. *J. Polym. Environ.* **2021,** *29* (2), 672–684.

Tjong, S. C. *Nanocrystalline Materials: Their Synthesis-Structure-Property Relationships and Applications*, 2nd ed.; Newnes, 2013; p 418.

Usman, A.; Hussain, Z.; Riaz, A.; Khan, A. N. Enhanced Mechanical, Thermal and Antimicrobial Properties of poly(vinyl alcohol)/Graphene Oxide/Starch/Silver Nanocomposites Films. *Carbohydr. Polym.* **2016,** *153,* 592–599.

Viguié, J.; Molina-Boisseau, S.; Dufresne, A. Processing and Characterization of Waxy Maize Starch Films Plasticized by Sorbitol and Reinforced with Starch Nanocrystals. *Macromol. Biosci.* **2007,** *7* (11), 1206–1216.

Wang, C.; Gao, X.; Chen, Z.; Chen, Y.; Chen, H. Preparation, Characterization and Application of Polysaccharide-Based Metallic Nanoparticles: A Review. *Polymers* **2017,** *9* (12), 689.

Wang, N.; Zhang, X.; Han, N.; Bai, S. Effect of Citric Acid and Processing on the Performance of Thermoplastic Starch/Montmorillonite Nanocomposites. *Carbohydr. Polym.* **2009,** *76* (1), 68–73.

Wang, Y.; Tian, H.; Zhang, L. Role of Starch Nanocrystals and Cellulose Whiskers in Synergistic Reinforcement of Waterborne Polyurethane. *Carbohydr. Polym.* **2010,** *80* (3), 665–671.

Woehl, M. A.; Canestraro, C. D.; Mikowski, A.; Sierakowski, M. R.; Ramos, L. P.; Wypych, F. Bionanocomposites of Thermoplastic Starch Reinforced with Bacterial Cellulose Nanofibres: Effect of Enzymatic Treatment on Mechanical Properties. *Carbohydr. Polym.* **2010,** *80* (3), 866–873.

Wu, J. G.; Wang, P. J.; Chen, S. C. Antioxidant and Antimicrobial Effectiveness of Catechin-Impregnated PVA—Starch Film on Red Meat. *J. Food Qual.* **2010,** *33* (6), 780–801.

Youssef, A. M.; El-Sayed, S. M. Bionanocomposites Materials for Food Packaging Applications: Concepts and Future Outlook. *Carbohydr. Polym.* **2018,** *193,* 19–27.

Zeppa, C.; Gouanvé, F.; Espuche, E. Effect of a Plasticizer on the Structure of Biodegradable Starch/Clay Nanocomposites: Thermal, Water, Sorption, and Oxygen-Barrier Properties. *J. Appl. Polym. Sci.* **2009,** *112* (4), 2044–2056.

Zhai, X.; Shi, J.; Zou, X.; Wang, S.; Jiang, C.; Zhang, J.; Huang, X.; Zhang, W.; Holmes, M. Novel Colorimetric Films based on Starch/Polyvinyl Alcohol Incorporated with Roselle Anthocyanins for Fish Freshness Monitoring. *Food Hydrocoll.* **2017,** *69,* 308–317.

CHAPTER 3

Release Studies to Improve the Mechanical Properties of the Biopolymer, Polylactic Acid (PLA) for Food Packaging Applications

Y. K. REYES ACOSTA[1], W. E. CRUZ MARTINEZ[1], A. V. REYES ACOSTA[2], L. SEPÚLVEDA TORRE[1], C. N. AGUILAR GONZALES[1], and R. I. NARRO CESPEDES[1]

[1]*Facultad de Ciencias Químicas de la Universidad Autónoma de Coahuila, Saltillo, Coahuila, México*

[2]*Facultad de Sistemas de la Universidad Autónoma de Coahuila, Zona Centro, Arteaga, Coahuila, México*

ABSTRACT

Polylactic acid (PLA) is a biodegradable thermoplastic and aliphatic polyester obtained from renewable and sustainable starch-rich materials. Due to its biodegradability, biocompatibility, and high mechanical properties, such as tensile strength and Young's modulus, comparable to those of polystyrene (PS) and polyethylene terephthalate (PET), PLA has been considered one of the most important biopolymers. However, due to its lower impact resistance, many materials have been used to improve its properties. Materials such as food and its derivatives, natural oils, organic material, different types of fibers, nanoparticles, and others have been incorporated into the PLA matrix with the aim of showing improved mechanical properties. This paper presents a review of the research on PLA and the use of enhancers, as well as the results for the improvement of mechanical properties and applications of food containers.

Engineering Principles for Food Processing Technology and Product Realization. Leonardo Sepúlveda Torre, Porteen Kannan, A. K. Haghi, PhD (Eds.)

3.1 INTRODUCTION

In recent years, there has been a growing scientific and industrial interest in the use of compounds obtained from renewable sources and with biodegradability properties that reduce the environmental effect (Scaffaro et al., 2018). One of the methods recently introduced to resolve this issue is the use of biopolymers that are naturally occurring in living organisms (Suryanegara et al., 2017). The use of biopolymers as raw materials would be of great benefit to environmental protection efforts compared to the use of fossil-based materials due to less adverse effects (Suryanegara et al., 2017).

Biopolymers have important physicochemical and biological properties. These properties are targeted for applications in a broad range of fields, especially in the packaging, textile, agricultural, pharmaceutical, electronics, and biomedical sectors (Hammiche et al., 2019). Among biodegradable polymers, poly(lactic acid) (PLA) derived from sugar feedstock, corn, etc. has gained growing attention as a sustainable and readily biodegradable biomaterial. This aliphatic polyester, known as GRAS (Generally Recognized As Safe), was used for food packaging applications (Nilsuwan et al., 2018).

PLA has an immense potential to replace petroleum-based plastics due to its high mechanical properties, which are comparable to polystyrene (PS) and polyethylene terephthalate (PET) polymers currently used in the packaging industry (Herrera et al., 2016; Miao et al., 2019; Suryanegara et al., 2017). PLA has gained a lot of attention due to its mechanical properties such as high tensile strength and high Young's modulus (Carbonell-Verdu et al., 2017) and its relatively strong biocompatibility (Ge et al., 2018). However, due to the inherent fragility and lower impact resistance of PLA, a number of efforts have been made to develop its properties in order to achieve compatibility with thermoplastic processing and manufacturing, thereby enabling it to compete with commodity polymers (Suryanegara et al., 2017).

Many materials have been used to increase PLA properties. Materials that can be considered food or derivates: fish gelatin (FG), (Nilsuwan et al., 2018), carvacrol (CRV) (Scaffaro et al., 2018), beeswax (BW) (Reis et al., 2018), sugarcane bagasse pulp (SCB) (Suryanegara et al., 2018), olive husk flour (OHF) (Hammiche et al., 2019). Natural oils like epoxidized cottonseed oil (ECSO) (Carbonell-Verdu et al., 2017), epoxidized (ELO, ESBO) and non-epoxidized linseed oil (LO) and soybean oil (SBO) (Orue et al., 2018), and fennel (FEN) oil (Miao et al., 2019). Organic material as leaves of Posidonia oceanica (PO) (Scaffaro, Lopresti, and Botta, 2018), wood flour (WF) (Ge et al., 2018; Holcapkova et al., 2017), wood fiber (F) (Asadi et al., 2018) and diatomaceous earth (DE) (Aggarwal et al., 2019) can also be considered.

Different kinds of fiber like bamboo fibers (BF) (Li et al., 2019; Long et al., 2019), Agave fibers (AF) (Cisneros-López et al., 2018), ramie fibers (Debeli et al., 2019), and cellulose fibers (Laadila et al., 2017) are used. Nanocomposites as carbon nanotubes (CNT) (Liu et al., 2019) and cellulose nanocrystals (CNC) (Mariano et al., 2017; Wei et al., 2018). Finally, materials such as carbon fiber, microfibrillated wood (MFW), microfibrillated lignocellulose (MFLC) (Winter et al., 2018), and diatomaceous earth (DE) (Aggarwal et al., 2019) have been used in 3D printing.

3.2 FOOD DERIVATIVES USED TO IMPROVE PLA MECHANICAL PROPERTIES

Polymeric materials for packaging applications have been cautiously considered due to the limitations and regulations relating to primary and post-consumer plastic waste management (Nilsuwan et al., 2018). Ecological concern and the challenge of recycling packaging materials have stimulated the production of biodegradable materials (Reis et al., 2018). Mechanical properties of biobased materials used in food packaging structures are critical criteria for their performance and functionality. The mechanical properties of PLA biopolymers can vary from soft plastics to rigid elastics and high strength materials, depending on a wide variety of factors, such as molecular weight, stereochemical spine makeup. As a result, different materials considered as food or food derivatives were used to improve the mechanical properties of PLA: FG and PLA bilayer films were developed by casting, exhibiting lower tensile strength but higher elongation at break than PLA film (Nilsuwan et al., 2018), PLA/CRV nanofiber membranes were also prepared by electrospinning, adding 28 wt% of CRV caused an increment of E by ~70%, σm by ~450% and εb by ~1350% (Scaffaro, Lopresti, and Botta, 2018). In addition, attempts have been made to achieve sufficient mechanical and barrier properties. Thermoplastic starch (TPS) with PLA trays coated with beeswax (BW) emulsion was prepared by Reis and coworkers. The trays coated with 1% BW had higher tensile strength than the trays with a higher BW emulsion content (Reis et al., 2018).

Another reinforcement agent, sugar bagasse pulp (SCB), was used. It was found that the tensile strength decreased due to an increase in the quality of the filler. 'Young's modulus, however, increased by up to 13–20 wt % (Suryanegara et al., 2018). Polylactic acid and olive husk flour (PLA/OHF) composites were prepared by extrusion, finding that Young's modulus increased by 18% , the tensile strength and the elongation at break decreased as compared to neat PLA (Hammiche et al., 2019).

TABLE 3.1 Reinforcement Used in Polylactic Acid Food Packing Material.

Reinforcement agent	Concentration	Methodology	Results
Fish Gelatin (FG) (Nilsuwan et al., 2018)	Layer thickness ratios of 9:1, 8:2, 7:3, 6:4 and 5:5 (FG/ LPA)	Casting	σ_m = 31.79 MPa (PLA) σ_m = 24.23 MPa (FG) σ_m = 27.23 MPa (FG/PLA5:5) ε_b = 8.89% (PLA) ε_b = 36.27% (FG) ε_b = 11.6% (FG/PLA 5:5)
Carvacrol (CRV) (Scaffaro et al., 2018)	28 wt %	Stirring solution/ electrospinning	σ_m = 1.25 MPa (PLA) σ_m = 6.81 MPa (PLA/CRV 28%) ε_b = 7.8% (PLA) ε_b = 115.2% (PLA/CRV 28%) E_b = 57.1 MPa (PLA) E = 98 MPa (PLA/CRV 28%)
Thermoplastic starch (TPS) using beeswax (BW) emulsion to coat (Reis et al., 2018)	1, 2 and 3 wt % of BW	Mixing/ extrusion	σ_m = 10.3 MPa (TPS/PLA) σ_m = 11.5 MPa (TPS/PLA–BW1) ε_b = 27% (TPS/PLA) ε_b = 22% (TPS/PLA – BW1) E = 294 MPa (TPS/PLA) E = 292 MPa (TPS/PLA – BW1)
Sugarcane bagasse pulp (SCB) (Suryanegara et al., 2018)	5, 10 and 20 wt %	Stirring solution	σ_m = 47.9 MPa (PLA) σ_m = 38.3 MPa (PLA/pulp5) ε_b = 15.6% (PLA) ε_b = 5.5% (PLA/pulp 5) E = 1.18 GPa (PLA) E = 1.32 GPa (PLA/pulp 5)
			σ_m = 75 MPa (PLA) σ_m = 64 MPa (PLA/untreated OHF) σ_m = 71 MPa (PLA/treated OHF) ε_b = 6.8% (PLA)

TABLE 3.1 *(Continued)*

Reinforcement agent	Concentration	Methodology	Results
Untreated and treated olive husk flour (OHF) (Hammiche et al., 2019)	20 wt %	Extrusion	ε_b= 3.2% (PLA/untreated OHF) ε_b = 3.5% (PLA/treated OHF) E = 3101 GPa (PLA) E = 3635 GPa (PLA/ untreated OHF) E = 3757 GPa (PLA/ treated OHF)
Fish Gelatin (FG) (Nilsuwan et al., 2018)	Layer thickness ratios of 9:1, 8:2, 7:3, 6:4 and 5:5 (FG/LPA)	Casting	σ_m = 31.79 MPa (PLA) σ_m = 24.23 MPa (FG) σ_m = 27.23 MPA (FG/PLA 5:5) ε_b = 8.89% (PLA) ε_b = 36.27% (FG) ε_b = 11.6% (FG/PLA 5:5)
Carvacrol (CRV) (Scaffaro et al., 2018)	28 wt %	Stirring solution/ electrospinning	σ_m = 1.25 MPa (PLA) σ_m = 6.81 MPa (PLA/CRV 28%) ε_b = 7.8% (PLA) ε_b = 115.2% (PLA/CRV 28%) E = 57.1 MPa (PLA) E = 98 MPa (PLA/CRV 28%)
Thermoplastic starch (TPS) using beeswax (BW) emulsion to coat (Reis et al., 2018)	1, 2 and 3 wt % of BW	Mixing/ extrusion	σ_m=10.3 MPa (TPS/PLA) σ_m = 11.5 MPa (TPS/PLA–BW1) ε_b = 27% (TPS/PLA) ε_b = 22% (TPS/PLA–BW1) E = 294 MPa (TPS/PLA) E = 292 MPa (TPS/PLA–BW1)
Sugarcane bagasse pulp (SCB) (Suryanegara et al., 2018)	5, 10 and 20 wt %	Stirring solution	σ_m = 47.9 MPa (PLA) σ_m = 38.3 MPa (PLA/pulp 5) ε_b = 15.6% (PLA) ε_b = 5.5% (PLA/pulp 5) E = 1.18 GPa (PLA) E = 1.32 GPa (PLA /pulp 5) σ_m = 75 MPa (PLA) σ_m = 64 MPa (PLA/untreated OHF) σ_m = 71 MPa (PLA/treated OHF) ε_b = 6.8% (PLA)

TABLE 3.1 *(Continued)*

Reinforcement agent	Concentration	Methodology	Results
Untreated and treated olive husk flour (OHF) (Hammiche et al., 2019)	20 wt %	Extrusion	ε_b = 3.2% (PLA/untreated OHF) ε_b = 3.5% (PLA/treated OHF) E = 3101 GPa (PLA) E = 3635 GPa (PLA/ untreated OHF) E = 3737 GPa (PLA/ treated OHF)

3.3 NATURAL OILS USED AS PLA REINFORCEMENT AGENT

The use of compatible plasticizers for PLA may be an effective solution to minimize the fragility of PLA, since plasticizers lead to a drop in the glass transfer temperature of PLA resulting in more durable materials. Increased interest in low migration and low toxicity plasticizers focus on bio-based plasticizers manufactured from vegetable oils, citrus, and sugar derivatives as an alternative to common plasticizers. Epoxidized oils are known as reactive plasticizers due to the reactivity between the epoxy functional groups and the –OH and –COOH PLA polymer groups. Epoxidized vegetable oils can be produced by epoxidizing C=C bonds of oils and plasticizers with zero volatility and exceptionally low migration patterns (Orue et al., 2018).

In order to boost its mechanical properties, certain oils have been applied to PLA: the melt extrusion was used to plasticize PLA formulations with various ECSO contents in the range of 0–10 wt. %. The 10 wt.% composition of PLA demonstrates a surprising improvement in mechanical ductile properties (Carbonell et al., 2017). PLA was also plasticized with epoxidized and non-epoxidized linseed and soybean oils, and the effect of different plasticizers was assessed. Epoxidized oils have been shown to have slightly higher tensile modulus values than non-epoxidized oils, particularly for those blends plasticized with ELO (Orue et al., 2018). In addition, a biodegradable film based on PLA and PHB was produced with the addition of FEN oil, finding that the complete mechanical properties of PLA-PHB film were comparable to those of EVOH film (Miao et al., 2019), which are summarized in Table 3.2.

TABLE 3.2 Organic Materials Addition in PLA: Effect in the Stability Thermal.

Reinforcement agent	Concentration	Methodology	Results
Exporized cottonseed oil (ECSO) (Carbonell et al., 2017)	2.5, 5, 7.5 and 10 wt %	Mixing/ extrusion	σ_m = 63.7 MPa (PLA) σ_m = 49.5 MPa (PLA-10wt. %ECSO) E = 3.592 GPa (PLA) E = 3.99 GPa (PLA-10wt. %ECSO) ε_b = 9.06% (PLA) ε_b = 110.5% (PLA-10 wt. % ECSO) FM = 3.296 GPa (PLA) FM = 3.361 GPa (PLA-10 wt. % ECSO) FS = 116.3 MPa (PLA) FS = 53.3 MPa (PLA-10 wt. % ECSO)
Exporized (ELO, ESBO) and non-exporized Linseed Oil (LO) and soybean oil (SBO) (Orue et al., 2018)	5, 10 y 20 wt %	Melt mixing	ε_b = increased 119% (PLA-20 wt. % oils) a increased 100% (PLA-20 wt. % oils)
Fennel (FEN) Oil and polyhydroxybutyrate (PBH). Using ethanol vinyl alcohol (EVOH) (Miao et al., 2019)	24.46 wt % PHB and 6 wt %FEN oil	Extrusion	σ_m = 30.38 MPa (EVOH) σ_m = 41.20 MPa (PLA-PBH) σ_m = 41.49 MPa (PLA-PBH-FEN) ε_b = 212.7% (EVOH) ε_b = 4.93% (PLA-PBH) ε_b = 6.90% (PLA-PBH-FEN)

3.4 ORGANIC MATERIAL IN PLA-BASED COMPOSITES

Recently, various functional groups or fillers have been introduced into polymeric networks to add new functionalities and boost efficiency and preserve reusability. In this sense, increasing attention has been paid to natural fibers and particles as a reinforcement or filler agent in PLA-based composites. These materials allowed the recovery of agricultural or marine waste, providing goods with final properties suitable for a range of applications (Scaffaro et al., 2018).

Many materials, which can be called organic, have been used as reinforcement or filler in PLA composites. Leaves of Posidonia oceanica (PO) have been used for the preparation of PLA-based biocomposites through melting. The findings revealed that the increase in PO material improved both the tensile and the flexural modulus (Scaffaro et al., 2018). Also, the PLA composites produce 20 wt% of the wood flour (WF) was prepared by melting.

The inclusion of ZIKA did not influence the final mechanical properties of the composite PLA/WF-ZIKA (Holcapkova et al., 2017). Wood flour, PLA, and other ingredients have also been combined. The effect of four factors: heat temperature, molding pressure, holding pressure time, and azodicarbonamide (AC) content was investigated. The optimal conditions were a heating temperature of 178°C, a molding pressure of 7 MPa, and a holding pressure of 25 s. In addition, the integration of nanographene (NG) into PLA/wood fiber composites strengthened the tensile and bending properties (Asadi et al., 2018). The findings showed that the composites made of PLA68.5/F30/NG1.5 have the best tensile and bending properties. And a hybrid electrospun membrane of electroactive PLLA polymer with glass-reinforced hydroxyapatite (Bonelike®) microparticles has been processed. It has been found that the addition of Bonelike® particles contributes to a decline in the Young modulus and to tension at breakage (Santos et al., 2017). As well as applying Diatomaceous Earth (DE) to the PLA, there has been a decrease in the mechanical properties of the filaments (Aggarwal et al., 2019).

TABLE 3.3 Organic Material in PLA-Based Composites.

Reinforcement agent	Concentration	Methodology	Results
Leaves of Posidonia oceanica (PO) (Scaffaro, Lopresti, and Botta, 2018)	10 and 20 wt % with PO particles size <150 μm and 150–300 μm	Melt mixing	E = 1.891 GPa (PLA) E = 2.071 GPa (PLA/PO150 – 10) ϵ_b = 3.66% (PLA) ϵ_b = 2.46% (PLA/PO150 – 10) σ_m = 51.1 MPa (PLA) σ_m = 39.8 MPa (PLA/PO150 – 10) *FM* = 3.39 GPa (PLA) *FM* = 3.97 GPa (PLA/PO150 – 10) FS = 82 MPa (PLA) FS = 65.6 MPa (PLA/ PO150 – 10)
Maple Wood Flour (WF) (Ge et al., 2018)	The effects of heating temperature, molding pressure, holding pressure time, and AC foaming agent	Mixing	Optimum conditions: heating temperature of 178°C, molding pressure of 7 MPa, and holding pressure time of 25 s.
Wood Fiber (F) and Nanographene (NG) (Asadi et al., 2018)	15 and 30 wt.% of F with 0.75, 1.5, and 3 wt.% of NG	Mixing	Highest σ_m for PLA68.5/F30/NG1.5 Highest FM for PLA68.5/F30/NG1.5 Highest for PLA83.5//F15/NG1.5

TABLE 3.3 *(Continued)*

Reinforcement agent	Concentration	Methodology	Results
Wood flour (WF) and ZIKA additive (Holcapkova et al., 2017)	20 wt.% of WF	Melt mixing	σ_m = 62.9 MPa (PLA) σ_m = 55.4 MPa (PLA/WF–ZIKA) ϵ_b = 1.02% (PLA) ϵ_b = 3.5% (PLA/WF–ZIKA) *E* = 2.48 GPa (PLA) *E* = 2.73 GPa (PLA/WF–ZIKA)
Diatomaceous Earth (DE) (Aggarwal et al., 2019)	5, 10 and 15 wt.%	Extrusion	σ_m = 56.14 MPa (PLA) σ_m = 44.46 MPa (PLA/DE – 5%) *E* = 2.25 GPa (PLA) *E* = 2.16 GPa (PLA /DE – 5%)
Bonelike® and PLLA (Santos et al., 2017)	80 wt.%	Stirring solution/ electrospin-ning	*E* = 18 MPa (PLLA) *E* = 5.4 MPa (Bonelike® PLLA) σ_m = 8 MPa (PLLA) σ_m = 0.57 MPa (Bonelike® PLLA) ϵ_b = 38% (PLLA) ϵ_b = 41% (Bonelike® PLLA)
Leaves of Posidonia oceanica (PO) (Scaffaro, Lopresti, and Botta, 2018)	10 and 20 wt.% with PO particles size <150 μm and 150–300 μm	Melt mixing	E = 1.891 GPa (PLA) E = 2.071 GPa (PLA/PO150 – 10) ϵ_b = 3.66% (PLA) ϵ_b = 2.46% (PLA/PO150 – 10) σ_m = 51.1 MPa (PLA) σ_m = 39.8 MPa (PLA/ PO150 – 10) *FM* = 3.39 GPa (PLA) *FM* = 3.97 GPa (PLA/PO150 – 10) *FS* = 82 GPa (PLA) *FS* = 65.6 MPA (PLA/PO150 – 10)

3.5 NATURAL FIBERS IN PLA-BASED COMPOSITES

In recent decades, the use of plant fibers in composite materials has drawn significant attention. Natural fibers can greatly enhance fracture strength, crack resistance, stress performance, flexural properties, impact strength, and fatigue behavior of composites (Long et al., 2019). However, the mechanical properties of any fiber-reinforced composite depend on the direction of the fibers, the quality of the fibers, and the form of matrix used to determine the adhesion strength of the fiber–matrix interface (Debeli et al., 2019).

Various types of fibers have been integrated into PLA, e.g. bamboo fiber (BF), which has been incorporated as reinforced by a twin-screw extruder and injection molding. Compared to UBF/PLA composites, the mechanical properties of ABF/PLA composites have improved (Long et al., 2019). In addition, PLA/agave fiber biocomposites were prepared by dry-molding and rotational molding to determine the suitability of this process for the manufacture of these materials. Tensile and flexural strengths decreased with fiber content for both manufacturing processes, while tensile and flexural moduli, as well as impact strength and stiffness, were above the matrix (PLA) value of 10 wt % agave for R-BCs (Cisneros-López et al., 2018). In addition, the properties of molded ramie fibers reinforced with PLA biocomposites have been investigated. DAP was added to the surface of ramie fibers with and without pretreatment prior to the preparation of composites. DAP-treated ramie fiber reinforced composites improved mechanical strength (Debeli et al., 2019).

In addition, the properties of PLA-g-BF/PLA composites developed by blending with a three-component plasticizer, glycerol/formamide/tributyl citrate, were investigated. The bending power and elongation at breakage of PLA-g-BF/PLA composites improved by 19.3% and 30.1% compared to ungreased composites (Li et al., 2019). And biocomposites from recycled PLA and pulp and paper solid waste processed cellulosic fibers. Mechanical properties of biocomposites with 2% (w/w) of treated cellulosic fiber have been improved relative to recycled PLA (Laadila et al., 2017).

Many nanoparticles have been used to strengthen PLA, for example by using electrospinning to produce PLA/CNTs/CS composite fibers, and it has been observed that, as the CS concentration rose from 1 to 5 wt.%, the tensile strength of the fiber steadily increased, while the elongation at breakage decreased (Liu et al., 2019). Nanocomposite films primed with melt compounding and film blowing have also been tested for packaging applications. Mechanical properties of nanocomposite and reference films are found to be higher in the machine direction (MD) compared to the transverse direction (TD) (Herrera et al., 2016). In addition, high lignin-containing cellulose nanocrystals (HLCNCs) and lignin-coated CNCs (BLCNCs)/PLA composites have been made, showing substantial improvements in both Young's modulus and brittle elongation but minor decreases in tensile strength (Wei et al., 2018). Two separate PLLA-based surfactants have also been prepared. It consists of either the PEG block PEG-b-PLLA or the Im-PLLA, which can communicate with the surface of the CNCs in various ways. The findings reveal that the addition of both surfactants to the PLA matrix induces a decrease in Young's modulus, intensity, and even elongation at split. The addition of CNCs has shown no significant improvement in the mechanical properties (Mariano et al., 2017). In addition to the tensile strength and elasticity modules of the PLA/CNC composite films, the addition of 0.1 wt.% and 0.3 wt.% CNCs was improved (Wang et al., 2019). And PLA/CNC bio-nanocomposite films have been developed using a twin-screw extruder. The tensile strength and Young's modulus increased at both 1 and 3% CNCs (Sung et al., 2017).

TABLE 3.4 Natural Fibers in PLA.

Reinforcement agent	Concentration	Methodology	Results
Untreated bamboo fibers (UBF) and Alkali-treated Bamboo Fibers (ABF) (Long et al.,2019)	20 wt %	Mixing and extrusion	σ_m = 50.83 MPa (PLA/UBF) σ_m = 56.46 MPa (PLA/ABF) ε_B= 2.21% (PLA/UBF) ε_B= 2.491% (PLA/ABF) FM = 4.75 GPa (PLA/UBF) FS = 4.85 GPa (PLA/ABF) FS = 83.38 MPa (PLA/UBF) FS = 82 MPa (PLA/ABF) α = 2.59 kJ/m^2 (PLA/UBF) α = 3.49 kJ/m^2 (PLA/ABF) σ_M = 59 MPa (PLA) σ_m = 6 MPa (PLA/AF40) E = 1,908 MPa (PLA) E = 352 MPa (PLA/AF40) ε_b= 3.5% (PLA) ε_b= 2.8% (PLA/AF10) FM = 3.51 GPa (PLA) FM = 3.59 GPa (PLA/AF40)

TABLE 3.4 *(Continued)*

Reinforcement agent	Concentration	Methodology	Results
Agave Fibers (AF) (Cisneros-López et al., 2018)	10,20,30 and 40 wt %	Dry-blending and compression-molded (CM)	FS = 93 MPa (PLA) FS = 4 MPa (PLA/AF40) σ_m = 60 MPa (PLA) σ_m = 29 MPa (PLA/AF40) E = 2.03 GPa (PLA) E = 1.77 GPa (PLA/AF40) ε_b = 3.5% (PLA) ε_b = 2.1% (PLA/AF10) FM = 3.73 GPa (PLA) FM = 3.93 GPa (PLA/AF10) FS = 92 MPa (PLA) FS = 36 MPa (PLA/AF40)
Alklai treated, silane treated, alkali-silane treated, and no treated ramie fiber (Debeli et al., 2019)	30 wt %	Mixing	σ_m = 44.5 MPa (PLA) σ_m = 58.0 MPa (Ram/PLA) σ_m = 72.7 MPa (Ram-Alk/PLA) σ_m = 57.2 MPa (Ram-Sil/PLA) σ_m = 63.7 MPa (Ram-Alk-Sil/PLA) FS = 46.6 MPa (PLA) FS = 35.9 MPa (Ram/PLA) FS = 64.1 MPa (Ram-Alk/PLA) FS = 49.2 MPa (Ram-Sil/PLA) FS = 57.8 MPa (Ram-Alk-Sil/PLA) α = 4.9 kJ/m^2 (PLA) α = 10.4 kJ/m^2 (Ram/PLA) α = 12.1 kJ/m^2 (Ram-Alk/PLA) α = 9.4 kJ/m^2 (Ram-Sil/PLA) α = 7.9 kJ/m^2 (Ram-Alk-Sil/PLA)
			σ_m = 44.5 MPa (PLA)

TABLE 3.4 *(Continued)*

Reinforcement agent	Concentration	Methodology	Results
Alkali treated, silane treated, alkali-silane treated, and no treated (Debeli et al., 2019)	30 wt %	Mixing	σ_m = 58.3 MPa (Ram-Alk-DAP/PLA) σ_m = 53.4 MPa (Ram-DAP/PLA) σ_m = 17.0 MPa (Ram-Sil-DAP/PLA) σ_m = 38.7 MPa (Ram-Alk-Sil-DAP/PLA) FS = 46.6 MPa (PLA) FS = 38.2 MPa (Ram-Alk-DAP/PLA) FS = 37.3 MPa (Ram-DAP/PLA) FS = 25.8 MPa (Ram-Sil-DAP/PLA) FS = 34.7 MPa (Ram-Alk-Sil-DAP/PLA) α = 8.1 kJ/m^2(Ram-Alk-DAP/PLA) α = 8.7 kJ/m^2 (Ram-DAP/PLA) α = 5.2 kJ/m^2 (Ram-Sil-DAP/PLA) α = 6.5 kJ/m^2 (Ram-Alk-Sil-DAP/PLA)
Bamboo Fiber (BF) to obtain PLA-g-BF/PLA and BF/PLA (Li et al., 2019)	-	Mixing/ extrusion	σ_m = 16.1 MPa (BF/PLA) σ_m = 19.43 MPa (PLA-g-BF/PLA) ε_b = 4.27 % (BF/PLA) ε_b = 5.59 % (PLA-g-BF/PLA) E = 4.5 GPa (BF/PLA) E = 5.1 GPa (PLA-g-BF/PLA)
Treated cellulose fibers, using recycled PLA (PLAr) as matrix (Laadila et al., 2017)	2, 5 and 10 wt %	Dry blending	E = 644.4 MPa (PLAr) E = 887.8 MPa (PLAr+2% fibers) E = 649.1 MPa (PLAr+5% fibers) E = 714.1 MPa (PLAr+10% fibers)

TABLE 3.5 Nanoparticles in PLA.

Reinforcement agent	Concentration	Methodology	Results
Carbon nanotubes, (CNTs), Chitosan (CS) (Liu et al., 2019)	1, 3, 5, 7 and 9 wt. % of CS	Stirring solution/ electrospinning	σ_m = 28.01 MPa (PLA/CNTs) σ_m = 38.45 MPa (PLA/CNTs/CS-5) ε_b = 83.41% (PLA/CNTs) ε_b = 73.57% (PLA/CNTs/CS-1)

TABLE 3.5 *(Continued)*

Reinforcement agent	Concentration	Methodology	Results
Chitin nanocomposite (ChNCs) (Herrera et al., 2016)	Reference compound: PLA/PBAT/GTA/Talc (60/24/10/6 wt %) and nanocomposite compound: PLA/PBAT/GTA/Talc/ChNCs (59/25/9/6/1 wt %)	Extrusion/blown film machine direction (MD)	σ_m = 33 MPa (PLA) σ_m = 35 MPa (PLA/ChNCs) E = 0.8 GPa (PLA) E = 0.7 GPa (PLA/ChNCs) ε_b = 282% (PLA) ε_b = 281% (PLA/ChNCs)
		Extrusion/blown film, transverse direction, (TD)	σ_m = 17 MPa (PLA) σ_m = 28 MPa (PLA/ChNCs) E = 0.4 GPa (PLA) E = 0.7 GPa (PLA/ChNCs) ε_b = 174% (PLA) ε_b = 220% (PLA/ChNCs)
High lignin containing cellulose nanocrystals (HLCNCs) and commercial lignin coated CNCs (BLCNCs) (Wei et al., 2018)	0.5, 1, 2 and 5 wt %	Mixing/extrusion	σ_m = 64.4 MPa (PLA) σ_m = 61.2 MPa (PLA/0.5%HLCNC) σ_m = 60.3 MPa (PLA/0.5%BLCNC) ε_b = 7.1% (PLA) ε_b = 11.4% (PLA/0.5%HLCNC) ε_b = 8.2% (PLA/0.5%BLCNC) E = 3.7 GPa (PLA) E = 4.3 GPa (PLA/0.5%HLCNC) E = 4.1 GPa (PLA/0.5%BLCNC)
Polyethylene glycol (PEG) block (PEG*b*-PLLA), Imidazolium group (Im-PLLA) and CNC (Mariano et al., 2017)	20 wt % of CNC	Extrusion	σ_m = 40 MPa (PLA) σ_m = 33.2 MPa (PLAPEG-*b*-PLLA) σ_m = 31.7 MPa (PLAIm-PLLA) σ_m = 33 MPa (PLAPEG-*b*-PLLA+20CNC) σ_m = 14.8 MPa (PLAIm-PLLA+20CNC) ε_b = 3.5% (PLA) ε_b = 2.6% (PLAPEG-*b*-PLLA) ε_b = 2.4% (PLAIm-PLLA) ε_b = 1.9% (PLAPEG-*b*-PLLA+20CNC) ε_b = 1.6% (PLAIm-PLLA+20CNC) E = 2 GPa (PLA) E = 1.7 GPa (PLAPEG-*b*-PLLA) E = 1.9 GPa (PLAIm-PLLA) E = 2.2 GPa (PLAPEG-*b*-PLLA+20CNC)

TABLE 3.5 *(Continued)*

Reinforcement agent	Concentration	Methodology	Results
CNCs from waste cotton cloth cellulose (Wang et al., 2019)	0.1, 0.3, 0.5 and 0.7 wt %	Mixing	E = 2.2 GPa (PLAIm-PLLA+20CNC) σ_m = 30 MPa (PLA) σ_m = 294 MPa (PLA-0.1% CNCs) ε_b = 20.1% (PLA) ε_b = 5.2% (PLA-0.1% CNCs) E = 2.8 GPa (PLA) E = 12.8 GPa (PLA-0.1% CNCs)
CNC (extracted form coffee silverskin) (Sung et al., 2017)	1, 3 and 5 wt %	Extrusion	σ_m = 54.6 MPa (PLA) σ_m = 56.2 MPa (PLA/CNC 1%) ε_b = 6.3% (PLA) ε_b = 6.4% (PLA/CNC 1%) E = 1166 MPa (PLA) E = 1343 MPa (PLA/CNC 5%)

TABLE 3.6 Materials en PLA.

Reinforcement agent	Concentration	Methodology	Results
Carbon fiber and modified carbon fiber (Li et al., 2016)	-	Extrusion	σ_m = 28 MPa (PLA) σ_m = 80 MPa (carbon fiber/PLA) σ_m = 91 MPa (modified carbon fiber/PLA) FS= 53 MPa (PLA) FS= 59 MPa (carbon fiber/ PLA) FS= 156 MPa (modified carbon fiber/PLA)
Microfibrillated wood (MFW) Microfibrillated lignocellulose (MFLC) (Winter et al., 2018)	1 wt %	Extrusion	E = 3.5 GPa (PLA) E = 3.6 GPa (PLA-MFW) E = 3.3 GPa (PLA-MFLC) σ_m = 107 MPa (PLA-MFW) σ_m = 91 MPa (PLA-MFLC) α = 17.3 kJ/m^2 (PLA) α = 18.0 kJ/m^2 (PLA-MFW) α = 19.1 kJ/m^2 (PLA-MFLC)
Diatomaceous Earth (DE) (Aggarwal et al., 2019)	5, 10 and 15 wt %	Extrusion	σ_m = 41.69 MPa (PLA) σ_m = 38.65 MPa (PLA/DE-5%) E = 941.61 MPa (PLA) E = 990.40 MPa (PLA/DE-5%)

3.6 3D PRINTING AS A METHOD OF INCORPORATING REINFORCEMENT AGENTS IN PLA MATRICES

The 3D printing method is extensively studied in the manufacturing of PLA thermoplastic resins due to the good operating characteristics, low cost, and no need for tooling or molding. But the quick procurement of fiber-reinforced PLA composites with outstanding mechanical properties by 3D printing is still a major challenge (Li et al., 2016).

Some materials were integrated into the PLA (matrix) and used in 3D printing, e.g. the carbon fiber reinforced PLA composite was developed through the rapid prototyping method to 3D printing (Li et al., 2016). The findings showed that the tensile strength and flexicurity of the improved carbon fiber reinforced composites are 13.8 and 164% higher than the initial carbon fiber reinforced samples (Li et al., 2016). In addition, MFW and MFLC filaments with PLA have been prepared. The bars printed with PLA-MFW behaved similar to pure PLA (no significant difference), although PLA-MFW demonstrated marginally increased impact strength (Winter et al., 2018). Diatomaceous earth has also been used as a possible part of PLA. Adding DE to the PLA created an improvement in the elasticity module (Aggarwal et al., 2019).

3.7 CONCLUSIONS

Polylactic acid is at the forefront of biopolymers and introduces itself as an alternative to traditional polymers. Despite its high mechanical properties, many materials have been integrated into PLA to improve properties such as impact resistance. This work provides a description of the use of various types of reinforcement materials to improve the mechanical properties of PLA and extrusion as a form of integration. It should be considered that the integration of such reinforcement materials into the PLA can positively increase the tensile strength and the Young's composite modulus. This study of the role of reinforcement agents in the mechanical properties of PLA opens up a vision of the vast field of operation for the application of this biopolymer by optimizing its properties in order to focus on a particular application.

KEYWORDS

- **mechanical properties**
- **food packaging**
- **biocompatibility**
- **biopolymers**
- **polylactic acid (PLA)**
- **organic materials**

REFERENCES

Aggarwal, S.; Johnson, S.; Saloni, D.; Hakovirta, M. Novel 3D Printing Filament Composite using Diatomaceous Earth and Polylactic Acid for Materials Properties and Cost Improvement. *Compos. B: Eng.*, **2019,** *177*. https://doi.org/10.1016/j.compositesb.2019.107310

Asadi, M.; Bazyar, B.; Hemmasi, A. H.; Ghsemi, I.; Talaeipoor, M. Assessment of Mechanical and Morphological Properties of New Poly Lactic Acid (PLA)/Wood Fibers/Nanographene Composite. *Drvna Industrija*, **2018,** *69* (2).

Carbonell-Verdu, A.; Samper, M. D.; Garcia-Garcia, D.; Sanchez-Nacher, L.; Balart, R. Plasticization Effect of Epoxidized Cottonseed Oil (ECSO) on Poly (Lactic Acid). *Indus. Crops Prod.* **2017,** *104*. https://doi.org/10.1016/j.indcrop.2017.04.050

Cisneros-López, E. O.; Pérez-Fonseca, A. A.; González-García, Y.; Ramírez-Arreola, D. E.; González-Núñez, R.; Rodrigue, D.; Robledo-Ortíz, J. R. Polylactic Acid–Agave Fiber Biocomposites Produced by Rotational Molding: A Comparative Study with Compression Molding. *Adv. Polym. Technol.* **2018,** *37* (7). https://doi.org/10.1002/adv.21928

Debeli, D. K.; Zhang, Z.; Jiao, F.; Guo, J. Diammonium Phosphate-Modified Ramie Fiber Reinforced Polylactic Acid Composite and Its Performances on Interfacial, Thermal, and Mechanical Properties. *J. Nat. Fibers* **2019,** *16* (3). https://doi.org/10.1080/15440478.2017.1423255

Ge, Z.; Qi, Z.; Si, D.; Yu, M. The Effects of Processing Parameters and ac Foaming Agent on the Mechanical Properties and Morphology of Foamed Wood-polylactic Acid (PLA) Composites. *BioRes.* **2018,** *13* (1). https://doi.org/10.15376/biores.13.1.1605-1618

Hammiche, D.; Boukerrou, A.; Azzeddine, B.; Guermazi, N.; Budtova, T. Characterization of Polylactic Acid Green Composites and Its Biodegradation in a Bacterial Environment. *Int. J. Polym. Anal. Charact.* **2019,** *24* (3). https://doi.org/10.1080/1023666X.2019.1567083

Herrera, N.; Roch, H.; Salaberria, A. M.; Pino-Orellana, M. A.; Labidi, J.; Fernandes, S. C. M.; Radic, D.; Leiva, A.; Oksman, K. Functionalized Blown Films of Plasticized Polylactic Acid/Chitin Nanocomposite: Preparation and Characterization. *Mater. Des.* **2016,** *92*. https://doi.org/10.1016/j.matdes.2015.12.083

Holcapkova, P.; Stloukal, P.; Kucharczyk, P.; Omastova, M.; Kovalcik, A. Anti-Hydrolysis Effect of Aromatic Carbodiimide in Poly(Lactic Acid)/Wood Flour Composites. *Compos. Part A: Appl. Sci. Manuf.* **2017,** *103*. https://doi.org/10.1016/j.compositesa.2017.10.003

Laadila, M. A.; Hegde, K.; Rouissi, T.; Brar, S. K.; Galvez, R.; Sorelli, L.; Cheikh, R. B., Paiva, M.; Abokitse, K. Green Synthesis of Novel Biocomposites from Treated Cellulosic Fibers and Recycled Bio-Plastic Polylactic Acid. *J. Clean. Prod.* **2017,** *164*. https://doi.org/10.1016/j.jclepro.2017.06.235

Li, W.; He, X.; Zuo, Y.; Wang, S.; Wu, Y. Study on the Compatible Interface of Bamboo Fiber/Polylactic Acid Composites by In-Situ Solid Phase Grafting. *Int. J. Biol. Macromol.* **2019,** *141*. https://doi.org/10.1016/j.ijbiomac.2019.09.005

Liu, Y.; Wang, S.; Lan, W.; Qin, W. Fabrication of Polylactic Acid/Carbon Nanotubes/Chitosan Composite Fibers by Electrospinning for Strawberry Preservation. *Int. J. Biol. Macromol.* **2019,** *121*. https://doi.org/10.1016/j.ijbiomac.2018.09.042

Long, H.; Wu, Z.; Dong, Q.; Shen, Y.; Zhou, W.; Luo, Y.; Zhang, C.; Dong, X. Effect of Polyethylene Glycol on Mechanical Properties of Bamboo Fiber-Reinforced Polylactic Acid Composites. *J. Appl. Polym. Sci.* **2019,** *136* (26). https://doi.org/10.1002/app.47709

Mariano, M.; Pilate, F.; de Oliveira, F. B.; Khelifa, F.; Dubois, P.; Raquez, J. M.; Dufresne, A. Preparation of Cellulose Nanocrystal-Reinforced Poly(Lactic Acid) Nanocomposites through Noncovalent Modification with PLLA-Based Surfactants. *ACS Omega* **2017,** *2* (6). https://doi.org/10.1021/acsomega.7b00387

Miao, L.; Walton, W. C.; Wang, L.; Li, L.; Wang, Y. Characterization of Polylactic Acids-Polyhydroxybutyrate Based Packaging Film With Fennel Oil, and Its Application on Oysters. *Food Pack. Shelf Life* **2019,** *22*. https://doi.org/10.1016/j.fpsl.2019.100388

Nilsuwan, K.; Benjakul, S.; Prodpran, T. Physical/Thermal Properties and Heat Seal Ability of Bilayer Films Based on Fish Gelatin and Poly(Lactic Acid). *Food Hydrocol.* **2018,** *77*. https://doi.org/10.1016/j.foodhyd.2017.10.001

Orue, A.; Eceiza, A.; Arbelaiz, A. Preparation and Characterization of Poly(Lactic Acid) Plasticized with Vegetable Oils and Reinforced with Sisal Fibers. *Indus. Crops Prod.* **2018,** *112*. https://doi.org/10.1016/j.indcrop.2017.11.011

Reis, M. O.; Olivato, J. B.; Bilck, A. P.; Zanela, J.; Grossmann, M. V. E.; Yamashita, F. Biodegradable Trays of Thermoplastic Starch/Poly (Lactic Acid) Coated with Beeswax. *Indus. Crops Prod.* **2018,** *112*. https://doi.org/10.1016/j.indcrop.2017.12.045

Santos, D.; Correia, C. O.; Silva, D. M.; Gomes, P. S.; Fernandes, M. H.; Santos, J. D.; Sencadas, V. Incorporation of Glass-Reinforced Hydroxyapatite Microparticles into Poly(Lactic Acid) Electrospun Fibre Mats for Biomedical Applications. *Mater. Sci. Eng. C* **2017,** *75*. https://doi.org/10.1016/j.msec.2017.03.007

Scaffaro, R.; Lopresti, F.; Botta, L. PLA Based Biocomposites Reinforced with Posidonia Oceanica Leaves. *Comp. B: Eng.* **2018,** *139*. https://doi.org/10.1016/j.compositesb.2017.11.048

Scaffaro, R.; Lopresti, F.; D'Arrigo, M.; Marino, A.; Nostro, A. Efficacy of Poly(Lactic Acid)/Carvacrol Electrospun Membranes Against Staphylococcus Aureus and Candida Albicans in Single and Mixed Cultures. *Appl. Microbiol. Biotechnol.* **2018,** *102* (9). https://doi.org/10.1007/s00253-018-8879-7

Sung, S. H.; Chang, Y.; Han, J. Development of Polylactic Acid Nanocomposite Films Reinforced with Cellulose Nanocrystals Derived from Coffee Silverskin. *Carbohydr. Polym.* **2017,** *169*. https://doi.org/10.1016/j.carbpol.2017.04.037

Suryanegara, L.; Kurniawan, Y. D.; Syamani, F. A.; Nurhamiyah, Y. Mechanical Properties of Composites based on Poly(Lactic Acid) and Soda-Treated Sugarcane Bagasse Pulp. In *Sustainable Future for Human Security: Society, Cities and Governance*; 2017. https://doi.org/10.1007/978-981-10-5433-4_19

Wang, Z.; Yao, Z.; Zhou, J.; He, M.; Jiang, Q.; Li, A.; Li, S.; Liu, M.; Luo, S.; Zhang, D. Improvement of Polylactic Acid Film Properties Through the Addition of Cellulose Nanocrystals Isolated from Waste Cotton Cloth. *Int. J. Biol. Macromol.* **2019,** *129.* https://doi.org/10.1016/j.ijbiomac.2019.02.021

Wei, L.; Agarwal, U. P.; Matuana, L.; Sabo, R. C.; Stark, N. M. Performance of High Lignin Content Cellulose Nanocrystals in Poly(Lactic Acid). *Polymer* **2018,** *135.* https://doi.org/10.1016/j.polymer.2017.12.039

Winter, A.; Mundigler, N.; Holzweber, J.; Veigel, S.; Müller, U.; Kovalcik, A.; Gindl-Altmutter, W. Residual Wood Polymers Facilitate Compounding of Microfibrillated Cellulose with Poly(Lactic Acid) for 3D Printer Filaments. *Philos. Transac. R. Soc. A: Math. Phys. Eng. Sci.* **2018,** *376* (2112). https://doi.org/10.1098/rsta.2017.0046

CHAPTER 4

Thermal Properties by Adding Natural Oils, Foods, Organic Materials, Fibers, and Nanocomposites in the PLA, and Applications in 3D Printing

Y. K. REYES ACOSTA[1], W. E. CRUZ MARTINEZ[1], A. V. REYES ACOSTA[2], V. A. CEPEDA TOVAR[1], J. C. CONTRERAS ESQUIVEL[1], C. N. AGUILAR GONZALES[1], R. I. NARRO CESPEDES[1], and R. REYNA MARTINEZ[3]

[1]*Facultad de Ciencias Químicas de la Universidad Autónoma de Coahuila, Blvd. V. Carranza s/n, Saltillo, Coahuila, México*

[2]*Facultad de Sistemas de la Universidad Autónoma de Coahuila, Zona Centro, Arteaga, Coahuila, México*

[3]*Colegio de Estudios Científicos y Tecnológicos del Estado de Coahuila Plantel San Antonio de las Alazanas, Arteaga, Coahuila, México*

ABSTRACT

Polylactic acid (PLA) is a biodegradable thermoplastic and aliphatic polyester derived from sustainable and sustainable starch rich materials with important physicochemical and biological properties. These properties are targeted for applications in a broad range of fields, especially in the packaging, textile, agricultural, pharmaceutical, electronics, and biomedical sectors. PLA is one of the most significant biopolymers due to its biodegradability, biocompatibility, and thermal properties with glass transition temperature degradation temperature, comparable to polystyrene (PS) and polyethylene terephthalate (PET). However, due to their lower thermal properties, many materials have

Engineering Principles for Food Processing Technology and Product Realization. Leonardo Sepúlveda Torre, Porteen Kannan, A. K. Haghi, PhD (Eds.)

been used to improve. By adding, such as food and its derivatives, natural oils, organic material, various types of fibers, nanoparticles, and others have been integrated into the PLA matrix with the objective of improved thermal properties. This study presents a review of the research on PLA and the use of enhancers, as well as the findings of the thermal property's improvement.

4.1 INTRODUCTION

Thermal properties are those properties of materials that are related to their heat conductivity, in other words, the properties that a material exhibits when it is heated.

The increment in molecular mobility by increasing temperature induces changes in properties such as density, heat capacity, electrical properties, thermal conductivity, optical or mechanical properties, and rheological behavior. In addition, phase transitions can occur associated with dramatic modifications within the structure of materials, as in the case of composite fabrication.

Consequently, thermal behavior is essential to understand fundamental aspects of the structure, kinetics of crystallization, and properties of a polymer in its final application.

Below are presented some studies in which the thermal properties of materials have been of great importance for the development of different technologies. In addition to this, many government organizations in all countries have focused their efforts in seeking environmental protection for decades, seeking in this way the friendliest processing of resources with ecosystems, which will achieve a better use of natural resources—fossil source or its elimination in the future; with this, countries are forced to obtain a more conscious environmental legislation that promotes the quality of the environment in the near future. As a result, the sustainable and recyclable use of resources plays an important role in solving the growing demand for wood resources and the scarcity of forest resources, thereby achieving a reduction in the pressure on the preservation of the environment. Plastic and wood source composites have become the most optimal and sustainable alternatives to replace green forest source wood, due to their very good capabilities and lower cost, since the materials emerged on the market (Nilsuwan et al., 2018). Some applications have used this improvement in obtaining materials from sustainable raw materials, such as boards, flooring, public and private infrastructure, components such as interior decorations, furniture material, automotive trim, and of course, engineering applications (Carbonell-Verdu et al., 2017; Reis et al., 2018)—due to properties such as high resistance to putrefaction, decomposition, contact with insects and corrosion, followed by lower moisture absorption, higher rigidity strength and thermal stability (Carbonell-Verdu et al., 2017; Orue et al., 2018; Reis et al., 2018). In the manufacturing process, it consists of mixing wood fibers with polymers considered thermoplastic (Aggarwal et al., 2019; Santos et al., 2017); this is how thermoplastic materials exert important influences on the capacity of the resulting compounds. This is where the use of polylactic acid (PLA) influences, as a precursor in the manufacture of new tangible goods considered important in the field of 3D printing engineering (Scaffaro et al., 2018).

In this sense, the different technologies that have been developed seek a significant improvement in the properties of materials in general, with thermal properties being those that directly influence the conformational structure of the final products. In this way, it is important to consider how their effect influences the finished products, specifically in the food, pharmaceutical, and, of course, 3D printing areas. 3D printing is becoming very popular due to its wide spectrum in the creation of new prototypes that seek to meet the needs at low manufacturing cost and really short lead times (Li et al., 2016; Sun et al., 2021; Wang et al., 2021). Likewise, the biopolymer considered PLA and its respective PLA-wood compounds have been used in the formulation of filaments for 3D printing with properties that have been considered green technologies. PLA is considered to be one of the biocomposites obtained from agricultural sources; it has a great potential to be an alternative as a self-sustaining material coupled or in substitution for petroleum-based polymers. Likewise, inorganic loads from wood fibers are considered renewable and biodegradable; due to this, the implementation of this compound on polymeric matrices has resulted in the production of new materials that have significantly improved the development of new commercial products (Huang et al., 2021; Jin et al., 2017; le Duigou et al., 2016; Sun et al., 2021; Z. Wang et al., 2021; Zhou et al., 2015). Therefore, the objective of this present review was to understand the behavior depending on the thermal decomposition of biopolymers formed by the natural addition of oils, organic materials, fibers, and PLA nanocomposites as fillers destined for the process and future commercial applications in the field of 3D printing.

4.2 FOODS AND NATURAL OILS AS ADDITAMENTS TO PLA

A study carried out by Nilsuwan et al. (2018) showed that when formulating bilayer films of fish gelatin (FG) and PLA in different concentrations, they presented a glass transition temperature (Tg) higher than those of PLA independently, demonstrating that the bilayer FG/PLA films could be used as a promising food packaging as they have a higher Tg producing a high resistance to an increment of temperature, and it is biodegradable.

Another study carried out by Altan et al. (2018) demonstrates an important change in the thermal properties of materials in which they designed fibrous films composed of zein and PLA by incorporating carvacrol in three different concentrations or weight percentage (5, 10, and 20% wt) using the electrospinning technique. The results show an effect on carvacrol incorporation and thermal stability of electrospun zein and PLA fibers. The thermogravimetric analysis (TGA) curve for pure zein fiber shows two stages of weight loss as a function of temperature. The first stage occurs in the range of 130–220 °C with a weight loss of around 3.4%, while the second stage is observed between 240 and 400 °C. Both weight losses are attributed to the main degradation of zein, which was observed to be relatively slow at first. Pure carvacrol completely evaporates at around 170°C. Zein fibers show a further stage of weight loss between 75 and 150 °C after carvacrol incorporation. Carvacrol weight loss was calculated as ~3.3, 6.1, and 9.2% for 5, 10, and 20% carvacrol, respectively, indicating indirect confirmation of the presence of carvacrol in zein fibers. The maximum temperatures at the highest rate of weight loss are 329, 330, and 326 °C.

These results show that the incorporation of carvacrol in zein does not significantly affect the thermal degradation profile of zein fibers. The addition of carvacrol could act as a natural plasticizer and disrupt the interactions between the zein molecules. Therefore, no significant improvement in thermal stability was observed for the carvacrol-loaded zein fibers.

Another case is the study carried out by Reis et al. (2018). Biodegradable trays of thermoplastic starch (TPS)/PLA blends were produced by flat extrusion, calendering, and thermopressing. The trays were coated by immersion in beeswax (BW) emulsion (1, 2, and 3 g BW/100 g ethyl alcohol + Tween 80 solution) to improve their barrier properties. The thermal stability of the coated and non-coated TPS/PLA trays was measured by TGA, and all the samples had two weight-loss steps. The first step was related to the starch and the second one corresponded to the thermal degradation of PLA. The BW content (1, 2, and 3% wt) affected the thermal stability of glycerol, although the thermal stability of starch and PLA remains unchanged in the coated samples, showing no interactions between the materials and different mechanisms of thermal degradation.

Some studies demonstrate the importance of thermal properties in polymers when making composite products or mixtures with natural oils; for example, Carbonell-Verdu et al. (2017) studied new formulations of environmentally friendly plasticized PLA incorporating epoxidized cottonseed oil (ECSO). The Tg dropped by about 5–6 °C for formulations with 7.5–10.0% wt ECSO. This decrease in Tg evidenced an increase in free volume with a positive effect on the mobility of the chain, so that both the Tg and the cold crystallization temperature (Tc) move to lower values. Therefore, the results obtained in this work suggest that highly environmentally friendly toughened PLA formulations can be obtained using ECSO in the range of 5–10% wt.

Another case is the study carried out by Orue et al. (2018) in which are prepared and characterized PLA plasticized with vegetable oils and reinforced with sisal fibers; their DSC study show that PLA plasticized with epoxidized oils have lower Tg values than PLA plasticized with non-epoxidized oils. Although the addition of plasticizer accelerates the crystallization process of PLA, the values of the degree of crystallinity increase slightly. After the addition of sisal fibers in PLA systems plasticized with 10% wt of vegetable oils, in addition to increase the degree of crystallinity of the samples at least twice due to the nucleating effect of the sisal fibers, the values of Tg and onset Tc decrease slightly. Regarding the cold Tc values, it is observed that after the addition of plasticizers, the Tc decrease considerably, and the crystallization process is accelerated due to the increase in the segmental mobility of the PLA chains by plasticization.

There are several cases, such as those presented in Table 4.1, we show some studies carried out and how the addition of various materials influences the thermal properties of polymers.

4.3 ORGANIC MATERIALS INCORPORATING IN PLA

In the organic materials field, there exists applications in which the thermal properties of the polymers vary depending on the complement in which they work; this undoubtedly represents a great area of opportunity for the development of new compounds. Below we mention some examples.

Aggarwal et al. (2019) used diatomaceous earth as a potential component for PLA-based 3D printing composites. DSC analyzes show that the Tc in the filament composites shows a clear shift toward a Tc lower for PLA-DE filaments compared to 100% PLA; the authors show that diatomaceous earth can be used as a filler for PLA to create a composite material for 3D printing filaments, showing no major degradation of material properties after 3D printing.

Santos et al. (2017) incorporated glass-reinforced hydroxyapatite macroparticles into PLA electrospun fiber meshes for biomedical

TABLE 4.1 Foods and Natural Oils Addition in PLA: Effect in the Stability Thermal.

Complement	Concentration	Method	Results
Fish gelatin (FG) (Nilsuwan et al., 2018)	Layer thickness ratios of 9:1, 8:2, 7:3, 6:4, and 5:5 (FG/PLA)	Casting	Tg = 33.25 °C (PLA) Tg = 43.09 °C (FG/PLA 6:4) Tc = 88.7 °C (PLA) Tc = 89.07 °C (FG/PLA 6:4) Tmax = 156.28 °C (PLA) Tmax = 145.11–156.28 °C (FG/PLA 6:4)
Carvacrol (CRV) (Altan et al., 2018)	5%, 10%, and 20% wt	Mixing/electros-pinning	T_{TGA} = 346 °C (PLA) T_{TGA} = 377 °C (PLA-5 CRV)
Thermoplastic starch (TPS) (Reis et al., 2018)	1, 2, and 3 g BW/100 g ethyl alcohol + Tween 80 solution	Flat extrusion	T_{TGA} = 360 °C (PLA) T_{TGA} = 360 °C (PLA-5 CRV)
Epoxidized cottonseed oil (ECSO) (Carbonell-Verdu et al., 2017)	2.5%, 5%, 7.5%, and 10% wt	Mixing	Tg = 66.7 °C (PLA) Tg = 63.76 °C (PLA-2 ECSO) Tc = 97.16 °C (PLA) Tc = 91.17 °C (PLA-2 ECSO) Tm = 169.94 °C (PLA) Tm = 172.94 °C (PLA-2 ECSO) T_{TGA} = 335.5 °C (PLA) T_{TGA} = 340 °C (PLA-10 ECSO)
Epoxidized (ELO, ESBO) and non-epoxidized linseed oil (LO) and soybean oil (SBO) (Orue et al., 2018)	5%, 10%, and 20% wt	Mixing	Tg = 66.2 °C (PLA) Tg = 53.46 °C (PLA-5 LO) Tc = 123.8 °C (PLA) Tc = 114.4 °C (PLA-5 SBO) Tm = 154–157.9 °C (PLA) Tm = 150.8–157.8 °C (PLA-5 SBO) T_{TGA} = 358.3 °C (PLA) T_{TGA} = 357.2 °C (PLA-10 SBO)
	10% wt of oils with 30% wt of fiber		Tg = 56 °C (PLA/sF) Tg = 52.4 °C (PLA/sBO/sF) Tc= 102.2 °C (PLA/sF) Tc= 115 °C (PLA/sBO/sF) Tm= 149.2–158.7 °C (PLA/sF) Tm= 148.4–154.9 °C (PLA/sBO/sF)

applications. In their study, differential scanning calorimetry was performed on the electrospun fibrous mats to study the thermal properties of the materials. Electrospun pure PLLA fibers show a large overshoot in the Tg, around 60 °C, which is related to the enthalpy recovery of the sample stored at room temperature, around 30 °C below Tg, and therefore subject to physical aging. As the temperature increases, a cold crystallization process is observed immediately after the glass transition, which shows up as a broad exothermic peak between 70 and 110 °C with a maximum (Tc) of around 81 °C. Melting begins at 130–160 °C. Microcomposite membranes present a glass transition of the polymer shifted toward low temperatures, with a maximum around 51 °C, and the cold crystallization process that occurs immediately after Tg shows a peak of ~73 °C. Finally, the authors concluded that the presence of the ceramic microparticles plays an important role in the thermal degradation of the polymeric matrix, and a decrease in the activation energy of the thermal degradation of 134 kJ.mol–[1].

There are many studies like these, in which the importance of thermal properties in materials is highlighted before generating a final product, which are summarized in Table 4.2.

TABLE 4.2 Organic Materials Addition in PLA: Effect in the Stability Thermal.

Complement	Concentration	Method	Results
Diatomaceous earth (DE) (Aggarwal et al., 2019)	5%, 10%, and 15% wt	Extrusion	Tc = 115.3 °C (PLA) Tc = 109.2 °C (PLA/DE-5) Tm = 148.5 °C (PLA) Tm = 148.3–157.9 °C (PLA/DE-5)
Bonelike® and PLLA (Santos et al., 2017)	80% wt	Add to solution and electrospinning	Tg = 60 °C (PLLA) Tg = 51 °C (Bonelike® PLLA) Tc = 81 °C (PLLA) Tc = 73 °C (Bonelike® PLLA)
Leaves of Posidonia oceanica (PO) (Scaffaro et al., 2018)	10% and 20% wt with PO particles size <150 μm and 150–300 μm	Mixing	Tg = 56 °C (PLA) Tg = 54.8 °C (PLA/PO300-10) Tc = 105.8 °C (PLA) Tc = 101.1 °C (PLA/PO150-10) Tm = 161.7–167.8 °C (PLA) Tm = 161.6–169.8 °C (PLA/PO300-10)

TABLE 4.2 *(Continued)*

Complement	Concentration	Method	Results
Wood flour (WF) and ZIKA additive (Holcapkova et al., 2017)	20% wt of WF	Mixing	Tg = 59.1 °C (PLA) Tg = 58.6 °C (PLA/WF) Tc = 99.6–155.8 °C (PLA) Tc = 98.6–154.8 °C (PLA/WF) Tm = 168.4 °C (PLA) Tm = 167.4 °C (PLA-ZIKA) $T_{TGA\,(50\%)}$ = 358 °C (PLA) $T_{TGA\,(50\%)}$ = 359 °C (PLA-ZIKA)
Corn starch and wood flour (WF) with artificial weathering (300 h, 600 h, 900 h, and 1200 h) (Lv et al., 2018)	30% wt	Extrusion	Tg = 59.4 °C (PLA) Tg = 54.8 °C (PLA-1200) Tg = 58.2 °C (PLA/starch) Tg = 54.4 °C (PLA/starch-600) Tg = 58.1 °C (PLA/WF) Tg = 56.9 °C (PLA/WF-900) Tc = 99.6 °C (PLA) Tc = 93.4 °C (PLA-900) Tc = 90.1 °C (PLA/starch) Tc = 89.5 °C (PLA/starch-1200) Tc = 93.3 °C (PLA/WF) Tc = 90.1 °C (PLA/WF-600) Tm = 156.2 °C (PLA) Tm = 154.7 °C (PLA-1200) Tm = 151 °C (PLA/starch) Tm = 154.3 °C (PLA/starch-900) Tm = 154.4 °C (PLA/WF) Tm = 153.4 °C (PLA/WF-1200)

4.4 FIBERS INCORPORATING IN PLA

In this field, there are recent studies, such as Long et al. (2019), who studied the effect of polyethylene glycol on the mechanical properties of PLA compounds reinforced with bamboo fiber (BF). The results regarding the DSC analysis show that the crystallinity of PLA in composites with low BF content is increased. The crystallinity of PLA was 44.19% when the BF content was 15%. However, PLA crystallinity decreased when the BF content continued to increase. It can be concluded that a small amount of BF acted as a nucleating agent and accelerated crystallization rates; on the other hand, the excess of fiber hindered the crystallization of the matrix and reduced the thermal stability of the composites.

Another study carried out by Cisneros-López et al. (2018) looked into the development of biocomposites of agave fiber and PLA produced by rotational molding; in the thermal analysis, a Tg of 60–61 °C was obtained and a Tm of 168–169 °C, which could, again, indicate that both rotomolding and compression molding do not produce significant degradation as optimal processing conditions were used. The crystallinity of the rotomolded biocomposites (R-BC) increases 22% (from 18% for pure PLA to 22% with 10% wt of agave fiber). However, higher fiber content decreased crystallinity by up to 11% (16–40% wt). Compression molded biocomposites (CM-BC) present higher crystallinity than R-BC for all fiber contents. The level of crystallinity increased from 20% for PLA to 26–30% wt.

W. Li et al. (2019) conducted a study on the compatible interface of BF/PLA composites by in situ solid phase grafting. The DSC thermograms of the BF/PLA and PLA-g-BF/PLA systems show that the relative changes in the crystal structure of the BF/PLA and PLA-g-BF/PLA compounds were detected by the DSC test. The PLA-g-BF/PLA compounds show 170.63 °C values, which were higher than those of the unmodified compound (163.25 °C). The Tc increased considerably, and the crystallization process slowed down due to the higher mobility of the PLA chains. On the contrary, the BF/PLA compounds show a double crystallization peak, low and high temperature endotherm. It is attributed to the melting mechanism, recrystallization, and melting behaviors of the samples. It was concluded that the nucleation effect of the PLA matrix is enhanced and that crosslinking of the interface is enhanced. The thermal degradation of the compounds was divided into three stages with temperature ranges of 50–90, 240–375, and 375–550 °C. The first stage was the evaporation of water and other BF extracts. In the second stage, hemicellulose, cellulose, a part of the lignin in the BF, and most of the PLA were thermally decomposed. In the third stage (370 °C), the remaining lignin and PLA were decomposed. In addition, the carbon residue content of the PLA-g-BF/PLA composites at 550 °C is found to be marginally lower than that of the unmodified composites. The terminal temperature of thermal decomposition of the second stage for the composites prepared by treatment with graft polymers is 278.4 °C, higher than the 252.3 °C of the unmodified composites. The second thermal decomposition phase explained that the BF grafting treatment strengthened the interfacial interactions between the filler BF and the PLA matrix and reduced the mobility of the PLA molecular chain. However, in contrast to the present results, it has also been inferred in the above text that the BF modification improved the thermal stability of BF/PLA. Table 4.3 shows some progress in this area.

TABLE 4.3 Fibers Incorporating in PLA: Effect in the Stability Thermal.

Complement	Concentration	Method	Results
Alkali-treated Bamboo Fibers (ABF) (Long et al., 2019)	20% wt	Mixing and extrusion	Tc = 110.76 °C (PLA) Tc= 110.93 °C (ABF/PLA (10/90)) Tc= 110.98 °C (ABF/PLA/PEG (19/76/5)) Tm = 173.08 °C (PLA) Tm = 173.19 °C (ABF/PLA (10/90)) Tm = 171.56 °C (ABF/PLA/PEG (19/76/5))
Agave fibers (AF) (Cisneros-López et al., 2018)	10%, 20%, 30%, and 40% wt	Dry-blending and roto-molded (RM)	Tg= 60 °C (PLA) Tg = 61 °C (PLA/AF10) Tc= 102 °C (PLA) Tc= 104 °C (PLA/AF30) Tm = 169 °C (PLA) Tm = 168 °C (PLA/AF20)
		Dry-blending and compression-molded (CM)	Tg= 60 °C (PLA) Tg = 61 °C (PLA/AF20) Tc= 104 °C (PLA) Tc= 103 °C (PLA/AF10) T m = 169 °C (PLA) T m = 168 °C (PLA/AF20)
Bamboo fiber (BF) to obtain PLA-g-BF/PLA and BF/PLA (W. Li et al., 2019)	-	Graft reaction/extrusion	Tc= 167.36 °C (PLA) Tc= 163.25 °C (BF/PLA) Tc= 170.63 °C (PLA-g-BF/PLA)
Alkali treated, silane treated, alkali-silane treated, and no treated ramie fiber, no soaked and soaked in DAP solution (Debeli et al., 2019)	30% wt	Mixing	Tg = 54.9 °C (PLA) Tg= 62.4 °C (Ram-Sil/PLA) Tg= 60.3 °C (Ram-Alk/PLA) Tc = -- °C (PLA) Tc= 99.9 °C (Ram-Sil/PLA) Tc= 93.1 °C (Ram-Alk/PLA) Tm= 164.76 °C (PLA) Tm= 163.54 °C (Ram/PLA) Tm= 163.48 °C (Ram-Alk-Sil-DAP/PLA)

4.5 INCORPORATING THE NANOCOMPOSITES IN PLA

Herrera et al. (2016) evaluated bionanocomposite films prepared with compounds in melt and blown film for packaging applications. Their thermal analyzes show that the Tg and melting temperature (Tm) of the nanocomposite were slightly higher (33 and 172 °C) compared to the values of the reference films, which were 29 and 171 °C, respectively. The results indicate that the addition of 1% wt of Chitin nanocrystals (ChNC) slightly improved the thermal stability and the interaction between the components of the nanocomposite, which hinders the mobility of the polymer chain. The degree of crystallinity (Xc) of the nanocomposite films was slightly higher (56%) compared to the reference (52%), indicating higher nucleation due to chitin nanocrystals, which is consistent with our previous results. The higher crystallinity results in a higher enthalpy of fusion (ΔHm) of the nanocomposite. Thus, they concluded that the addition of ChNC to the compound increased its viscosity, modified the melt strength, and improved the thermal stability and crystallinity of the compound. The tear and puncture resistance of the reference films was increased by the addition of the nanocrystals and thus the polymer was reinforced.

Wei et al. (2018) evaluated the performance of cellulose nanocrystals with high lignin content as a filler in PLA; DSC analysis showed that the Tg was not affected by the incorporation of cellulose nanocrystals with high lignin content. lignin content (HLCNC), or commercial cellulose nanocrystals (BLCNC) in PLA. This indicated that the relatively small number and large sizes of CNCs are not enough to change the polymer chain mobility within the glass transition region. In the current study, the degree of crystallinity (Xc) was found to increase by adding HLCNC or BLCNC, because CNCs act as a nucleating agent for PLA compounds and facilitate PLA crystallization. When concentrations of CNC were increased from 2% to 5%, the Xc was slightly reduced. It is likely that this was due to a relatively poorer dispersion of the CNCs in the polymeric matrix and to the higher fraction of large agglomerates, which slightly inhibited the nucleation effect of the CNCs. It is interesting the representation of the double melting peaks (representing low to high temperature peaks) of the PLA/HLCNC compounds. It is speculated that there could be two types of crystal structures present in PLA/HLCNC nanocomposites. The lower temperature endotherm could be attributed to the melting of the original PLA crystals, while the higher temperature could be due to the melting of secondary crystallization resulting from the rearrangement of the crystallizable amorphous polymer segments around the nanolignin particles. This further confirmed the higher cold crystallization enthalpies, ΔHcc (J/g), of PLA/HLCNCs compared to pure PLA and the corresponding PLA/BLCNC system.

Incorporation of BLCNC or HLCNC resulted in significant improvements in both Young's modulus and elongation at break, but small decreases in tensile strength. When BLCNC (or HLCNC) increased from 2% to 5%, no additional increase in Young's modulus was observed. Similarly, the degree of PLA crystallinity (χc%) increased for both BLCNC and HLCNC. Dynamic mechanical analysis showed that the storage modulus of the nanocomposites increased by adding CNC, especially HLCNC. A broader tanδ peak indicated a better interaction between the HLCNCs and the polymeric matrix. It can be concluded that HLCNC derived from hydrothermally treated wood can be used as excellent fillers and/or reinforcing agents to improve the physical and thermal properties of polymeric nanocomposites. However, approaches to improve the dispersion of cellulose nanomaterials in PLA are still needed. Table 4.4 shows a summary of the area.

TABLE 4.4 Addition the Nanocomposites in PLA and the Stability Thermal.

Complement	Concentration	Method	Results
Chitin nanocrystals (ChNCs) (Herrera et al., 2016)	Reference compound: PLA/PBAT/GTA/Talc (60%/24%/10%/6% wt) and nanocomposite compound: PLA/PBAT/GTA/Talc/ChNCs (59%/25%/9%/6%/1% wt)	Extrusion	Tg = 29 °C (reference) Tg = 33 °C (nanocomposite) Tm = 171 °C (reference) Tm = 172 °C (nanocomposite) T_{TGA} = 327 °C (reference) T_{TGA} = 326 °C (nanocomposite)
High lignin containing cellulose nanocrystals (HLCNCs) and commercial lignin coated CNCs (BLCNCs) (Wei et al., 2018)	0.5%, 1%, 2%, and 5% wt	Extrusion	Tg = 60.8 °C (PLA) Tg = 60.9 °C (PLA/2 HLCNC) Tg = 60.8 °C (PLA/0.5 BLCNC) Tc = 117.4 °C (PLA) Tc = 115.8 °C (PLA/0.5 HLCNC) Tc = 115.8 °C (PLA/0.5 BLCNC) Tm = 149 °C (PLA) Tm = 148–154 °C (PLA/2 HLCNC) Tm = 148.4 °C (PLA/0.5 BLCNC)

TABLE 4.4 *(Continued)*

Complement	Concentration	Method	Results
Polyethylene glycol (PEG) block (PEG-*b*-PLLA), Imidazolium group (Im-PLLA), and CNC (Mariano et al., 2017)	20% wt of CNC	Extrusion	Tg = 61.5 °C (PLA) Tg = 58.6 °C (PLA/PEG-*b*-PLLA) Tg =56.4 °C (PLA/Im-PLLA) Tg = 59.9 °C (PLA/PEG-b-PLLA+10CNC) Tg = 56.3 °C (PLA/Im-PLLA+10CNC) Tm = 111.9 °C (PLA) Tm = 97.4 °C (PLA/PEG - b -PLLA) Tm = 93.1 °C (PLA/Im-PLLA) Tm = 99.2 °C (PLA/PEG - *b* -PLLA+10CNC) Tm = 94.9 °C (PLA/Im-PLLA+10CNC)
CNCs from waste cotton cloth cellulose (Z. Wang et al., 2019)	0.1%, 0.3%, 0.5%, and 0.7% wt	Mixing	Tg = 60.1 °C (PLA) Tg = 63.6 °C (PLA-0.5 CNCs) Tc = 114 °C (PLA) Tc = 134 °C (PLA-0.7 CNCs) Tm = 167.8 °C (PLA) Tm = 166.5 °C (PLA-0.7 CNCs)
CNC (extracted form coffee silverskin) (Sung et al., 2017)	1%, 3%, and 5% wt	Extrusion	Tg = 61.4 °C (PLA) Tg = 61.1 °C (PLA/CNC 1) Tc = 127.8 °C (PLA) Tc = 126.2 °C (PLA/CNC 1) Tm = 148.7 °C (PLA) Tm = 149.6 °C (PLA/CNC 1)
TiO_2 nanoparticles (González et al., 2018)	1%, 5%, 10%, and 20% wt	Mixing	$T_{TGA(5\%)}$ = 331.9 °C (PLA) $T_{TGA(5\%)}$ = 341.9 °C (PLA/TiO_2-21-10)

4.6 3D PRINTING AS A METHOD OF INCORPORATING REINFORCEMENT AGENTS IN PLA MATRICES

Today, in the scientific and technological field, many wood-based materials have been used, which are commonly valid for their various applications, to mention construction. However, some related to the properties of wood such as dimensional instability, biodegradability by microbial agents (fungi), and/or insects, restrict the use of materials in certain applications.

These undesirable properties could be improved by chemical treatments, as reported in the literature (Temiz et al., 2010, 2016). The general applications of composites structured by wood source fillers are decks, docks, design of garden items, and fences increase the wettability of composites, as cited in the literature (Morrell et al., 2006; Sommerhuber et al., 2016).

Bio-based polymers are susceptible to biodegradation under long periods of exposure in outdoor conditions, citing wood as a sustainable organic source (Khavkine et al., 2000). Kamdem et al. (2004) studied the feasibility of wood-structured composites (WPC) filled with CCA-treated wood particles mixed with raw high-density polyethylene (HDPE) and another from a recycled source, of which they reported an increase in physical properties, specifically resistance, biological durability, and photo protection. It is in this sense that scientific studies have been carried out that focus on the use of waste from wood sources as organic loads within polymer matrices from a sustainable source, such is the case of PLA.

As mentioned above, the use and exploitation of biodegradable polymers and biomaterials in the synthesis of plastics as an option for the protection and preservation of the environment, as well as the reduction of the use of petrochemical polymers, has attracted special attention in the scientific and industrial field. To exemplify, poly(lactic acid) or commonly known as PLA has become an important biodegradable and bio-based thermoplastic that is fully biodegradable, biocompatible, and produced from renewable source resources, mainly corn kernel starch or cob husks and of course sugar cane through highly sustainable bioconversion processes and polymerization routes (Carmona et al., 2015; Chaiwutthinan et al., 2015; Chuayjuljit et al., 2017; Ibrahim et al., 2017; Moura et al., 2014; van Cong et al., 2012; Zhang et al., 2019).

Currently, this biopolymer has been marketed at an increasing level in many countries to solve environmental problems caused by waste from synthetic sources (plastics). PLA disintegrates into practically harmless components for the environment, such as carbon dioxide (CO_2), water (H_2O), and other non-toxic small molecules that are safely released in landfills (Carmona et al., 2015; Petersson et al., 2007). In addition to this, it can be used in numerous applications, which provides advantages such as high tensile strength, elastic modulus and transparency, good processing and reaction performance when it is synthesized, recyclability, and of course its good availability in the market sustainable environment (Jiang et al., 2011; Mihai and Ton-That, 2019; Mohapatra et al., 2014; Syazeven Effatin Azma Mohd Asri et al., 2021). In addition, PLA is manufactured by many standard plastic processes, including injection, compression and blow molding, extrusion, thermoforming, and film formation (Makhetha, 2016). Next, the most common processes for the synthesis of this biopolymer are described, as well as an explanation of this compound charged by species that cause modifications in some of its physical, chemical, mechanical, and thermal properties.

In this concept, certain researchers modified PLA by incorporating it in a copolymerization process other polymers of synthetic and semi-synthetic origin, as a second polymeric phase, for example: poly-caprolactone (Chaiwutthinan et al., 2019), poly(butylenesuccinate) (Chaiwutthinan et al., 2015), thermoplastic polyurethane (Ibrahim et al., 2017), ethylene vinyl acetate (EVA) copolymer (Makhetha, 2016), poly(adipate-co-butylene terephthalate) (Chaiwutthinan et al., 2019), epoxidized natural rubber (Syazeven Effatin Azma Mohd Asri et al., 2021), and linear low density polyethylene, LDPE (Jiang et al., 2011), which manage to improve both the physical and mechanical properties and therefore its high processability. Among these polymers, EVA is a good candidate to blend with PLA due to its high flexibility and excellent impact resistance (Salehi et al., 2017). The wide range of applications that its use involves has interested in the scientific and technological field, one of them is 3D printing, mainly due to its easy availability and great use as an economical, viable, and sustainable product. It is commonly known as a mainly additive or sequential manufacturing technique or for the formation of products or prototypes (X. Wang et al., 2017).

The three-dimensional entities that can be rapidly elaborated under the control of the computer and software capable of producing a formed or preformed product (Xie et al., 2017). Fused deposition modeling (FDM) is the fastest, most widely used and easily accessible diffusion prototyping technology for the manufacture of next-generation polymer products (Tian et al., 2016), widely used for the manufacture of prototypes with certain complex structures and shapes using thermoplastic polymers, among these polymers, research has been carried out on acrylonitrile butadiene styrene (ABS) and, of course, PLA (Jin et al., 2017; le Duigou et al., 2016). In this sense, adding certain components that provide an improvement in properties benefits the polymer that is intended to be printed in 3D, thus giving it the best conditions before and after obtaining it as a finished product. With increasing environmental change and low availability of natural source resources, biodegradable materials have become an important topic in current research. PLA is considered a highly biodegradable and environmentally friendly biopolymer, which is why its use and disposal as a material for 3D printing has been considered (Bhattacharjee and Bajwa, 2017; Zhou et al., 2015). The

PLA monomer is lactic acid (LA), produced by fermentation of corn and other crops (Dong et al., 2014; Lai and Lan, 2013).

Increasing the use of PLA reduces the use of resources or fossil source species in an impressive proportion, according to the latest reports cited in the literature (Guo et al., 2018). The combination of PLA allows to improve the relatively low performance in question at its cost (Lee and Wang, 2006; X. Liu et al., 2017). Likewise, the by-products obtained from conventional wood are considered as they are quite cheap and easily available, useful for the formulation of compounds of sustainable origin, added as a by-product of the conventional tree processing industry (Pilla et al., 2009). These residues have advantages such as: low density, high rigidity, and, of course, degradability, as well as low cost (Xia et al., 2015). As reported, they have been designed to improve the toughness of PLA composites, including polymer blending and grafting (Rasal et al., 2010). Polyurethane-derived polymers (TPU) are the most widely used thermoplastic elastomers, ideal as a recyclable material and respectful of the ecosystem, allowing the formation of materials with a certain degree of crosslinking, thus providing good mechanical properties, biocompatibility, and biodegradability, respectively (Dogan et al., 2017; Ferri et al., 2016).

Just as the mixing of PLA with other polymers is very efficient, fillers as reinforcement provide an improvement in these systems of organic origin. However, the hydrophilicity that wood powders present, the compatibility when they are mixed mechanically, is poor, the polymeric matrices of hydrophobic character, which results in a somewhat deficient interfacial adhesion and therefore mechanical properties with little efficiency (D. Li et al., 2011; N. Wang et al., 2007). One of the most outstanding experimental developments in research to improve the compatibility between the polymer matrix and the reinforcements from wood residue is the use of compatibilizers, especially PLA grafted with glycidyl methacrylate (GMA), in the composites is relatively efficient and environmentally friendly, compared to conventional chemical treatments where only and exclusively these reinforcements are added (Hill et al., 1998; Mishra et al., 2003; Ray et al., 2001).

Among the compatibilizers, the compounds that are used found in the literature handle GMA, since it provides a relatively better compatibility between PLA and reinforcements such as wood dust or sawdust (Juntuek et al., 2012; J. Liu et al., 2012; Y. N. Wang et al., 2014). Guo et al. (2018) concluded that this addition of both functional groups in GMA is due to the acrylic group and the epoxy group, present in the polymers and the reinforcing fillers. The acrylic group can couple with the polymer chain and form a structure with good chemical stability. Meanwhile, the epoxy group can react with the hydroxyl group that is in fact present in the wood residue, since in some cases it presents a certain humidity typical of conventional wood.

Complementing the work carried out by Z. Wang et al. (2021) they designed four varieties of wood flour compounds (powder considered sawdust or WF) in the presence of the PLA biopolymer, modified together with organomontmorillonite (OMMT) as reinforcement and moisture precursor, through different modification process. These distributions were characterized by testing the mechanical capabilities of these resulting compounds. When analyzing the different processes modifying the OMMT, the mechanical capacities of the compounds were clearly different, as reported by the authors in the manuscript. However, an exception was presented in the toughness, almost in its entirety the mechanical capacities showed improvements after the modification by OMMT; the compound made of pure PLA (homopolymer) together with the wood filler modified with OMMT behaved with the best mechanical properties, while the compound made of pure WF and PLA modified with OMMT showed a low performance in terms of mechanical properties; Whether for OMMT-modified PLA or OMMT-modified WF-co-PLA compounds, the existence of OMMT aggregates could cause damage to mechanical properties. However, the addition of OMMT intercalated in the content of the WF or the WF-co-PLA compound was beneficial for the resulting physical-mechanical properties in all the tests.

Sun et al. (2021) carried out a dynamic thermogravimetric study (DMA) to understand the thermal decomposition process of 3D printing filaments, made from a mixture of PLA and starch, as a polymeric matrix adding wood filler as reinforcement. With increasing starch content in the filament, the onset of thermal decomposition for the filaments presented by the authors of the manuscript gradually decreased from 272.4 to 155.1 °C, respectively. The degree of thermal degradation became smaller, and the transition temperature range became larger as the proportion of polysaccharide (starch) content increased. For a better understanding of the thermal decomposition behavior of starch-PLA wood composites, thermal stability calculations were performed, employing more detailed studies that allowed the development of starch/PLA-based filaments to achieve material suitable for 3D printing. The boom that 3D printing shows today, reaches high levels leading to its development and exponential growth, seeking a gradual improvement in the design of new materials that replace conventional polymers, making use of self-sustaining polymeric materials or obtained from waste of vegetable sources, etc. Additive manufacturing is considered one of the new technologies that give a key point in the field of research for future high precision manufacturing, benefiting in the first instance the industry, in the foundation of buildings, development of finished products, innovation, and invention in medicine and biomedicine, etc. Its growing applications mainly depend on its significant merits of reduced weight, minimal materials used, high precision, and shorter production time (Sun et al., 2021; Z. Wang et al., 2021).

Furthermore, it is very crucial that such 3D printed components can maintain the same or even better material performance and product quality as those obtained by conventional manufacturing methods. In the present work (Dong et al., 2018), successfully fabricated 3D-printed mechanical components in structured PLA trials of wood-fiber-based composites. The parameters that stood out in the work were the density of the applied fill, the height of the layer and the number of treatments investigated, this study was carried out through a design of experiments, among which the number of tests was determined as the most significant factor. to maximize the physical properties, especially the tensile strength of PLA samples in conjunction with wood fibers. In addition, the experimental design evaluated the effect of the type of material (i.e., the content of pure PLA and the content of PLA with wood fibers), resistance to bending and impact of the samples. Based on the results supported in this work, it was suggested that the type of material is the only predominant factor to maximize all the mechanical forces that were present in all the tests; however, they are consistently lower for the wood fiber/PLA compared to pure PLA—this is due to the mechanical resistance provided by the PLA homopolymer.

4.7 CONCLUSIONS

PLA is at the forefront of biopolymers and is presented as an alternative to traditional polymers. Despite its high thermal properties, many materials have been integrated into PLA to improve thermal properties such as TGA (thermal stability), Tg, Tm (melting point) and TC. This paper describes the use of various types of reinforcing materials to improve the thermal properties of PLA and the application in 3D printer. It should be noted that the integration of these reinforcing materials into PLA can positively increase the thermal stability, Tg, Melting Point, Tc. This study about the role of reinforcing agents in the thermal properties of PLA opens an insight into the vast field of operation for the application of this biopolymer by optimizing its properties to focus on a particular application with 3D printing.

KEYWORDS

- **food, organic materials**
- **thermal properties**
- **nanoparticles**
- **nanofibers**
- **natural oils**

REFERENCES

Aggarwal, S.; Johnson, S.; Saloni, D.; Hakovirta, M. Novel 3D Printing Filament Composite Using Diatomaceous Earth and Polylactic Acid for Materials Properties and Cost Improvement. *Compos. B: Eng.* **2019,** *177*. https://doi.org/10.1016/j.compositesb.2019.107310

Altan, A.; Aytac, Z.; Uyar, T. Carvacrol Loaded Electrospun Fibrous Films from Zein and Poly(Lactic Acid) for Active Food Packaging. *Food Hydrocol.* **2018,** *81*. https://doi.org/10.1016/j.foodhyd.2018.02.028

Bhattacharjee, S.; Bajwa, D. S. Feasibility of Reprocessing Natural Fiber Filled Poly(Lactic Acid) Composites: An In-depth Investigation. *Adv. Mater. Sci. Eng.* **2017,** *2017*. https://doi.org/10.1155/2017/1430892

Carbonell-Verdu, A.; Samper, M. D.; Garcia-Garcia, D.; Sanchez-Nacher, L.; Balart, R. Plasticization Effect of Epoxidized Cottonseed Oil (ECSO) on Poly(Lactic Acid). *Indus. CropsProd.* **2017,** *104*. https://doi.org/10.1016/j.indcrop.2017.04.050

Carmona, V. B.; Corrêa, A. C.; Marconcini, J. M.; Mattoso, L. H. C. Properties of A Biodegradable Ternary Blend of Thermoplastic Starch (TPS), Poly(ε-Caprolactone) (PCL) and Poly(Lactic Acid) (PLA). *J. Polym. Env.* **2015,** *23*(1). https://doi.org/10.1007/s10924-014-0666-7

Chaiwutthinan, P.; Chauyjuljit, S.; Thipkham, N.; Kowalski, C. P.; Boonmahitthisud, A. Poly(Lactic Acid)/Ethylene Vinyl Acetate Copolymer Blend Composites With Wood Flour and Wollastonite: Physical Properties, Morphology, and Biodegradability. *J. Vinyl Additive Technol.* **2019,** *25* (4). https://doi.org/10.1002/vnl.21697

Chaiwutthinan, P.; Pimpan, V.; Chuayjuljit, S.; Leejarkpai, T. Biodegradable Plastics Prepared from Poly(Lactic Acid), Poly(Butylene Succinate) and Microcrystalline Cellulose Extracted from Waste-Cotton Fabric With A Chain Extender. *J. Polym. Environ.* **2015,** *23* (1). https://doi.org/10.1007/s10924-014-0689-0

Chuayjuljit, S.; Wongwaiwattanakul, C.; Chaiwutthinan, P.; Prasassarakich, P. Biodegradable Poly(Lactic Acid)/Poly(Butylene Succinate)/Wood Flour Composites: Physical and Morphological Properties. *Polym. Compos.* **2017,** *38* (12). https://doi.org/10.1002/pc.23886

Cisneros-López, E. O.; Pérez-Fonseca, A. A.; González-García, Y.; Ramírez-Arreola, D. E.; González-Núñez, R.; Rodrigue, D.; Robledo-Ortíz, J. R. Polylactic Acid–Agave Fiber Biocomposites Produced by Rotational Molding: A Comparative Study with Compression Molding. *Adv. Polym. Technol.* **2018,** *37* (7). https://doi.org/10.1002/adv.21928

Debeli, D. K.; Zhang, Z.; Jiao, F.; Guo, J. Diammonium Phosphate-Modified Ramie Fiber Reinforced Polylactic Acid Composite and Its Performances on Interfacial, Thermal, and Mechanical Properties. *J. Nat. Fibers* **2019,** *16* (3). https://doi.org/10.1080/15440478.2017.1423255

Dogan, S. K.; Boyacioglu, S.; Kodal, M.; Gokce, O.; Ozkoc, G. Thermally Induced Shape Memory Behavior, Enzymatic Degradation and Biocompatibility of PLA/TPU Blends: "Effects of Compatibilization." *J. Mech. Behav. Biomed. Mater.* **2017,** *71*. https://doi.org/10.1016/j.jmbbm.2017.04.001

Dong, Y.; Ghataura, A.; Takagi, H.; Haroosh, H. J.; Nakagaito, A. N.; Lau, K. T. Polylactic Acid (PLA) Biocomposites Reinforced with Coir Fibres: Evaluation of Mechanical Performance and Multifunctional Properties. *Compos. Part A: Appl. Sci. Manufac.* **2014,** *63*. https://doi.org/10.1016/j.compositesa.2014.04.003

Dong, Y.; Milentis, J.; Pramanik, A. Additive Manufacturing of Mechanical Testing Samples Based on Virgin Poly (Lactic Acid) (PLA) and PLA/Wood Fibre Composites. *Adv. Manufac.* **2018,** *6* (1). https://doi.org/10.1007/s40436-018-0211-3

Ferri, J. M.; Fenollar, O.; Jorda-Vilaplana, A.; García-Sanoguera, D.; Balart, R. Effect of Miscibility on Mechanical and Thermal Properties of Poly(Lactic Acid)/Polycaprolactone Blends. *Polym. Int.* **2016,** *65* (4). https://doi.org/10.1002/Pi.5079

González, E. A. S.; Olmos, D.; Lorente, M. Ángel, Vélaz, I.; González-Benito, J. Preparation and Characterization of Polymer Composite Materials Based on PLA/TiO_2 for Antibacterial Packaging. *Polymers* **2018,** *10* (12). https://doi.org/10.3390/polym10121365

Guo, R.; Ren, Z.; Bi, H.; Song, Y.; Xu, M. Effect of Toughening Agents on the Properties of Poplar Wood Flour/Poly (Lactic Acid) Composites Fabricated with Fused Deposition Modeling. *Eur. Polym. J.* **2018,** *107*. https://doi.org/10.1016/j.eurpolymj.2018.07.035

Herrera, N.; Roch, H.; Salaberria, A. M.; Pino-Orellana, M. A.; Labidi, J.; Fernandes, S. C. M.; Radic, D.; Leiva, A.; Oksman, K. Functionalized Blown Films of Plasticized Polylactic Acid/Chitin Nanocomposite: Preparation and Characterization. *Mater. Design* **2016,** *92*. https://doi.org/10.1016/j.matdes.2015.12.083

Hill, C. A. S.; Khalil, H. P. S. A.; Hale, M. D. A Study of the Potential of Acetylation to Improve the Properties of Plant Fibres. *Indus. Crops Prod.* **1998,** *8* (1). https://doi.org/10.1016/s0926-6690(97)10012-7

Holcapkova, P.; Stloukal, P.; Kucharczyk, P.; Omastova, M.; Kovalcik, A. Anti-Hydrolysis Effect of Aromatic Carbodiimide in Poly(Lactic Acid)/Wood Flour Composites. *Compos. Part A: Appl. Sci. and Manufac.* **2017,** *103*. https://doi.org/10.1016/j.compositesa.2017.10.003

Huang, Y.; Löschke, S.; Proust, G. In the Mix: The Effect of Wood Composition on The 3D Printability and Mechanical Performance of Wood-Plastic Composites. *Comp. Part C: Open Access* **2021,** *5*. https://doi.org/10.1016/j.jcomc.2021.100140

Ibrahim, N.; Wahab, M. K. A.; Uylan, D. N.; Ismail, H. Physical and Degradation Properties of Polylactic Acid and Thermoplastic Starch Blends—Effect of Citric Acid Treatment on Starch Structures. *Bioresources* **2017,** *12* (2). https://doi.org/10.15376/Biores.12.2.3076-3087

Jiang, G.; Huang, H. X.; Chen, Z. K. Rheological Responses and Morphology of Polylactide/Linear Low Density Polyethylene Blends Produced by Different Mixing Type. *Polym. Plast. Technol. Eng.* **2011,** *50* (10). https://doi.org/10.1080/03602559.2011.557822

Jin, Y.; Wan, Y.; Zhang, B.; Liu, Z. Modeling of the Chemical Finishing Process for Polylactic acid Parts in Fused Deposition Modeling and Investigation of Its Tensile Properties. *J. Mater. Proc. Technol.* **2017,** *240*. https://doi.org/10.1016/j.jmatprotec.2016.10.003

Juntuek, P.; Ruksakulpiwat, C.; Chumsamrong, P.; Ruksakulpiwat, Y. Effect of Glycidyl Methacrylate-Grafted Natural Rubber on Physical Properties of Polylactic Acid and Natural Rubber Blends. *J. Appl. Polym. Sci.* **2012,** *125* (1). https://doi.org/10.1002/app.36263

Kamdem, D. P.; Jiang, H.; Cui, W.; Freed, J.; Matuana, L. M. Properties of Wood Plastic Composites Made of Recycled HDPE and Wood Flour from CCA-Treated Wood Removed from Service. *Compos. Part A: Appl. Sci. Manufac.* **2004,** *35* (3). https://doi.org/10.1016/j.compositesa.2003.09.013

Khavkine, M.; Kazayawoko, M.; Law, S.; Balatinecz, J. J. Durability of Wood Flour-Thermoplastic Composites Under Extreme Environmental Conditions and Fungal Exposure. *Int. J. Polym. Mater. Polym. Biomater.* **2000,** *46* (1–2), 255–269. https://doi.org/10.1080/00914030008054859

Lai, S. M.; Lan, Y. C. Shape Memory Properties of Melt-Blended Polylactic Acid (PLA)/Thermoplastic Polyurethane (TPU) Bio-Based Blends. *J. Polym. Res.* **2013,** *20* (5). https://doi.org/10.1007/s10965-013-0140-6

Le Duigou, A.; Castro, M.; Bevan, R.; Martin, N. 3D Printing of Wood Fibre Biocomposites: From Mechanical to Actuation Functionality. *Mater. Des.* **2016,** *96*. https://doi.org/10.1016/j.matdes.2016.02.018

Lee, S. H.; Wang, S. Biodegradable Polymers/Bamboo Fiber Biocomposite with Bio-Based Coupling Agent. *Compos. A: Appl. Sci. Manufac.* **2006,** *37* (1). https://doi.org/10.1016/j.compositesa.2005.04.015

Li, D.; Shentu, B.; Weng, Z. Morphology, Rheology, and Mechanical Properties of Polylactide/Poly(Ethylene-Co-Octene) Blends. *J. Macromol. Sci. Part B: Phys.* **2011,** *50* (10). https://doi.org/10.1080/00222348.2011.557617

Li, W.; He, X.; Zuo, Y.; Wang, S.; Wu, Y. Study on the Compatible Interface of Bamboo Fiber/Polylactic Acid Composites by In-Situ Solid Phase Grafting. *Int. J. Biol. Macromol.* **2019,** *141*. https://doi.org/10.1016/J.Ijbiomac.2019.09.005

Liu, J.; Jiang, H.; Chen, L. Grafting of Glycidyl Methacrylate onto Poly(Lactide) and Properties of PLA/starch Blends Compatibilized by the Grafted Copolymer. *J. Polym. Environ.* **2012,** *20* (3). https://doi.org/10.1007/s10924-012-0438-1

Liu, X.; Lv, S.; Jiang, Y.; Shi, J.; Tan, H.; Gu, J.; Zhang, Y. Effects of Alkali Treatment on the Properties of WF/PLA Composites. *J. Adhes. Sci. Technol.* **2017,** *31* (10). https://doi.org/10.1080/01694243.2016.1248703

Long, H.; Wu, Z.; Dong, Q.; Shen, Y.; Zhou, W.; Luo, Y.; Zhang, C.; Dong, X. Effect of Polyethylene Glycol on Mechanical Properties of Bamboo Fiber-Reinforced Polylactic Acid Composites. *J. Appl. Polym. Sci.* **2019,** *136* (26). https://doi.org/10.1002/app.47709

Lv, S.; Gu, J.; Tan, H.; Zhang, Y. Enhanced Durability of Sustainable Poly (Lactic Acid)-Based Composites with Renewable Starch and Wood Flour. *J. Clean. Prod.* **2018,** *203*. https://doi.org/10.1016/j.jclepro.2018.08.266

Makhetha, T. A. *Preparation and Characterization of Ethylene Vinyl Acetate Copolymer/Poly(Lactic Acid)/sugarcane Bagasse Composites for Water Purification*; University of The Free State, 2016. http://scholar.ufs.ac.za/xmlui/handle/11660/7811

Mariano, M.; Pilate, F.; De Oliveira, F. B.; Khelifa, F.; Dubois, P.; Raquez, J. M.; Dufresne, A. Preparation of Cellulose Nanocrystal-Reinforced Poly(Lactic Acid) Nanocomposites Through Noncovalent Modification with PLLA-Based Surfactants. *ACS Omega* **2017,** *2* (6). https://doi.org/10.1021/acsomega.7b00387

Mihai, M.; Ton-That, M. T. Novel Bio-Nanocomposite Hybrids Made from Polylactide/Nanoclay Nanocomposites and Short Flax Fibers. *J. Thermoplast. Compos. Mater.* **2019,** *32* (1). https://doi.org/10.1177/0892705717743293

Mishra, S.; Mohanty, A. K.; Drzal, L. T.; Misra, M.; Parija, S.; Nayak, S. K.; Tripathy, S. S. Studies on Mechanical Performance of Biofibre/Glass Reinforced Polyester Hybrid Composites. *Compos. Sci. Technol.* **2003,** *63* (10). https://doi.org/10.1016/s0266-3538(03)00084-8

Mohapatra, A. K.; Mohanty, S.; Nayak, S. K. Study of Thermo-Mechanical and Morphological Behaviour of Biodegradable PLA/PBAT/Layered Silicate Blend Nanocomposites. *J. Polym. Environ.* **2014,** *22* (3). https://doi.org/10.1007/s10924-014-0639-X

Morrell, J. J.; Stark, N. M.; Pendleton, D. E.; Mcdonald, A. G. Durability of Wood-Plastic Composites. *Wood Design Focus* **2006,** *16* (3).

Moura, I.; Botelho, G.; Machado, A. V. Characterization of EVA/PLA Blends When Exposed to Different Environments. *J. Polym. Environ.* **2014,** *22* (1). https://doi.org/10.1007/s10924-013-0614-y

Nilsuwan, K.; Benjakul, S.; Prodpran, T. Physical/Thermal Properties and Heat Seal Ability of Bilayer Films Based on Fish Gelatin and Poly(Lactic Acid). *Food Hydrocol.* **2018,** *77*. https://doi.org/10.1016/j.foodhyd.2017.10.001

Petersson, L.; Kvien, I.; Oksman, K. Structure and Thermal Properties of Poly(Lactic Acid)/Cellulose Whiskers Nanocomposite Materials. *Compos. Sci. Technol.* **2007,** *67* (11–12). https://doi.org/10.1016/j.compscitech.2006.12.012

Pilla, S.; Gong, S.; O'Neill, E.; Yang, L.; Rowell, R. M. Polylactide-Recycled Wood Fiber Composites. *J. Appl. Polym. Sci.* **2009,** *111* (1). https://doi.org/10.1002/app.28860

Rasal, R. M.; Janorkar, A. V.; Hirt, D. E. Poly(Lactic Acid) Modifications. In *Progress In Polymer Science*; Oxford, Vol. 35 (3), 2010. https://doi.org/10.1016/j.progpolymsci.2009.12.003

Ray, D.; Sarkar, B. K.; Rana, A. K.; Bose, N. R. Effect of Alkali Treated Jute Fibres on Composite Properties. *Bull. Mater. Sci.* **2001,** *24* (2). https://doi.org/10.1007/BF02710089

Reis, M. O.; Olivato, J. B.; Bilck, A. P.; Zanela, J.; Grossmann, M. V. E.; Yamashita, F. Biodegradable Trays of Thermoplastic Starch/Poly (Lactic Acid) Coated with Beeswax. *Indus. Crops Prod.* **2018,** *112*. https://doi.org/10.1016/j.indcrop.2017.12.045

Salehi, S.; Ehsani, M.; Khonakdar, H. A. Assessment of Thermal, Morphological, and Mechanical Properties of Poly(Methyl Methacrylate)/Glass Flake Composites. *J. Vinyl Addit. Technol.* **2017,** *23* (1). https://doi.org/10.1002/vnl.21476

Santos, D.; Correia, C. O.; Silva, D. M.; Gomes, P. S.; Fernandes, M. H.; Santos, J. D.; Sencadas, V. Incorporation of Glass-Reinforced Hydroxyapatite Microparticles into Poly(Lactic Acid) Electrospun Fibre Mats for Biomedical Applications. *Mater. Sci. Eng. C* **2017,** *75*. https://doi.org/10.1016/j.msec.2017.03.007

Scaffaro, R.; Lopresti, F.; Botta, L. PLA Based Biocomposites Reinforced with Posidonia Oceanica Leaves. *Compos. B: Eng.* **2018,** *139*. https://doi.org/10.1016/j.compositesb.2017.11.048

Sommerhuber, P. F.; Wang, T.; Krause, A. Wood-Plastic Composites as Potential Applications of Recycled Plastics of Electronic Waste and Recycled Particleboard. *J. Clean. Prod.* **2016,** *121*. https://doi.org/10.1016/j.jclepro.2016.02.036

Sun, Y.; Lee, D.; Wang, Y.; Li, S.; Ying, J.; Liu, X.; Xu, G.; Gwon, J.; Wu, Q. Thermal decomposition behavior of 3D printing filaments made of wood-filled polylactic acid/starch blend. *J. Appl. Polym. Sci.* **2021,** *138* (9). https://doi.org/10.1002/app.49944

Sung, S. H.; Chang, Y.; Han, J. Development of Polylactic Acid Nanocomposite Films Reinforced with Cellulose Nanocrystals Derived from Coffee Silverskin. *Carbohydr. Polym.* **2017,** *169*. https://doi.org/10.1016/j.carbpol.2017.04.037

Syazeven Effatin Azma Mohd Asri, Zakaria, Z.; Hassan, A.; Kassim, M. H. M.; Arjmandi, R. Exploring the Effects of Fermented Chitin Nanowhiskers on Tensile and Thermal Properties of Poly(Ethylene Glycol) Modified Polylactic Acid Nanocomposites. *Mal. J. Fundam. Appl. Sci.* **2021,** *17* (2). https://doi.org/10.11113/mjfas.v17n2.2002

Temiz, A.; Akbas, S.; Aydin, I.; Demirkir, C. The Effect of Plasma Treatment on Mechanical Properties, Surface Roughness and Durability of Plywood Treated with Copper-Based Wood Preservatives. *Wood Sci. Technol.* **2016,** *50* (1). https://doi.org/10.1007/s00226-015-0773-1

Temiz, A.; Alma, M. H.; Terziev, N.; Palanti, S.; Feci, E. Efficiency of Bio-Oil Against Wood Destroying Organisms. *J. Biobased Mater. Bioenergy* **2010,** *4* (4). https://doi.org/10.1166/jbmb.2010.1092

Tian, X.; Liu, T.; Yang, C.; Wang, Q.; Li, D. Interface and Performance of 3D Printed Continuous Carbon Fiber Reinforced PLA Composites. *Compos. A: Appl. Sci. Manufac.* **2016,** *88*. https://doi.org/10.1016/j.compositesa.2016.05.032

van Cong, D.; Hoang, T.; Giang, N. V.; Ha, N. T.; Lam, T. D.; Sumita, M. A Novel Enzymatic Biodegradable Route for PLA/EVA Blends Under Agricultural Soil of Vietnam. *Mater. Sci. Eng. C* **2012,** *32* (3). https://doi.org/10.1016/j.msec.2011.12.012

Wang, N.; Yu, J.; Ma, X. Preparation and Characterization of Thermoplastic Starch/PLA Blends by One-Step Reactive Extrusion. *Polym. Int.* **2007,** *56* (11). https://doi.org/10.1002/pi.2302

Wang, X.; Jiang, M.; Zhou, Z.; Gou, J.; Hui, D. (2017). 3D Printing of Polymer Matrix Composites: A Review and Prospective. In *Composites Part B: Engineering* (Vol. 110). https://doi.org/10.1016/J.Compositesb.2016.11.034

Wang, Y. N.; Weng, Y. X.; Wang, L. Characterization of Interfacial Compatibility of Polylactic Acid and Bamboo Flour (PLA/BF) in Biocomposites. *Polym. Test.* **2014,** *36*. https://doi.org/10.1016/j.polymertesting.2014.04.001

Wang, Z.; Wang, C.; Qin, Y.; Huang, A.; Liu, R. Organo-Montmorillonite Modified Wood Flour/Poly (Lactic Acid) Composites via Different Modification Process. *Polym. Compos.* **2021,** *42* (2). https://doi.org/10.1002/pc.25880

Wang, Z.; Yao, Z.; Zhou, J.; He, M.; Jiang, Q.; Li, A.; Li, S.; Liu, M.; Luo, S.; Zhang, D. Improvement of Polylactic Acid Film Properties Through the Addition of Cellulose Nanocrystals Isolated from Waste Cotton Cloth. *Int. J. Biol. Macromol.* **2019,** *129*. https://doi.org/10.1016/J.Ijbiomac.2019.02.021

Wei, L.; Agarwal, U. P.; Matuana, L.; Sabo, R. C.; Stark, N. M. Performance of High Lignin Content Cellulose Nanocrystals in Poly(Lactic Acid). *Polymer* **2018,** *135*. https://doi.org/10.1016/j.polymer.2017.12.039

Xia, X.; Liu, W.; Zhou, L.; Liu, H.; He, S.; Zhu, C. Study on Flax Fiber Toughened Poly (Lactic Acid) Composites. *J. Appl. Polym. Sci.* **2015,** *132* (38). https://doi.org/10.1002/app.42573

Xie, G.; Zhang, Y.; Lin, W. Plasticizer Combinations and Performance of Wood Flour-Poly(Lactic Acid) 3D Printing Filaments. *Bioresources* **2017,** *12* (3). https://doi.org/10.15376/biores.12.3.6736-6748

Zhang, M.; Jia, H.; Weng, Y.; Li, C. Biodegradable PLA/PBAT Mulch on Microbial Community Structure In Different Soils. *Int. Biodeteriorat. Biodegrad.* **2019,** *145*. https://doi.org/10.1016/J.Ibiod.2019.104817

Zhou, Y.; Luo, L.; Liu, W.; Zeng, G.; Chen, Y. Preparation and Characteristic of PC/PLA/TPU Blends by Reactive Extrusion. *Adv. Mater. Sci. Eng.* **2015,** *2015*. https://doi.org/10.1155/2015/393582

CHAPTER 5

Antioxidant and Antimicrobial Properties of *Moringa oleifera* in Food Preservation

ALAÍN MARTÍNEZ-PÉREZ[1], ELISEO SANCHEZ-LOREDO[2], DIANA B. MUÑIZ MÁRQUEZ[3], and JORGE E. WONG-PAZ[3]

[1]*Tecnológico Nacional de México/Instituto Tecnológico de Durango, Durango, México*

[2]*Bioprocesses and Bioproducts Research Group, Food Research Department, School of Chemistry, Universidad Autónoma de Coahuila, Saltillo, Coahuila, Mexico*

[3]*Universidad Autónoma de San Luis Potosí, Facultad de Estudios Profesionales de la Zona Huasteca, San Luis Potosí, México*

ABSTRACT

The use of preservatives has become indispensable to extend the shelf life of foods. However, uncertainty about the toxicity and safety of the use of synthetic preservatives has led the population to demand more natural ingredients for food preservation. There are naturally occurring substances in plants with high antioxidant and antimicrobial capacities that could be used as natural preservatives. Thus, extending the shelf life of perishable foods and in turn providing phytochemical compounds, phenolics, and nutrients. *Moringa oleifera* is a plant rich in phenolic compounds with antioxidant and antimicrobial capacities that have been shown to inhibit the growth of microorganisms such as *Staphylococcus aureus*, *Escherichia coli,* and *Listeria monocytogenes*. The purpose of this chapter is to provide detailed

Engineering Principles for Food Processing Technology and Product Realization. Leonardo Sepúlveda Torre, Porteen Kannan, A. K. Haghi, PhD (Eds.)

information on the use of *M. oleifera* extracts in the preservation of different types of foods due to their antioxidant and antimicrobial capacity.

5.1 INTRODUCTION

The current trend in the population is the consumption of healthy and minimally processed foods, coupled with the negative perception of food additives such as preservatives, the demand for natural extracts as food preservatives is increasing. Natural plant extracts contain phytochemicals and essential oils consisting of a wide variety of bioactive compounds with antioxidant and antimicrobial properties. Studies focused on the development of new preservatives based on plant extracts are based on polyphenols and essential oils with antimicrobial and antioxidant activity to be applied to perishable foods such as meat, dairy, seafood, fruits, and vegetables (Debonne et al., 2018; Efenberger-Szmechtyk et al., 2021; Gutiérrez-del-Río et al., 2018; El-Saber Batiha et al., 2021). However, it is not known with certainty what the mechanisms of action are, since, in most research, crude extracts are used without purifying and/or identifying the compounds present in the extracts. In general, extracts obtained from plants are a complex mixture of substances such as carbohydrates, proteins, polyphenols, terpenoids, and others.

Among the plants most studied for their antioxidant and antimicrobial properties is *Moringa oleifera*. This is a plant with a high nutritional value due to its vitamin and mineral content. It also contains beneficial phytochemicals with antioxidant, anticarcinogenic, antimicrobial, and anti-inflammatory properties (Abd El-Hack et al., 2022; Sultana, 2020). Moreover, it has a high protein content, especially the seeds (~52%), which in addition to their nutritional value have multiple biological activities such as antioxidant, antimicrobial, and antidiabetic (Kumar et al., 2022). *M. oleifera* polysaccharides also possess biological activity such as antihyperlipidemic, antidiabetic, immunomodulatory, antihypertensive, and gastrointestinal protection (Sharma et al., 2022). The high nutritional value and content of bioactive compounds make the moringa plant an important source of ingredients for the development of functional foods, and its antimicrobial capacity makes it important for food preservation (Saucedo-Pompa et al., 2018).

The uses of moringa are varied, which range from its consumption as food to wastewater treatment due to the variety of bioactive compounds present in the leaves, seeds, flowers, and root of moringa. Extraction plays an important role in antimicrobial and antioxidant capacity since; depending on the method and solvents used, different compounds with different activities will be extracted. Therefore, traditional methods to the use of emerging technologies will be addressed. Identifying the compounds present in moringa extracts is important to know the mechanisms of action, as well as possible synergistic interactions between the extracted compounds. Finally, the use of moringa extracts (seeds, leaves, flowers, and roots.) applied in perishable foods such as dairy products, meat, seafood, edible oils, and others, as a preservative against microbial and oxidative deterioration will be addressed.

5.2 MORINGA USES

M. oleifera is a native plant mainly in India. It is a very large plant having nutritional, medicinal, and many other outstanding uses both in the preparation of other food and industrial products. It is one of the most nutrient-rich plants discovered today and the use of all parts is well known, from the use as fodder for animals to water purification, including their use for treating infant malnutrition (Oyeyinka and Oyeyinka, 2018; Terefe et al., 2017). Due to the number of uses and properties, it has been called the "miraculous tree" among cultures that commonly use it (Badmos et al., 2014). *M. oleifera* has been reported to have many beneficial compounds, from vitamins such as vitamins A, C, and E, to minerals such as calcium, magnesium, potassium, copper, iron, and manganese. Also, *M oleifera* have carotenoid, acid ascorbic, and tocopherols among others (Oyeyinka and Oyeyinka, 2018). For its nutritional, medicinal, and technological properties, *M. oleifera* is used in different areas such as food, food additives, pharmaceuticals, wastewater treatment, and more (Table 5.1).

TABLE 5.1 Uses of *M. oleifera* in Different Areas.

Study area	Application
Food	All parts of the plant are edible, seeds oil is used to cook, and the leaves are used as livestock feed
Food additive	Thickener in yogurt, milk coagulant for cheese elaboration, clarifier in fruit juices, and food preservative
Medicine	Treatment of disease such as inflammation, hepatorenal disorders, hematological, cardiovascular, gastrointestinal, and as wound healing
Treatment of wastewater	Coagulant and clarifier
	Adsorbent of organic pollutants
Others	Polysaccharides from trunk for gum formation, biofilm inhibition caused by *S. aureus*

M. oleifera is used in a variety of foods because all its parts are edible from the pods, leaves, flowers, seeds, and even the roots (Benitez and Benítez Wilberth, 2012). Among important nutritional uses include its use as food for humans mainly alleviating malnutrition for babies and children, and livestock due is considered a great food to increase weight and milk production in animals due to its high fiber content, proteins, vitamins, and minerals, leaves being the main sources of forage for animals even more than other leaves (Oyeyinka and Oyeyinka, 2018; Terefe et al., 2017). On the other hand, studies highlight the great importance of its edible use in humans, which is something very interesting, because important studies for application in food start from there; *M. oleifera* has been used in various countries over the years as a source of the human diet, as in the case of the Philippines, in which the use of moringa flowers stands out consuming the leaves of various firms, including cooked, fresh leaves, even as a dry powder (Lee et al., 2016; Pontual et al., 2012). *M. oleifera* flour has been evaluated as a food supplement in snacks, broiler breast meat, rice crackers, yogurt, and cheese. Moreover, moringa seed flour has also been used in soups, bread, and cookies to increase the fiber dietary and protein content and to reduce levels of lipid oxidation as well as improve fatty acid profiles (Feitosa et al., 2020). Also, moringa seeds have the power to compete against other oils for use in the kitchen, as well as for the lubrication of mechanisms, as well as for the manufacture of soaps and cosmetics (Benitez and Benítez Wilberth, 2012; Rodríguez-Pérez et al., 2016). Due to its great antioxidant potential with extracts and oils from all parts of *M. oleifera,* it has been able to demonstrate its antioxidant effect in food matrices such as in meats. Moringa leaves have demonstrated their effectiveness in the inhibition of lipid oxidation in meats through packaging technologies (Shah et al., 2015).

Moringa oleifera has been used as food additive, highlighting the properties of the seed, since seed extracts of moringa seeds have been evaluated as a thickener agent in yogurt, where moringa seeds demonstrated their effect as a thickening agent and higher protein content (Cardines et al., 2018). Since moringa leaves have shown their potential as a milk coagulant due to their protease content and the action of proteins on the casein micelle (Pontual et al., 2012). However, although moringa leaves are effective in coagulating milk, the best properties for coagulation of milk have been seen in the seeds of moringa, since this property has been low in other parts such as flowers and leaves of this plant (Tajalsir et al., 2014; Sánchez-Muñoz et al.,2017). This property of coagulation of milk has been verified for the elaboration of food, being in the coagulation of milk for the formulation of cheeses with different types of milk such as skim milk and soy milk (Sánchez-Muñoz et al., 2017). Also, it was been used for clarification in honey as well as in the clarification of sugar cane juice, both important for food production (Benitez and Benítez Wilberth, 2012).

Its another important use is its application in cosmetics and pharmaceuticals (Cardines et al., 2018), that lies in the bioactive compounds it contains, from phenolic acids, alkaloids, phytosterols, flavonoids, as well as vitamins and mineral content (Gautier et al., 2022). *M. oleifera* is also rich in phenolic acids such as gallic acid, chlorogenic acid, ellagic acid, ferulic acid, and flavanols, among which kaempferol and quercetin stand out (Feitosa et al., 2020). Extracts from *M. oleifera* have been used in traditional medicine to treat diseases such as inflammation, hepatorenal disorders, hematological, cardiovascular, gastrointestinal diseases, treatment for gastric ulcers as well as for the relief of fatigue and bronchitis, due to their high content of antioxidants and anti-inflammatory compounds (Zhang et al., 2019; Vongsak et al., 2013). Multiple in vitro and in vivo investigations in mice and humans have studied the anti-obesity mechanisms of the different compounds from moringa isolated or as crude extracts obtained from maceration, ultrasound-assisted extraction, microwave-assisted extraction, supercritical fluid extraction, among others, in addition to the use of different solvents such as ethanol, petroleum ether, acetone, and their aqueous mixtures. The main mechanisms suggested are the improvement of the lipid profile, regulation of genes associated with adipogenesis, glucose uptake, insulin resistance, and satiety-related hormones (Ali Redha et al. 2021). One of the properties of the hydroalcoholic extracts of *M. oleifera* seeds recently studied is their potential for wound healing. Seed extracts were incorporated into a 5% and 10% hydrogel and applied to 1 cm wounds in male Swiss albino mice. After 4 days, a wound closure rate of up to 81% was observed and total healing at 13 days, while the control and standard (povidone) groups were 62% and 82% at the end of treatment (14 days). In addition, mice to which the hydrogel with 10% *M. oleifera* seed extract was applied showed higher resistance to skin breaking of the healed wound, which was 243 g (Ali et al., 2022). Also, its beneficial properties against asthma-related inflammatory responses have been studied in Wistar rats. The ethanol extract of *M. oleifera* seeds was given to rats at 100 and 200 mg/kg body weight, a similar effect to that of dexamethasone (2.5 mg/kg body weight) was found since asthmatic symptoms only occurred in the control group (Mahajan et al., 2007). Oil-free *M. oleifera* seed extract can alleviate arsenic toxicity, mitigate ROS generation, decrease Hcy, IL-6, and TNF-α levels, and increase vitamin B9 and B12 levels (Jana et al., 2020).

Their use as a food source has already been demonstrated, but now industrially the use of *M. oleifera* also stands out. One of the uses that have been highlighted for *M. oleifera* is its potential as a coagulant for the treatment of wastewater from the dairy industry and the slaughterhouse as a coagulant, to decrease turbidity, and adsorbent of organic pollutants (Formentini-Schmitt et al., 2013; del Real-Olvera et al., 2015). Among all the uses of *M. oleifera* that has already been discussed about the leaves, seeds, and flowers, another use is the trunk—this part of the plant has been used for the formation of gums (Oyeyinka and Oyeyinka, 2018). One application, although in the food or water industry, is for the inhibition of biofilms caused by *S. aureus*, being the antibacterial property of extracts of *M. oleifera* the one that acts on this microorganism on surfaces with *S. aureus* (de Oliveira et al., 2020). Also, it has wide use both industrially, in wastewater treatment, as well as in food use again both in nutritionally, fortifying foods, as well as acting as a potential new food additive.

5.3 BIOACTIVE COMPOUNDS IN MORINGA

Plant extracts that have bioactive compounds can be obtained from different parts of the plant such as roots, leaves, seeds, flowers, stems, and fruits. These extracts are mainly composed of polyphenols and aromatic hydrocarbons with antioxidant activity (Ong et al., 2021). In addition to plant extracts, essential oils are another group of volatile compounds with a characteristic aroma of the plant and possess antimicrobial and antioxidant properties, consisting mainly of terpenes, terpenoids, phenylpropenes, and to a lesser extent other compounds (Reis et al., 2022; El Khetabi et al., 2022). The presence of bioactive compounds in moringa extracts has been reported mainly in leaves, seeds, and flowers (Fig. 5.1).

In a study for the optimization of the extraction of moringa leaves by microwave and pressurized liquid-assisted extraction using aqueous ethanol, a total of 62 compounds were identified by HPLC-ESI-QTOF-MS. These include essential amino acids, organic acids, phenolic acids, flavonoids, flavonoids, glucosinolates, and lignans. The most important bioactive compounds in moringa leaves are the flavonoids kaempferol and quercetin as well as all their derivatives, in addition to their antioxidant properties, which are greater than those of traditional vitamins, they have anti-inflammatory effects (Rodríguez-Pérez et al. 2016). In addition to the conventional flavonoids reported in *M. oleifera* leaves, four flavonoids (vitexin, robinetin,

FIGURE 5.1 Bioactive compounds in *M. oleifera*.

nictoflorin, and taxifolin) found in other plants such as *Pigeonpea*, *Cayratia trifolia*, *Lycium* berries, and *Pinus pinaster* were recently reported for the first time (Wang et al., 2022). Another type of bioactive compound identified in the leaves of *M. oleifera* is carotenoids, with the highest concentrations of β-carotene, lutein, violaxanthin, and neoxanthin, the concentrations found were higher than those reported in other foods such as broccoli, chicory, cabbage, and chard (Teixeira et al., 2014). This shows that in addition to their nutritional value, the leaves of *M. oleifera* have a wide variety of bioactive compounds that can have health benefits for people.

In the case of *M. oleifera* seeds, they have higher protein (~50.8%) and oil (~11.2%) content than leaves and other vegetable sources such as yellow peas, red lentils, and chickpeas. The essential amino acid content is also higher in seeds. However, the contents of polyphenols and flavonoids are considerably higher in moringa leaves than in seeds (Cattan et al., 2022). In addition to phenolic compounds, in the seeds of *M. oleifera,* we can find isothiocyanates with bioactive properties obtained by enzymatic conversion from glucosinolates. These compounds can modulate the lipid profile in diabetic patients, decreasing total cholesterol, triglycerides, and low and very low-density lipoproteins, and at the same time increasing the levels of high-density lipoproteins (Ali Redha et al., 2021).

Edible flowers are recognized for their nutritional, bioactive, and preservative properties, including *M. oleifera* flowers (Purohit et al., 2021). *M. oleifera* flowers extract containing polyphenols, flavonoids, terpenoids, and tannins has a promising anti-arthritic effect on Wistar rats with *Mycobacterium* tuberculosis-induced arthritis injected in the hind paw. The results indicate that adverse physical changes such as weight loss, increased paw volume, and arthritis index was reversed, and serum levels of rheumatoid factor and proinflammatory cytokine levels were decreased (Mahajan et al., 2009). In another study, one *M. oleifera* flower trypsin inhibitor (MoFTI) was evaluated as a potential inflammatory agent in two rat models of acute inflammation (paw edema and peritonitis) induced by carrageenan. The results showed that a dose of 15–30 mg/kg decreased edema size, and reduced leukocyte migration, nitric oxide levels, and proinflammatory cytokines (Patriota et al., 2021).

A polysaccharide isolated from the roots of *M. oleifera* MRP-1 is composed mainly of rhamnose, arabinose, fructose, xylose, mannose, and galactose. It possesses anti-inflammatory activity by preventing the LPS-induced increase in nitric oxide and TNF-a production in RAW264.7 macrophage cells. In another study, the methanol extract of *M. oleifera* roots was tested in two rat models of acute and chronic inflammation induced by carrageenan (Cui et al., 2019). Moringa root extract (660 mg/kg) showed inhibition of rat edema (acute model) as well as cellular accumulation and fluid exudation (chronic inflammation), while indomethacin (5 mg/kg) treatments only inhibited the acute but not the chronic form (Ezeamuzie et al., 1996).

5.4 EXTRACTION METHODS

The extraction is the most important step in obtaining bioactive compounds from vegetable sources (González-González et al., 2022). The extraction of *M. oleifera* has been used by different methods, such as percolation, ultrasound-assisted extraction (UAE), microwave-assisted extraction (MAE), and maceration (Fig. 5.2) (Zhang et al., 2019). The percolation extraction has been used for the extraction of bioactive compounds from *M. oleifera* seeds, allowing them to heat to 100 °C and later filtered (Zhang et al., 2019). Other alternatives to the extraction of bioactive compounds from *M. oleifera* are solid-state fermentation (SSF) (Feitosa et al., 2020). The extraction for SSF is an alternative method for the extraction of bioactive compounds, which involves the action enzymatic of microorganisms to release the compounds from the solid substrate. Moreover, other bioactive compounds can be formed with a high capacity antioxidant as reported by Feitosa et al. (2020), that showed a high content of total phenolic (134%) and an increase in total flavonoid contents (783.1%) in extracts of *M. oleifera* seeds of SSF compared with unfermented moringa seeds extracts compared with flour unfermented. Furthermore, SSF showed an improved antimicrobial activity on *Bacillus subtilis*. As result of fermentation, 14 new compounds were obtained, such as, 2-hydroxycinnamic acid, kaempferol 3-glycoside, p-coumaric acid, trans-cinnamic acid, luteolin, caffeic acid, isorhamnetin, 6,2 dihydroxyflavone, chlorogenic acid, epicatechin, epigallocatechin gallate, quercetin, taxifolin, and ferulic acid, which they have been reported with antimicrobial and antioxidant properties.

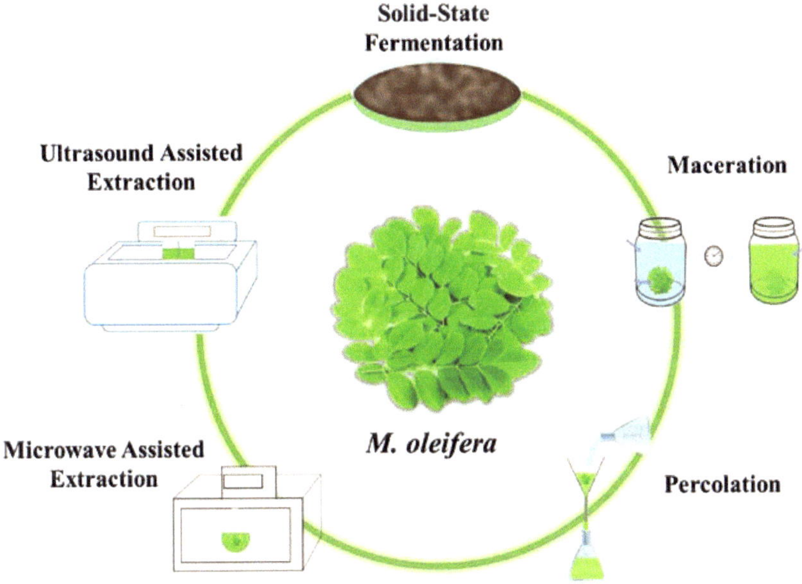

FIGURE 5.2 Extraction methods used for the bioactive compounds *M. oleifera* extraction.

Among solvents used for the extraction of bioactive compounds, ethanol is more used, since it has submitted extractions of bioactive compounds on best conditions of ethanol to 60%. However, extraction has been reported as the best condition extract to 70% of ethanol in maceration (Vongsak et al., 2013; Valdés-Hernández et al., 2015). The extraction of *M. oleifera* seeds has been reported by the obtention of bioactive compounds with high-value antioxidant, antimicrobial, and other properties through the use of solvents such as ethanol, acetone, water, and/or a combination of these (Feitosa et al., 2020; Zhang et al., 2019; Ali et al. 2022). The maceration with heating has been used for the extraction of bioactive compounds for employed as nanoparticles with antibacterial activity (Ao et al., 2022).

5.5 USE IN DAIRY FOODS

Due to consumer demands for dairy products that contain bioactive properties, fortified dairy products such as fortified yogurts, cheeses, and others with preservation potential have been developed, such as yogurt fortified with extracts of lyophilized *M. oleifera* seeds with various concentrations from 0.05%, 0.1%, and 0.2% (Zhang et al., 2019). Likewise, the microbial properties of yogurt during fermentation have been put under study, because the microbiological analysis is the second important quality criterion in the acceptability as well as the shelf-life of dairy products (Salem et al., 2015). However, color, pH viscosity measurements, as well as determination of radical scavenging activity and phenolic content are important parameters. It has been shown that the addition of *M. oleifera* accelerates the fermentation rate of yogurt and allows the elimination of free radicals and that it can even be a food that provides health benefits, contributing to the development of yogurt with bioactive properties (Zhang et al., 2019).

Some dairy products from other mammals besides cows have been the subject of research to improve their conservation as sour cream from buffalo milk due to *M. oleifera* as an oil or extract can improve microbial stability, therefore an increase sensorial activity and they can improve the shelf life of sour cream. In addition, the extracts and oils do not affect the punctuation of the attributes such as flavor, texture, body, acidity, and appearance. Furthermore, the *M. oleifera* extracts were permitted that the acidity increased; however, the increase of *M. oleifera* oils decreased. *M. oleifera* extracts and oils showed a decrease in the content of microorganisms such as proteolytic bacteria, lipolytic bacteria, and mold and yeast to 4 weeks of storage at 5 °C. (Salem et al., 2015). Badmos et al. (2014) applied extracts from *M. oleifera* in soft cheese and evaluated the antibacterial effect. The results showed that *M. oleifera* has antibacterial activity in soft cheese. The extracts with a 3% ethanol condition had a decrease in bacterial content, among them are enterococcus, *Bacillus pasteuri, Staphylococcus spp., S aureus, and Corynebacterium spp*. Furthermore, was shown that moisture content was low in the condition of extraction with 1 and 2% ether, which helps to avoid contamination by microorganisms. Among the dairy products which has presented positive effects of preservation with *M. oleifera* is the cheese, as reported by Cui et al. (2020), who evaluated the preservative effect of moringa oils applied by spray in mozzarella cheese, cheddar cheese, parmesan cheese, and camembert cheese against *L. monocytogenes* and evaluated a sensorial analysis. Moringa oils showed positive effects on the inhibition of *L. monocytogenes* in cheese and demonstrated no changes in the textural characteristics of the cheese during storage. The use of *M. oleifera* has not only been oriented to preservation, since it has been used as an additive for the coagulation of milk in cheese processing (Abdeen et al., 2021). Since the coagulation of milk is a very important step in the manufacture of various types of cheese, a new source of enzymes for cheese production has been sought, in addition to calf rennet, which has been widely used for cheese production (Terefe et al., 2017). Among the application in cheeses, *M. oleifera* is important in addition to the conservation against microorganisms, it has lengthened the shelf life in gouda cheese through a film delaying the oxidation of lipids in the cheese (Lee et al., 2016). Interestingly, two of the main microorganisms among dairy products are *L. monocytogenes* and *S. aureus*. A study has achieved to inhibit growing these two microorganisms in cheeses through nanoparticles (Lin et al., 2019).

5.6 USE IN MEAT FOOS

Meat and meat by-products are very susceptible to lipid oxidation and microbial spoilage resulting in undesirable physicochemical changes such as loss of color, texture, and flavor. Bioactive compounds derived from plants have been investigated for the preservation of meat products, in addition, they provide functional properties (Munekata et al., 2020; Falowo et al., 2014).

M. oleifera extracts have been studied in beef preservation. A study by Manzoor et al. evaluated the effect of aqueous extract of moringa leaves on the preservation of raw beef packaged with modified atmospheres (80% O_2 and 20% CO_2). The results showed significant effects on pH, thiobarbituric acid reactive substances (TBARS) and color, but no effect on microbiological quality and water-holding capacity (Shah et al., 2015). Similarly, direct application of moringa leaf powder at concentrations of 0.2–0.4% in ground beef has the same effects on lipid oxidation without affecting the sensory profile; in addition, the total microbial load, coliforms, fungi, and yeasts were reduced (Mashau et al., 2021). On the other hand, the application of moringa leaf extracts in combination with *Bidens pilosa* leaves in veal minced meat showed a higher antioxidant activity, obtaining lower pH and TBARS values. In addition, the antimicrobial properties of the extracts showed inhibition against pathogens of interest such as *S. aureus*, *E. coli*, and *Shigella* (Falowo et al., 2016, 2017). Furthermore, moringa seed meal applied in beef burgers improves cooking properties as the concentration of moringa seed meal increases, reduces thiobarbituric acid and pH values, while higher concentration of moringa seed in meals reduces thiobarbituric acid, pH values, and aerobic plate counts during storage. However, sensory acceptability was lower in beef burger with moringa seed flour and reduced at longer times, but that of beef burger without moringa seed flour decreased significantly at longer storage times (Al-Juhaimi et al., 2015).

The addition of moringa leaf powder in the processing of chicken patties decreases their moisture content, while increasing the fat, protein, and ash content. As for the total microbial load, the patties with moringa leaf powder remained at 3.78 Log CFU/g, while the control had 7.87 Log CFU/g, MLP maintained the absence of *E. coli* and *S. aureus* at 2.52 Log CFU/g, and the control showed 2.91 and 3.92 Log CFU/g, respectively. However, sensory acceptability was significantly lower than the control (Elhadi et al., 2016). Similarly, moringa flower extracts were added to the formulation of nuggets, which were shown to improve oxidative stability decrease TBARS values without demonstrating changes in sensory acceptability as occurs with moringa leaves (Madane et al., 2019). On the other hand, the extract of defatted moringa seeds applied in a chitosan-based coating was able to inhibit the growth of *B. cereus* and *E. coli* until day 6 of refrigerated storage in chicken sausages inoculated with both bacteria at a concentration of 4 Log CFU/g, while after 15 days of storage the microbial load increased to 7.3 Log CFU/g but still lower than the control without treatment 8 Log CFU/g (Sharma et al., 2020). In another study, it was reported that moringa leaf powder in 2% added to irradiated chicken meat packaged under aerobic and vacuum conditions decreased peroxide and TBARS values, and additionally improved the amino acid and fatty acid content (Nisar et al., 2019). The same treatment was evaluated in the preservation of chicken meatballs, reducing the total volatile basic nitrogen values. As for the microbial load, *M. oleifera* alone reduces more than 1 Log CFU/g of coliforms but in combination with radiation it inhibits them completely (Nisar et al., 2020).

Sheep meat is very susceptible to damage by pathogenic microorganisms and spoilage, having a very short shelf life. The use of natural antimicrobials such as plant extracts and essential oils in combination with refrigeration have been investigated for the preservation of mutton (Purgatorio et al. 2022). In this sense, the influence of moringa leaf extracts on the nutritional, physicochemical, and microbiological properties of sheep burgers was studied. The values of protein, ash, total phenolics, and total flavonoids increased, while moisture and fat decreased. The values of protein, ash, total phenolics and total flavonoids increased, while moisture and fat decreased. Cooking parameters such as yield, water and fat retention were also improved. Regarding the microbial load, the additional hamburgers with 5% moringa leaf extract had 4.30 Log CFU/g, while the control hamburgers had 6.63 Log CFU/g for the total plate count and 2.06 and 6.20 Log CFU/g for total coliforms (Mashau et al., 2021).

These studies demonstrate the potential of *M. oleifera* as a preservative in meat products. It has antimicrobial activity against pathogenic bacteria such as E. coli and S. aureus that are regularly present in this type of food. On the other hand, the antimicrobial properties of *M. oleifera* help against the oxidation of lipids and proteins, decreasing the values of TBARS and peroxide. In addition, they add functional characteristics due to bioactive compounds, mainly polyphenols.

5.7 USE IN SEAFOOD AND FISH

Seafood and fish are one of the most perishable foods and are an important part of the food supply in many countries, mainly on the coasts, and loss and waste in value chains represent losses in the millions of dollars (Kruijssen et al., 2020). One of the causes of spoilage of this type of food is proteases that degrade the structural proteins of fish and shellfish. This process is generally undesirable due to the changes in the physicochemical properties of fish as loss of texture, and changes in color, moisture distribution, and water-holding capability (Zhuang et al., 2021).

One solution to this problem is the use of protease inhibitors, which are usually chemical preservatives and antibiotics that inhibit both the proteolytic activity and the growth of the microorganisms that produce these enzymes. However, the negative perception of the use of chemically synthesized antimicrobials has increased the demand for natural preservatives, which would also be beneficial by decreasing antibiotic resistance of pathogenic microorganisms and increasing shelf life (Cotabarren et al., 2020). The protease inhibitor from *M. oleifera* leaves has been investigated and has shown inhibition on serine proteases thrombin, elastase, chymotrypsin, and cysteine proteases cathepsin B and papain, which are important in the industry. Their application in shrimp (*Penaeus monodon*) as a natural preservative (0.2 mg/mL) inhibited protein degradation and maintained a microbial load of 4×10^2 CFU/mL while the control registered 8×10^{10} CFU/mL, which shows that the proteolytic activity is closely related to the microbial load and therefore the deterioration of shrimp (Bijina et al., 2011). On the other hand, the application of moringa and curry leaf extracts is effective in delaying melanosis in Pacific white shrimp applied by immersion for 15 min, in addition to delaying microbial deterioration (Soni et al., 2021).

Biological selenium nanoparticles (Bio-SeNPs) synthesized using selenite and *M. oleifera* extracts are an effective antimicrobial to eliminate *L. monocytogenes* and *Corynebacterium diphtheriae* two pathogens of importance in food safety, 0.7 mg/mL of Bio-SeNPs significantly inhibit the growth of these two strains. the antimicrobial capacity of Bio-SeNPs was higher than when using moringa extract or chemical-Se(0) (formed by the reduction of selenite with cysteine). Furthermore, their application in raw salmon effectively eliminates *L. monocytogenes* without causing cytotoxicity (Ao et al., 2022). The incorporation of moringa seedmeal in a brine solution for the marinade of smoked African mud catfish (*Clarias gariepinus*) at a concentration of 15 g/kg achieves a significant decrease in microbial load, especially fungi and yeasts related to fish spoilage (Jimoh et al., 2021). In another study, moringa leaf extract was added to an edible gelatin-based coating of gray triggerfish (*Balistes capriscus*) skin and applied to smooth-hound shark (*Mulestus mulestus*) fillets. The moringa leaf extract was mainly composed of phenolic acids that presented high antioxidant and antimicrobial activity, thus decreasing spoilage indices such as malondialdehyde formation, total volatile basic nitrogen, microbial load, as well as physical parameters such as weight loss and texture profile (Mezhoudi et al., 2022). However, the use of moringa extracts does not always improve food preservation. Delfino et al. (2021) applied extracts of *M. oleifera* leaves and *Lavandula sp.* foliage to the preservation of tilapia fish burgers, which had an adverse effect by increasing the production of TBARS and negatively affecting the overall impression of tilapia fish burgers at the end of storage.

5.8 OXIDATIVE STABILIZER IN OILS

It is widely known that phenolic compounds have antioxidant properties and that they are part of many foods in greater or lesser abundance, and their incorporation in cooking oils can increase their stability in frying foods without affecting their sensory quality, through the elimination of free radicals and the elimination of carbonyls (Wu et al., 2019). In addition, polyphenols from *Eucalyptus citriodora* leaf extracts increase the stability of mixed vegetable oils during storage (Ali et al., 2016). However, there are few studies related to the stability of cooking oils during food frying. As reported by Dhibi et al. (2022), methanolic extracts of *M. oleifera* and *Zizipus Lotus* contained phenolic compounds (phenolic acids and flavonoids) which were able to increase the antioxidant activity (DPPH radical inhibition), decrease the free fatty acid content and peroxide value, and increase the oxidative stability of corn oil, after being heated at 180 °C for 8 h. Moringa leaves extract has a significant antioxidant effect in accelerated storage of soybean oil (12 days/60 °C), however, in the heating test its antioxidant activity decreased as the temperature increased (105 to 180 °C) but always had lower values than the control (Zhao et al., 2019). The mixture of moringa oil (20%) with sunflower oil (80%) improves stability during potato frying, decreasing the formation of polymers, and oxidized triglycerides reduces the value of p-anisidine and total volatile aldehydes, and preserves vitamin E and antioxidant activity due to a possible interaction with the components of moringa oil, which gives it a protective effect (Boukandoul et al., 2019). Moreover, the incorporation of moringa leaves extract in margarine increases oxidation resistance compared to vitamin E, it also improves physical aspects such as water droplet size, which is desirable since from a microbiological point of view it would help to prolong shelf life. However, one characteristic that may not be acceptable to the consumer is the change in color of the margarine for moringa leaves extract addition (Ouahrani et al., 2022).

The use of natural antioxidants obtained from plants helps to extend the shelf life of fish oils by retarding the oxidation of omega-3 unsaturated fatty acids, known not only for their health benefits but also for their instability and easy oxidation during storage, thus developing an undesirable flavor and potentially toxic compounds (Hrebień-Filisińska, 2021). Commercially available flavonoids have been used to increase the stability of fish oil by decreasing the formation of peroxides and thiobarbituric acid reactive substances (Nieto et al., 1993). These compounds are present in moringa extracts. The application of moringa leaf, flower, and seed extracts in fish oil retarded the oxidation of fish oil; moringa extracts showed lower peroxide and hydroperoxide values compared to the synthetic antioxidants tertiary butylated hydroquinone (TBHQ) and butylated hydroxytoluene (BHT) in fish oil in accelerated storage for 16 days at 60 °C and showed a greater protective effect on polyunsaturated fatty acids (Nascimento et al., 2016).

5.9 USE IN OTHER FOODS

Another microorganism against which plants have antimicrobial effects is *Aspergillus flavus*, allowing its application in nuts such as peanuts, which are often attacked by this aflatoxigenic fungus (Razzaghi-Abyaneh et al., 2010). The extracts of the leaves of the medicinal plants *Cymbopogon citratus*, *M.*

oleifera, *Ocimum gratissimum*, and *Clerodendrum volubile* have antifungal activity against the toxigenic fungus *A. flavus*, phytochemical screening showed the presence of saponins, tannins, glycoside, and volatile oils (Jeff-Agboola et al., 2016). A study by Abu-Sree et al. (2021) demonstrated the ability of M. oleifera leaves extracts in combination with oil of seeds to inhibit *A. flavus* and aflatoxin production in peanuts. An emulsion consisting of the methanol extract of moringa leaves, seed oil, and a commercial emulsifier was applied. After three months, the results show a reduction from 6.8×10^4 to 9 CFU/kg for peanuts with red skin, while for peanuts without red skin the reduction was 9.4×10^5 to 14 CFU/kg. As for aflatoxin B1 production, it remained undetectable for peanuts with the moringa extract coating and controls with a content of 480 and 945 ng/kg for peanuts with and without the red skin, respectively.

In active films based on chitosan, guar gum, and polyvinyl alcohol, *M. oleifera* extract improves characteristics such as tensile strength, thermal stability, decrease in glass transition temperature and melting temperature, and effectively inhibits the growth of *E. coli* and *S. aureus*. These characteristics make possible its incorporation in this type of film that can be used in food preservation due to its antimicrobial capacity (Bhat et al., 2022). The application of *M. oleifera* coatings with chitosan nanoparticles increases the shelf life of bananas during storage. This is achieved by reducing respiration rate, ethylene production, weight loss, firmness, and delaying color changes, thus increasing the sensory acceptability of the fruit after 30 days of storage (Odetayo et al., 2022). On the other hand, papaya-based coatings with moringa extract and/or ascorbic acid increased antioxidant activity and shelf life of cut pears for more than 9 days; however, sensory acceptability was affected by moringa extract (Rodríguez et al., 2020).

5.10 CONCLUSIONS

Food preservatives are one of the most widely used additives in the food industry in order to keep food quality for a longer period. However, they are closely related to different pathologies, so the demand for natural preservatives is increasing and medicinal plants such as *M. oleifera* are an important source of bioactive compounds with antioxidant and antimicrobial activity which can be used to extend shelf life. The use of emerging technologies for the extraction of bioactive compounds helps to improve yields, recovering most of the phytochemicals of interest and preserving their properties. The bioactive compounds of moringa have been identified in leaves, seeds, flowers, and even roots, among which there are phenolic acids, polyphenols, flavonoids, isothiocyanates, terpenoids, and others. Currently, many studies support their potential use in the preservation of foods such as dairy products, meats, seafood, and edible oils, among others. It is important to establish clear mechanisms of action and application of moringa compounds in food preservation so that they can be used safely and efficiently to ensure safe food for consumers.

KEYWORDS

- **antioxidant**
- **food preservation**
- **antimicrobial capacity**
- **healthy foods**
- ***M. Oleifera* extracts**
- **bioactive compounds**

REFERENCES

Abd El-Hack, M. E.; Alqhtani, A. H.; Swelum, A. A.; El-Saadony, M. T.; Salem, H. M.; Babalghith, A. O.; Taha, A. E.; Ahmed, O.; Abdo, M.; El-Tarabily, K. A. Pharmacological, Nutritional and Antimicrobial Uses of *Moringa Oleifera* Lam. Leaves in Poultry Nutrition: An Updated Knowledge. *Poult. Sci.* **2022,** *101* (9), 102031. https://doi.org/10.1016/j.psj.2022.102031.

Abdeen, E.-S. M. M.; Ibrahim, O. A.; Kholif, A. M. M. Utility of *Moringa Oleifera* Waste as a Coagulant in Goat Soft Cheese Production. *Heliyon* **2021,** *7* (7), e07536. https://doi.org/10.1016/j.heliyon.2021.e07536.

Abu-Sree, Y. H.; Abdel-Fattah, S. M.; Abdel-Razek, A. G.; Badr, A. N. Neoteric Approach for Peanuts Biofilm Using the Merits of Moringa Extracts to Control Aflatoxin Contamination. *Toxicol. Rep.* **2021,** *8*, 1685–1692. https://doi.org/10.1016/j.toxrep.2021.08.006.

Ali Redha, A.; Perna, S.; Riva, A.; Petrangolini, G.; Peroni, G.; Nichetti, M.; Iannello, G.; Naso, M.; Faliva, M. A.; Rondanelli, M. Novel Insights on Anti-Obesity Potential of the Miracle Tree, *Moringa Oleifera*: A Systematic Review. *J. Funct. Foods* **2021,** *84*, 104600. https://doi.org/10.1016/j.jff.2021.104600.

Ali, A.; Garg, P.; Goyal, R.; Khan, A.; Negi, P.; Li, X.; Kulshrestha, S. An Efficient Wound Healing Hydrogel Based on a Hydroalcoholic Extract of *Moringa Oleifera* Seeds. *SA J. Bot.* **2022,** *145*, 192–198. https://doi.org/10.1016/j.sajb.2021.05.003.

Ali, S.; Chatha, S. A. S.; Ali, Q.; Hussain, A. I.; Hussain, S. M.; Perveen, R. Oxidative Stability of Cooking Oil Blend Stabilized with Leaf Extract of Eucalyptus Citriodora. *Int. J. Food Prop.* **2016,** *19* (7), 1556–1565. https://doi.org/10.1080/10942912.2015.1047514.

Al-Juhaimi, F.; Ghafoor, K.; Hawashin, M. D.; Alsawmahi, O. N.; Babiker, E. E. Effects of Different Levels of Moringa (*Moringa Oleifera*) Seed Flour on Quality Attributes of Beef Burgers. *CyTA–J Food* **2015,** *14* (1), 1–9. https://doi.org/10.1080/19476337.2015.1034784.

Ao, B.; Lv, J.; Yang, H.; He, F.; Hu, Y.; Hu, B.; Jiang, H.; Huo, X.; Tu, J.; Xia, X. *Moringa Oleifera* Extract Mediated the Synthesis of Bio-SeNPs with Antibacterial Activity against *Listeria monocytogenes* and *Corynebacterium Diphtheriae*. *LWT* **2022,** *165*, 113751. https://doi.org/10.1016/j.lwt.2022.113751.

Badmos, A. H. A. et al. The Effect of Crude Leaf Extracts of *Moringa Oleifera* on the. *Wayamba J. Anim. Sci.* **2014,** No. 140000**2484,** 939–946.

Benitez, M.; Benítez Wilberth, M. Aprovechamiento Poscosecha De La Moringa (Moringa Oleífera). *Rev. Iber. Tecnol. Postcosecha* **2012,** *13* (2), 171–174.

Bhat, V. G.; Narasagoudr, S. S.; Masti, S. P.; Chougale, R. B.; Vantamuri, A. B.; Kasai, D. Development and Evaluation of Moringa Extract Incorporated Chitosan/Guar Gum/Poly (Vinyl Alcohol) Active Films for Food Packaging Applications. *Int. J. Biol. Macromol.* **2022,** *200*, 50–60. https://doi.org/10.1016/j.ijbiomac.2021.12.116.

Bijina, B.; Chellappan, S.; Krishna, J. G.; Basheer, S. M.; Elyas, K. K.; Bahkali, A. H.; Chandrasekaran, M. Protease Inhibitor from *Moringa Oleifera* with Potential for Use as Therapeutic Drug and as Seafood Preservative. *Saudi J. Biol. Sci.* **2011,** *18* (3), 273–281. https://doi.org/10.1016/j.sjbs.2011.04.002.

Boukandoul, S.; Santos, C. S. P.; Casal, S.; Zaidi, F. Oxidation Delay of Sunflower Oil under Frying by Moringa Oil Addition: More Than Just a Blend. *J. Sci. Food Agric.* **2019,** *99* (12), 5483–5490. https://doi.org/10.1002/jsfa.9809.

Cardines, P. H. F.; Baptista, A. T. A.; Gomes, R. G.; Bergamasco, R.; Vieira, A. M. S. *Moringa Oleifera* Seed Extracts as Promising Natural Thickening Agents for Food Industry: Study of the Thickening Action in Yogurt Production. *LWT* **2018,** *97*, 39–44. https://doi.org/10.1016/j.lwt.2018.06.028.

Cattan, Y.; Patil, D.; Vaknin, Y.; Rytwo, G.; Lakemond, C.; Benjamin, O. Characterization of *Moringa Oleifera* Leaf and Seed Protein Extract Functionality in Emulsion Model System. *Innov. Food Sci. Emerg. Technol.* **2022,** *75*, 102903. https://doi.org/10.1016/j.ifset.2021.102903.

Cotabarren, J.; Lufrano, D.; Parisi, M. G.; Obregón, W. D. Biotechnological, Biomedical, and Agronomical Applications of Plant Protease Inhibitors with High Stability: A Systematic Review. *Plant Sci.* **2020,** *292*, 110398. https://doi.org/10.1016/j.plantsci.2019.110398.

Cui, C.; Chen, S.; Wang, X.; Yuan, G.; Jiang, F.; Chen, X.; Wang, L. Characterization of *Moringa Oleifera* Roots Polysaccharide MRP-1 with Anti-Inflammatory Effect. *Int. J. Biol. Macromol.* **2019,** *132*, 844–851. https://doi.org/10.1016/j.ijbiomac.2019.03.210.

Cui, H.; Li, H.; Li, C.; Abdel-Samie, M. A.; Lin, L. Inhibition Effect of Moringa Oil on the Cheese Preservation and Its Impact on the Viability, Virulence and Genes Expression of *Listeria monocytogenes*. *LWT* **2020,** *134* (September), 110163. https://doi.org/10.1016/j.lwt.2020.110163.

de Oliveira, A. M.; de Abreu Filho, B. A.; de Jesus Bassetti, F.; Bergamasco, R.; Gomes, R. G. Natural Extract of *Moringa Oleifera* Leaves Promoting Control of *Staphylococcus aureus* Strains Biofilm on PVC Surface. *Food Bioprocess Technol.* **2020,** *13* (10), 1817–1832. https://doi.org/10.1007/s11947-020-02521-x.

Debonne, E.; Van Bockstaele, F.; Samapundo, S.; Eeckhout, M.; Devlieghere, F. The Use of Essential Oils as Natural Antifungal Preservatives in Bread Products. *J. Essent. Oil Res.* **2018,** *30* (5), 309–318. https://doi.org/10.1080/10412905.2018.1486239.

del Real-Olvera, J.; Rustrian-Portilla, E.; Houbron, E.; Landa-Huerta, F. J. Adsorption of Organic Pollutants from Slaughterhouse Wastewater Using Powder of *Moringa Oleifera* Seeds as a Natural Coagulant. *New pub Balaban* **2015,** *57* (21), 9971–9981. https://doi.org/10.1080/19443994.2015.1033479.

Delfino, L. A.; Mattje, L. G. B.; Silva, M.; Araujo, M. C.; Tormen, L.; Bainy, E. M. Evaluation of Moringa and Lavandula Extracts as Natural Antioxidants in Tilapia Fish Burger. *J. Culin. Sci. Technol.* **2021,** 1–16. https://doi.org/10.1080/15428052.2021.1883494.

Dhibi, M.; Amri, Z.; Bhouri, A. M.; Hammami, S.; Hammami, M. Comparative Study of the Phenolic Profile and Antioxidant Activities of Moringa (*Moringa Oleifera* Lam.) and Jujube (Ziziphus Lotus Linn.) Leaf Extracts and Their Protective Effects in Frying Stability of Corn Oil. *Meas. Food* **2022,** *7*, 100045. https://doi.org/10.1016/j.meafoo.2022.100045.

Efenberger-Szmechtyk, M.; Nowak, A.; Czyzowska, A. Plant Extracts Rich in Polyphenols: Antibacterial Agents and Natural Preservatives for Meat and Meat Products. *Crit. Rev. Food Sci. Nutr.* **2021,** *61* (1), 149–178. https://doi.org/10.1080/10408398.2020.1722060.

El Khetabi, A.; Lahlali, R.; Ezrari, S.; Radouane, N.; Lyousfi, N.; Banani, H.; Askarne, L.; Tahiri, A.; El Ghadraoui, L.; Belmalha, S.; Barka, E. A. Role of Plant Extracts and Essential Oils in Fighting against Postharvest Fruit Pathogens and Extending Fruit Shelf Life: A Review. *Trends Food Sci. Technol.* **2022,** *120*, 402–417. https://doi.org/10.1016/j.tifs.2022.01.009.

Elhadi, D. A. E.; Elgasim, E. A.; Mohamed Ahmed, I. A. Microbial and Oxidation Characteristics of Refrigerated Chicken Patty Incorporated with Moringa (*Moringa Oleifera*) Leaf Powder. *CyTA–J Food* **2016,** *15* (2), 234–240. https://doi.org/10.1080/19476337.2016.1242157.

El-Saber Batiha, G.; Hussein, D. E.; Algammal, A. M.; George, T. T.; Jeandet, P.; Al-Snafi, A. E.; Tiwari, A.; Pagnossa, J. P.; Lima, C. M.; Thorat, N. D.; Zahoor, M.; El-Esawi, M.; Dey, A.; Alghamdi, S.; Hetta, H. F.; Cruz-Martins, N. Application of Natural Antimicrobials in Food Preservation: Recent Views. *Food Control* **2021,** *126*, 108066. https://doi.org/10.1016/j.foodcont.2021.108066.

Ezeamuzie, I. C.; Ambakederemo, A. W.; Shode, F. O.; Ekwebelem, S. C. Antiinflammatory Effects of *Moringa Oleifera* Root Extract. *Int. J. Pharmacogn.* **1996,** *34* (3), 207–212. https://doi.org/10.1076/phbi.34.3.207.13211.

Falowo, A. B.; Fayemi, P. O.; Muchenje, V. Natural Antioxidants against Lipid–Protein Oxidative Deterioration in Meat and Meat Products: A Review. *Food Res. Int.* **2014,** *64*, 171–181. https://doi.org/10.1016/J.FOODRES.2014.06.022.

Falowo, A. B.; Muchenje, V.; Hugo, A.; Aiyegoro, O. A.; Fayemi, P. O. Antioxidant Activities of *Moringa Oleifera* L. and Bidens Pilosa L. Leaf Extracts and Their Effects on Oxidative Stability of Ground Raw Beef during Refrigeration Storage. *CyTA - J. Food* **2017,** *15* (2), 249–256. https://doi.org/10.1080/19476337.2016.1243587.

Falowo, A. B.; Muchenje, V.; Hugo, C. J.; Charimba, G. In Vitro Antimicrobial Activities of Bidens Pilosa and *Moringa Oleifera* Leaf Extracts and Their Effects on Ground Beef Quality during Cold Storage. *CyTA - J. Food* **2016,** *14* (4), 541–546. https://doi.org/10.1080/19476337.2016.1162847.

Feitosa, P. R. B.; Santos, T. R. J.; Gualberto, N. C.; Narain, N.; de Aquino Santana, L. C. L. Solid-State Fermentation with Aspergillus Niger for the Bio-Enrichment of Bioactive

Compounds in *Moringa Oleifera* (Moringa) Leaves. *Biocatal. Agric. Biotechnol.* **2020,** *27*, 101709. https://doi.org/10.1016/j.bcab.2020.101709.

Formentini-Schmitt, D. M.; Alves, Á. C. D.; Veit, M. T.; Bergamasco, R.; Vieira, A. M. S.; Fagundes-Klen, M. R. Ultrafiltration Combined with Coagulation/Flocculation/Sedimentation Using *Moringa Oleifera* as Coagulant to Treat Dairy Industry Wastewater. *Water, Air, Soil Pollut.* **2013,** *224* (9), 1682. https://doi.org/10.1007/s11270-013-1682-2.

Gautier, A.; Duarte, C. M.; Sousa, I. *Moringa Oleifera* Seeds Characterization and Potential Uses as Food. *Foods* **2022,** *11* (11), 1629. https://doi.org/10.3390/foods11111629.

González-González, G. M.; Palomo-Ligas, L.; Nery-Flores, S. D.; Ascacio-Valdés, J. A.; Sáenz-Galindo, A.; Flores-Gallegos, A. C.; Zakaria, Z. A.; Aguilar, C. N.; Rodríguez-Herrera, R. Coffee Pulp as a Source for Polyphenols Extraction Using Ultrasound, Microwave, and Green Solvents. *Environ. Qual. Manag.* **2022,** *32* (1), 451–461. https://doi.org/10.1002/tqem.21903.

Gutiérrez-del-Río, I.; Fernández, J.; Lombó, F. Plant Nutraceuticals as Antimicrobial Agents in Food Preservation: Terpenoids, Polyphenols and Thiols. *Int. J. Antimicrob. Agents* **2018,** *52* (3), 309–315. https://doi.org/10.1016/j.ijantimicag.2018.04.024.

Hrebień-Filisińska, A. Application of Natural Antioxidants in the Oxidative Stabilization of Fish Oils: A Mini-review. *J. Food Process. Preserv.* **2021,** *45* (4), e15342. https://doi.org/10.1111/jfpp.15342.

Jana, S.; Chattopadhyay, S.; Dey, A.; Perveen, H.; Dolai, D. Involvement of Metallothionein, Homocysteine and B-Vitamins in the Attenuation of Arsenic-Induced Uterine Disorders in Response to the Oral Application of Hydro-Ethanolic Extract of *Moringa Oleifera* Seed: A Preliminary Study. *Drug Chem. Toxicol.* **2020,** *43* (1), 1–12. https://doi.org/10.1080/01480545.2018.1508296.

Jeff-Agboola, Y. A.; Awe, L. B. Antifungal and Phytochemical Screening of Some Nigerian Medicinal Plant Extracts against Toxigenic Aspergillus Flavus. *Cogent Food Agric.* **2016,** *2* (1). https://doi.org/10.1080/23311932.2016.1210556.

Jimoh, W. A.; Ayeloja, A. A.; Badmus, G. O.; Olateju, K. O. Antibacterial and Antifungal Effect of Moringa (*Moringa Oleifera*) Seedmeal on Marinated Smoked African Mud Catfish (*Clarias Gariepinus*). *J. Food Saf.* **2020,** *40* (3), e12772. https://doi.org/10.1111/jfs.12772.

Kruijssen, F.; Tedesco, I.; Ward, A.; Pincus, L.; Love, D.; Thorne-Lyman, A. L. Loss and Waste in Fish Value Chains: A Review of the Evidence from Low and Middle-Income Countries. *Glob. Food Sec.* **2020,** *26*, 100434. https://doi.org/10.1016/j.gfs.2020.100434.

Kumar, M.; Selvasekaran, P.; Kapoor, S.; Barbhai, M. D.; Lorenzo, J. M.; Saurabh, V.; Potkule, J.; Changan, S.; ElKelish, A.; Selim, S.; Sayed, A. A. S.; Radha; Singh, S.; Senapathy, M.; Pandiselvam, R.; Dey, A.; Dhumal, S.; Natta, S.; Amarowicz, R.; Kennedy, J. F. *Moringa Oleifera* Lam. Seed Proteins: Extraction, Preparation of Protein Hydrolysates, Bioactivities, Functional Food Properties, and Industrial Application. *Food Hydrocoll.* **2022,** *131*, 107791. https://doi.org/10.1016/j.foodhyd.2022.107791.

Lee, K.-Y.; Yang, H.-J.; Song, K. Bin. Application of a Puffer Fish Skin Gelatin Film Containing *Moringa Oleifera* Lam. Leaf Extract to the Packaging of Gouda Cheese. *J. Food Sci. Technol.* **2016,** *53* (11), 3876–3883. https://doi.org/10.1007/s13197-016-2367-9.

Lin, L.; Gu, Y.; Cui, H. Moringa Oil/Chitosan Nanoparticles Embedded Gelatin Nanofibers for Food Packaging against *Listeria monocytogenes* and *Staphylococcus aureus* on Cheese. *Food Packag. Shelf Life* **2019,** *19* (Dec 2018), 86–93. https://doi.org/10.1016/j.fpsl.2018.12.005.

Madane, P.; Das, A. K.; Pateiro, M.; Nanda, P. K.; Bandyopadhyay, S.; Jagtap, P.; Barba, F. J.; Shewalkar, A.; Maity, B.; Lorenzo, J. M. Drumstick (*Moringa Oleifera*) Flower as an Antioxidant Dietary Fibre in Chicken Meat Nuggets. *Foods* **2019,** *8* (8), 307. https://doi.org/10.3390/FOODS8080307.

Mahajan, S. G.; Mali, R. G.; Mehta, A. A. Effect of *Moringa Oleifera* Lam. Seed Extract on Toluene Diisocyanate-Induced Immune-Mediated Inflammatory Responses in Rats. *J. Immunotoxicol.* **2007,** *4* (2), 85–96. https://doi.org/10.1080/15476910701337472.

Mahajan, S. G.; Mehta, A. A. Anti-Arthritic Activity of Hydroalcoholic Extract of Flowers of *Moringa Oleifera* Lam. in Wistar Rats. *J. Herbs. Spices Med. Plants* **2009,** *15* (2), 149–163. https://doi.org/10.1080/10496470903139363.

Mashau, M. E.; Munandi, M.; Ramashia, S. E. Exploring the Influence of *Moringa Oleifera* Leaves Extract on the Nutritional Properties and Shelf Life of Mutton Patties during Refrigerated Storage. *CyTA–J. Food* **2021,** *19* (1), 389–398. https://doi.org/10.1080/19476337.2021.1910732.

Mashau, M. E.; Ramatsetse, K. E.; Ramashia, S. E. Effects of Adding *Moringa Oleifera* Leaves Powder on the Nutritional Properties, Lipid Oxidation and Microbial Growth in Ground Beef during Cold Storage. *Appl. Sci.* 2021, *Vol. 11, Page 2944* **2021,** *11* (7), 2944. https://doi.org/10.3390/APP11072944.

Mezhoudi, M.; Salem, A.; Abdelhedi, O.; Fakhfakh, N.; Mabrouk, M.; Khorchani, T.; Debeaufort, F.; Jridi, M.; Zouari, N. Development of Active Edible Coatings Based on Fish Gelatin Enriched with *Moringa Oleifera* Extract: Application in Fish (Mustelus Mustelus) Fillet Preservation. *Food Sci. Nutr.* **2022,** *10* (11), 3979–3992. https://doi.org/10.1002/fsn3.2993.

Munekata, P. E. S.; Rocchetti, G.; Pateiro, M.; Lucini, L.; Domínguez, R.; Lorenzo, J. M. Addition of Plant Extracts to Meat and Meat Products to Extend Shelf-Life and Health-Promoting Attributes: An Overview. *Curr. Opin. Food Sci.* **2020,** *31*, 81–87. https://doi.org/10.1016/J.COFS.2020.03.003.

Nascimento, J. A.; Magnani, M.; Sousa, J. M. B.; Araújo, K. L. G. V.; Epaminondas, P. S.; Souza, A. S.; Souza, A. L.; Silva, M. C. D.; Souza, A. G. Assessment of the Antioxidant Effects of M Oringa Oleifera Lam. Extracts in Fish Oil during Storage. *J. Food Process. Preserv.* **2016,** *40* (1), 29–36. https://doi.org/10.1111/jfpp.12580.

Nieto, S.; Garrido, A.; Sanhueza, J.; Loyola, L. A.; Morales, G.; Leighton, F.; Valenzuela, A. Flavonoids as Stabilizers of Fish Oil: An Alternative to Synthetic Antioxidants. *J. Am. Oil Chem. Soc.* **1993,** *70* (8), 773–778. https://doi.org/10.1007/BF02542599.

Nisar, M. F.; Arshad, M. S.; Yasin, M.; Arshad, M. U.; Nadeem, M. T. Influence of Irradiation and Moringa Leaf Powder on the Amino Acid and Fatty Acid Profiles of Chicken Meat Stored under Various Packaging Materials. *J. Food Process. Preserv.* **2019,** *43* (9), e14166. https://doi.org/10.1111/jfpp.14166.

Nisar, M. F.; Arshad, M. S.; Yasin, M.; Khan, M. K.; Afzaal, M.; Sattar, S.; Suleria, H. A. R. Evaluation of Gamma Irradiation and Moringa Leaf Powder on Quality Characteristics of Meat Balls under Different Packaging Materials. *J. Food Process. Preserv.* **2020,** *44* (10), e14748. https://doi.org/10.1111/JFPP.14748.

Odetayo, T.; Sithole, L.; Shezi, S.; Nomngongo, P.; Tesfay, S.; Ngobese, N. Z. Effect of Nanoparticle-Enriched Coatings on the Shelf Life of Cavendish Bananas. *Sci. Hortic. (Amsterdam).* **2022,** *304*, 111312. https://doi.org/10.1016/j.scienta.2022.111312.

Ong, G.; Kasi, R.; Subramaniam, R. A Review on Plant Extracts as Natural Additives in Coating Applications. *Prog. Org. Coatings* **2021,** *151*, 106091. https://doi.org/10.1016/j.porgcoat.2020.106091.

Ouahrani, S.; Tzompa-Sosa, D. A.; Dewettinck, K.; Zaidi, F. Oxidative Stability, Structural, and Textural Properties of Margarine Enriched with *Moringa Oleifera* Leaves Extract. *J. Am. Oil Chem. Soc.* **2022,** *99* (6), 485–499. https://doi.org/10.1002/aocs.12586.

Oyeyinka, A. T.; Oyeyinka, S. A. *Moringa Oleifera* as a Food Fortificant: Recent Trends and Prospects. *J. Saudi Soc. Agric. Sci.* **2018,** *17* (2), 127–136. https://doi.org/10.1016/j.jssas.2016.02.002.

Patriota, L. L. de S.; Ramos, D. de B. M.; Gama e Silva, M.; dos Santos, A. C. L. A.; Silva, Y. A.; Marinho, A. de O.; Coelho, L. C. B. B.; Paiva, P. M. G.; Pontual, E. V.; Mendes, R. L.; Napoleão, T. H. The Trypsin Inhibitor from *Moringa Oleifera* Flowers (MoFTI) Inhibits Acute Inflammation in Mice by Reducing Cytokine and Nitric Oxide Levels. *South African J. Bot.* **2021,** *143*, 474–481. https://doi.org/10.1016/j.sajb.2021.09.032.

Pontual, E. V.; Carvalho, B. E. A.; Bezerra, R. S.; Coelho, L. C. B. B.; Napoleão, T. H.; Paiva, P. M. G. Caseinolytic and Milk-Clotting Activities from *Moringa Oleifera* Flowers. *Food Chem.* **2012,** *135* (3), 1848–1854. https://doi.org/10.1016/j.foodchem.2012.06.087.

Purgatorio, C.; Serio, A.; Chaves-López, C.; Rossi, C.; Paparella, A. An Overview of the Natural Antimicrobial Alternatives for Sheep Meat Preservation. *Compr. Rev. Food Sci. Food Saf.* **2022,** *21* (5), 4210–4250. https://doi.org/10.1111/1541-4337.13004.

Purohit, S. R.; Rana, S. S.; Idrishi, R.; Sharma, V.; Ghosh, P. A Review on Nutritional, Bioactive, Toxicological Properties and Preservation of Edible Flowers. *Futur. Foods* **2021,** *4*, 100078. https://doi.org/10.1016/j.fufo.2021.100078.

Razzaghi-Abyaneh, M.; Shams-Ghahfarokhi, M.; Rezaee, M. B.; Sakuda, S. Natural Aflatoxin Inhibitors from Medicinal Plants. *Mycotoxins Food, Feed Bioweapons* **2010,** 329–352. https://doi.org/10.1007/978-3-642-00725-5_19/COVER.

Reis, D. R.; Ambrosi, A.; Luccio, M. Di. Encapsulated Essential Oils: A Perspective in Food Preservation. *Futur. Foods* **2022,** *5*, 100126. https://doi.org/10.1016/j.fufo.2022.100126.

Rodríguez, G. M.; Sibaja, J. C.; Espitia, P. J. P.; Otoni, C. G. Antioxidant Active Packaging Based on Papaya Edible Films Incorporated with *Moringa Oleifera* and Ascorbic Acid for Food Preservation. *Food Hydrocoll.* **2020,** *103*, 105630. https://doi.org/10.1016/j.foodhyd.2019.105630.

Rodríguez-Pérez, C.; Gilbert-López, B.; Mendiola, J. A.; Quirantes-Piné, R.; Segura-Carretero, A.; Ibáñez, E. Optimization of Microwave-Assisted Extraction and Pressurized Liquid Extraction of Phenolic Compounds from *Moringa Oleifera* Leaves by Multiresponse Surface Methodology. *Electrophoresis* **2016,** *37* (13), 1938–1946. https://doi.org/10.1002/elps.201600071.

Salem, A. S.; Salama, W. M.; Ragab, W. A. Prolonged Shelf Life of Sour Cream by Adding *Moringa Oleifera* Leaves Extract (MOLE) or *Moringa Oleifera* Oil (MOO). *Am. J. Food Technol.* **2015,** *10* (2), 58–67. https://doi.org/10.3923/ajft.2015.58.67.

Sánchez-Muñoz, M.; Valdez-Solana, M.; Avitia-Domínguez, C.; Ramírez-Baca, P.; Candelas-Cadillo, M.; Aguilera-Ortíz, M.; Meza-Velázquez, J.; Téllez-Valencia, A.; Sierra-Campos, E. Utility of Milk Coagulant Enzyme of *Moringa Oleifera* Seed in Cheese Production from Soy and Skim Milks. *Foods* **2017,** *6* (8), 62. https://doi.org/10.3390/foods6080062.

Saucedo-Pompa, S.; Torres-Castillo, J. A.; Castro-López, C.; Rojas, R.; Sánchez-Alejo, E. J.; Ngangyo-Heya, M.; Martínez-Ávila, G. C. G. Moringa Plants: Bioactive Compounds and

Promising Applications in Food Products. *Food Res. Int.* **2018,** *111*, 438–450. https://doi.org/10.1016/j.foodres.2018.05.062.

Shah, M. A.; Bosco, S. J. D.; Mir, S. A. Effect of *Moringa Oleifera* Leaf Extract on the Physicochemical Properties of Modified Atmosphere Packaged Raw Beef. *Food Packag. Shelf Life* **2015,** *3*, 31–38. https://doi.org/10.1016/j.fpsl.2014.10.001.

Sharma, K.; Kumar, M.; Waghmare, R.; Suhag, R.; Gupta, O. P.; Lorenzo, J. M.; Prakash, S.; Radha; Rais, N.; Sampathrajan, V.; Thappa, C.; Anitha, T.; Sayed, A. A. S.; Abdel-Wahab, B. A.; Senapathy, M.; Pandiselvam, R.; Dey, A.; Dhumal, S.; Amarowicz, R.; Kennedy, J. F. Moringa (*Moringa Oleifera* Lam.) Polysaccharides: Extraction, Characterization, Bioactivities, and Industrial Application. *Int. J. Biol. Macromol.* **2022,** *209*, 763–778. https://doi.org/10.1016/j.ijbiomac.2022.04.047.

Sharma, P.; Wichaphon, J.; Klangpetch, W. Antimicrobial and Antioxidant Activities of Defatted *Moringa Oleifera* Seed Meal Extract Obtained by Ultrasound-Assisted Extraction and Application as a Natural Antimicrobial Coating for Raw Chicken Sausages. *Int. J. Food Microbiol.* **2020,** *332*, 108770. https://doi.org/10.1016/J.IJFOODMICRO.2020.108770.

Soni, A.; Pandiyan, P.; Elumalai, P. Effective Treatment of Curry (Murraya Koenigii) and Moringa (*Moringa Oleifera*) Leaves Extracts on Melanosis of Pacific White Shrimp (Litopenaeus Vannamei) During Chilled Storage. *J. Aquat. Food Prod. Technol.* **2021,** *30* (10), 1304–1314. https://doi.org/10.1080/10498850.2021.1988792.

Sultana, S. Nutritional and Functional Properties of *Moringa Oleifera*. *Metab. Open* **2020,** *8*, 100061. https://doi.org/10.1016/j.metop.2020.100061.

Tajalsir, A. E.; Ebraheem, A. S. A.; Abdallah, A. M.; Khider, F. J.; Elsamani, M. O.; Mohamed Ahmed, I. A. Partial Purification of Milk-Clotting Enzyme from the Seeds of *Moringa Oleifera*. *J. Microbiol. Biotechnol. Food Sci.* **2014,** *4* (1), 58–62. https://doi.org/10.15414/jmbfs.2014.4.1.58-62.

Teixeira, E. M. B.; Carvalho, M. R. B.; Neves, V. A.; Silva, M. A.; Arantes-Pereira, L. Chemical Characteristics and Fractionation of Proteins from *Moringa Oleifera* Lam. Leaves. *Food Chem.* **2014,** *147*, 51–54. https://doi.org/10.1016/j.foodchem.2013.09.135.

Terefe, M. A.; Kebede, A.; Kebede, M.; Sciences, C. Clotting Activities of Partially Purified Extracts of *Moringa Oleifera* L. on Dromedary Camel Milk. *East African J. Sci.* **2017,** *11* (2), 117–128.

Valdés-Hernández, Geidy Vivian; Cruz-Viera, Luis; Comet-Rodríguez, R. Influencia de Las Condiciones de Operación En La Extracción de Polifenoles a Partir de Hojas de *Moringa Oleifera* Lam. *CENIC. Ciencias Químicas ISSN* **2015,** *46*, 135–145.

Vongsak, B.; Sithisarn, P.; Mangmool, S.; Thongpraditchote, S.; Wongkrajang, Y.; Gritsanapan, W. Maximizing Total Phenolics, Total Flavonoids Contents and Antioxidant Activity of *Moringa Oleifera* Leaf Extract by the Appropriate Extraction Method. *Ind. Crops Prod.* **2013,** *44* (November 2017), 566–571. https://doi.org/10.1016/j.indcrop.2012.09.021.

Wang, Y.-Y.; Peng, C.; Zhang, Y.; Wang, Z.-R.; Chen, Y.-M.; Dong, J.-F.; Xiao, M.-L.; Li, D.-L.; Li, W.; Zou, Q.-J.; Zhang, K.; Wei, P. Optimization, Identification and Bioactivity of Flavonoids Extracted from *Moringa Oleifera* Leaves by Deep Eutectic Solvent. *Food Biosci.* **2022,** *47*, 101687. https://doi.org/10.1016/j.fbio.2022.101687.

Wu, G.; Chang, C.; Hong, C.; Zhang, H.; Huang, J.; Jin, Q.; Wang, X. Phenolic Compounds as Stabilizers of Oils and Antioxidative Mechanisms under Frying Conditions: A Comprehensive Review. *Trends Food Sci. Technol.* **2019,** *92*, 33–45. https://doi.org/10.1016/j.tifs.2019.07.043.

Zhang, T.; Jeong, C. H.; Cheng, W. N.; Bae, H.; Seo, H. G.; Petriello, M. C.; Han, S. G. Moringa Extract Enhances the Fermentative, Textural, and Bioactive Properties of Yogurt. *LWT* **2019,** *101* (June 2018), 276–284. https://doi.org/10.1016/j.lwt.2018.11.010.

Zhao, B.; Lan, T.; Li, H.; He, Y.; Wu, D.; Chen, Z. Antioxidation Activity of *Moringa Oleifera* Lam. Leaves Extract on Soybean Oil during Both Storage and Thermal Treatment. *J. Food Process. Preserv.* **2019,** *43* (8), e13975. https://doi.org/10.1111/jfpp.13975.

Zhuang, S.; Hong, H.; Zhang, L.; Luo, Y. Spoilage-related Microbiota in Fish and Crustaceans during Storage: Research Progress and Future Trends. *Compr. Rev. Food Sci. Food Saf.* **2021,** *20* (1), 252–288. https://doi.org/10.1111/1541-4337.12659.

CHAPTER 6

Storage of Fruits and Vegetables in Controlled Atmospheres and Its Application in the Preservation of White Onion Quality

CELIA CHÁVEZ-MENDOZA and ALEXANDRO GUEVARA-AGUILAR

Centro de Investigación en Alimentación y Desarrollo A C, Coordinación Tecnología de Productos Hortofrutícolas y Lácteos, Delicias, Chihuahua, México

ABSTRACT

Controlled atmosphere (CA) technology consists of storing fruits and vegetables in a refrigeration chamber in which the initial or normal atmosphere is replaced by another one that is poorer in oxygen and richer in CO_2. Through this technology, it is possible to control various metabolic processes of fruits and vegetables, thus delaying the process of maturation and senescence and maintaining their nutritional quality and optimal state of maturation until consumption. However, if it is not applied correctly, it can cause various quality problems. Changing the normal atmosphere to a controlled one requires the use of gases such as oxygen, carbon dioxide, and nitrogen, which are mixed to achieve the desired storage concentration. There are several types of AC systems, which depend on the way of generating the atmosphere and the concentrations of O_2 and CO_2 used. CAs are currently used commercially almost exclusively for the storage of apple, pear, kiwi, and cabbage. In Mexico, they are mainly used to preserve apples and pears in the states of Chihuahua and Coahuila, but they have not been used in onions despite being a seasonal product and widely consumed. Experimental studies have shown that this vegetable, being a nonclimacteric product, responds

Engineering Principles for Food Processing Technology and Product Realization. Leonardo Sepúlveda Torre, Porteen Kannan, A. K. Haghi, PhD (Eds.)

well to this technology, managing to preserve its quality for up to 7 months. The use of CA, its characteristics, and its application, with emphasis on the conservation of onion quality, is briefly described in this chapter.

6.1 INTRODUCTION

The recent concern of consumers and health organizations to maintain a healthy diet has increased the demand for fresh fruits and vegetables to be consumed on a daily basis; this is because they represent an important source of antioxidant compounds that are beneficial against chronic diseases, including cancer and cardiovascular diseases. This scenario has created new opportunities for the horticultural industry to improve the quality of fruits and vegetables through their nutritional value (López et al., 2011). However, in many of these products, the storage period is short, damaging their quality and limiting their commercialization, thus missing out on market niches. An alternative to this drawback is the use of controlled atmospheres (CAs), a technology that involves storage in low O_2 and/or high CO_2 concentrations to delay the rate of respiration, ethylene production, and weight loss, and therefore the fruit deterioration (Crisosto et al., 2002) The use of CAs represents one of the most important advances in postharvest technology, as the gas composition in storage has an impact on the shelf life of the stored product, and is used to prevent damage to harvested fruits and vegetables during storage, transport and distribution (Arshad et al., 2023).

There are many variables that affect the fruit degradation during postharvest storage. Some of these variables are damage by microorganisms, the storage environment, and the maturity at which they were harvested, among others. Therefore, storage requirements may be different for each type of fruit, and they tend to behave differently after harvest; for this reason, they are classified as climacteric or nonclimacteric, and there are fruits with intermediate behavior. Climacteric fruits show a growth in respiration rate during their decomposition stage that is associated with ethylene production, and in nonclimacteric fruits, their chemical composition is maintained over time, except for sugars that are decomposed for metabolic processes. This leads to a greater number of physiological changes in climacteric fruits that can be reduced by manipulating the atmosphere in which they are stored (Thompson, 2016).

The scientific basis for the application of CA in fruit and vegetable storage has been the subject of several research studies and has evolved tremendously, some of the science dates back 200 years, but it was first refined and commercially applied in the mid-20th century (Thompson, 2010). Commercially, it has been applied in apples and pears due to its special adaptation to CA conditions. But its effect has been evaluated in other fruits such as pitahaya (Magaña-Benitez et al., 2010); papaya (Martins and De Resende, 2013); blackberry (Brackmann et al., 2016); lychee (Ali et al., 2021); avocado (Chirinos et al., 2021; Fuentealba et al., 2022); banana, mango, pineapple, cherimoya and rambutan (Kader, 1994). In vegetables, it has been applied in tomato storage (Dilmaçünal et al., 2022); in the preservation and shelf-life extension of spinach (Martínez-Damina and Cantwell de Trejo, 2002), and in a more limited way in onion (Rios-Gonzalez et al., 2018; Smittle, 1988; Yamashita et al., 2009).

Onion is one of the oldest and most widely consumed crops in the world, is highly valued for its characteristic pungent flavor, and is an essential ingredient for cooking worldwide. It is a seasonal vegetable and is sometimes overproduced (Nega et al., 2015). Produced in many countries around the world, it is the second most important crop after tomato. China is the leading producer worldwide, followed by India and the United States. It is also the most important income-generating crop in small, medium, and large-scale production in many developing countries (Muhie, 2022).

On the contrary, one of the characteristics of this vegetable crop is the susceptibility to postharvest losses due to its high water and low dry matter content, although it is less perishable than other crops, losses are unacceptable during medium and long storage periods; it has been reported that 40–50% of the product never reaches the ultimate user/consumer due to postharvest losses (Gorreapti et al., 2017).

Quality losses of this vegetable are due to several factors because, during their collection, handling, transport, packaging, and storage, they are exposed to several treatments and environmental conditions which not only can affect their quality attributes and physiological characteristics but could also be responsible for several reactions and stresses, causing important biochemical changes in the bulb tissues (Benkeblia and Varoquaux, 2003).

Onion is classified as a low perishability product because it maintains a low respiration rate (Kadder, 2002). In addition, after harvest, it enters a natural state of dormancy that favors its conservation. This phenomenon is controlled by several endogenous hormones that vary depending on the genetic composition of each variety, environmental factors, and in particular temperature, which can extend or shorten it (Sharma et al., 2016). However, during storage, especially for prolonged periods, there are losses of marketable bulbs mainly due to sprouting, rooting, decay, weight loss, or phytopathological damage (González et al., 2005). Prolonged storage can also significantly affect the appearance and chemical composition of bulbs (Sharma et al., 2014).

CA has been proven that it could be used as an environmentally friendly technology for prolonged storage of fruits and vegetables including onions

(Poldma et al., 2012). This technology helps to preserve its quality allowing to offer quality products to the market in times of low supply and high prices, satisfying the demand that exists throughout the year, some researchers have used this technology and obtained positive results (Chávez-Mendoza et al., 2018; Garba et al., 2014; Poldma et al., 2012).

Therefore, characteristics and importance of the technology of crop storage in CAs are discussed in this chapter, as well as its application in onion preservation.

6.2 CONTROLLED ATMOSPHERE TECHNOLOGY

Controlled atmosphere technology consists of storing fruits and vegetables in a refrigeration chamber in which the initial or normal atmosphere is characterized by a concentration of 20.95% oxygen (O_2), 0.03% carbon dioxide (CO_2), and 78.09% nitrogen (N_2) is replaced by an atmosphere poorer in oxygen and richer in CO_2, maintaining precise control of the concentrations of these gases during storage or transport (Graell and Ortiz, 2003). Air composition is adjusted based on product requirements and the change of the atmosphere is achieved by certain devices such as nitrogen generators, CO_2, ethylene absorbers, etc. Inside the chamber, the environmental physical conditions such as temperature, humidity, and air circulation must also be controlled.

CA is successful in fruit and vegetable storage only when combined with refrigeration or low temperatures. Therefore, standard refrigeration units are integral components of CA storage systems.

This technique associated with cold, reiterates the effect of refrigeration on the vital activity of the tissues, avoiding certain physiological problems and reducing losses due to rotting. The action of the atmosphere on fruit respiration is much more important than the action of low temperatures. A controlled atmosphere delays biochemical reactions causing slower respiration and a delay in ripening, the fruit being in a dormant condition, with the possibility of vegetative reactivation once the fruit is placed in normal atmospheric air.

It is desirable that the created atmosphere composition remains constant over time. However, internal changes (metabolic reactions) of certain products consume some gases (oxygen) and generate others (carbon dioxide, ethylene) that alter this initial composition. These variations are detected by monitoring devices and compensated by various gas production/removal mechanisms. These systems cannot be implemented in small packages for retail sales.

CA technology is relatively expensive compared to regular storage, but it adds value to the crop by extending the marketability of the product, which is important for those fruits or varieties that have a short shelf life such as sweet onions (Poldma *et al* 2012).

This technique has been improved and is currently very important in the preservation of foods of both vegetable and animal origin. Currently, large quantities of fruits and vegetables, such as apples, pears, kiwis, sweet onions, and cabbage, are stored in CA. Furthermore, this storage technology is also widely used in several countries during the maritime transport of various fruits, such as bananas, mangoes, avocados, and apples (Yahia et al., 2019).

6.3 GASES EMPLOYED IN CONTROLLED ATMOSPHERES

Changing the normal atmosphere to a CA basically requires three types of gases: oxygen, carbon dioxide, and nitrogen, which are mixed to achieve the desired concentration.

6.3.1 OXYGEN

A decrease in the amount of oxygen in the atmosphere results in reduced respiratory intensity and lower ethylene biosynthesis in fruits and vegetables (Graell and Ortiz, 2003). It is a necessary compound for the metabolism of living organisms and is involved in oxidation processes. It is usually reduced in CA chambers, which may favor the growth of anaerobic pathogens (Yahia et al., 2019).

6.3.2 NITROGEN

It is employed to displace oxygen, thereby decreasing the growth of aerobic spoilage microorganisms and protecting against food oxidation (Parry, 1993). It is also used to balance the atmosphere and achieve the desired concentration inside the chamber.

6.3.3 CARBON DIOXIDE

The increase in the level of CO_2 decreases the respiratory intensity and inhibits the action of ethylene, this reduces the metabolic activity of plant tissues (in medium concentrations) which enter a state of "dormancy." Also inhibits the activity of microorganisms (in high concentrations) (Graell and Ortiz, 2003; Yahia, 1995b). CO_2 dissolves in water to form carbonic acid and thus lowers the cellular pH. It also modifies the enzymatic activity of several metabolic pathways. Greater solubility at low temperatures and therefore more effective in controlling microorganisms under these conditions (Yahia et al., 2019).

6.4 CA STORAGE SYSTEMS

There are several types of CA storage systems, some of the most commonly used are described in the following:

a. **Conventional CA (CCA)**

This system stores the fruit under hermetic conditions and allows the fruit to change the surrounding atmosphere due to the process of respiration, which causes the amount of oxygen to decrease and the amount of CO_2 to increase until the necessary atmosphere is established. Gas levels should be monitored daily and when the desired concentrations are achieved, gas control begins. O_2 levels are maintained by the introduction of air from outside the room and CO_2 levels are maintained by removal. In this system, the time in which the desired concentrations are achieved can be very long. In apples, for example, it takes about 10 days and during this time the fruit can continue to mature (Yahia, 1995b). In apples, a concentration of 3kPa O_2 and 3kPa CO_2 has been used (Yahia et al., 2019).

b. **Fast conventional CA (FCCA)**

In this system, the atmosphere is generated immediately after the chamber is sealed. Oxygen concentration is reduced to desired levels in less than two days using nitrogen or CA generators. A precise control of the gas concentration, temperature, and relative humidity of the chamber is carried out by means of certain analyzing devices. It is currently the most commercially used system for the storage of apples, pears, and cabbage (Yahia, 1995c).

c. **Ultra-low oxygen CA (ULOCA)**

In this system, the oxygen concentration is reduced to the minimum levels tolerable by the fruit. It is a type of FCCA that uses very low O_2 concentrations (less than 1%) and without using high levels of CO_2. Such low O_2 levels are very critical, and a highly efficient monitoring and control system is needed to prevent the amount of oxygen from dropping to dangerous levels for the fruit (fermentation).

Preservation under such O_2 conditions is called ultralow oxygen (ULO). ULO preservation is carried out in airtight preservation chambers and is used for the long-term preservation of apples, pears, blueberries, and kiwi.

In Golden Delicious apples, for example, it was observed that the use of this system preserved firmness and achieved good acidity after storage for 6–10 months, especially when ULOCA was applied in a rapid manner. It also decreased some important physiological disorders such as darkening of the pulp core, reduced ethylene biosynthesis, decreased respiration rate, and susceptibility to decay. Special care must be taken when using this atmosphere as it may favor the development of anaerobic processes in the fruit causing the accumulation of fermentation-derived compounds such as ethanol, ethyl acetate, and acetaldehyde (Yahia, 1995b).

The main advantages achieved with low oxygen atmospheres, compared to standard CA, are longer shelf life and subsequent product shelf life, better quality retention during storage and upon release, and a reduction of certain physiological alterations (Graell et al., 1998).

d. High carbon dioxide CA

In this system, high amounts of CO_2 (10–15% in the case of apple) are used before starting with the ideal atmosphere. This CA has been used for two decades in apples, but its commercial use is limited to the northwestern United States for Golden Delicious apples. When applied to this fruit, it was observed that using a CO_2 concentration of 10–15% for 2–4 weeks and a temperature between 0 and 5°C, it was possible to reduce the texture and quality loss and prolong the shelf life of the fruit (Yahia,1995b).

e. Low-ethylene CA (LECA)

This system began to be used commercially in 1983 in the Northwest USA and consists of a fast CA with low O_2 levels, high CO_2 levels, and strict control of ethylene levels in pre-harvest and postharvest (Yahia, 1995c).

Levels of this hormone in this system are maintained at a concentration of 1 ppm or less. By storing "Mclntosh" apples in this system for 8 months, the fruit only lost 3.5% of its initial firmness compared to a 24.2% loss of firmness of apples stored in CCA (Yahia, 1995b).

f. Insecticidal CA

In this system, very low amounts of O_2 and high amounts of CO_2 are used to control insects as an alternative to the use of chemicals.

Atmosphere levels needed for insect control in very short periods are 0.5% O_2 or less and 50% CO_2 or more. These can control insects and have the potential for use in postharvest, especially as quarantine treatments. Insect response depends on their stage of development, but many insects can be controlled within 2–4 days at room temperature.

CA use for insect control is very advantageous because it does not leave toxic residues in the food and the cost is competitive compared to the application of agrochemicals. However, insecticidal atmospheres are very extreme and can cause fruit problems such as anaerobiosis and fermentation. Therefore, it is very important to use them only on those fruits that can tolerate these levels of atmosphere and only for a period shorter than the tolerated period (Yahia, 1995a).

g. **Dynamic control system CA (DCS)**

This system consists of maintaining oxygen levels within the minimum limits tolerated by the fruit, for which it is monitored by sensors, and its levels are periodically adjusted.

CA conditions are dynamically regulated during storage (instead of being set in advance). The CA dynamic conservation system equipment reacts to the physiological circumstances of the fruit inside the chamber, adjusting the conservation conditions to them automatically. Achieving oxygen concentration percentages below 1% is not a problem, as the oxygen concentration is adjusted directly as soon as there is a risk of it being too low.

So far, the commercial use of DCS has only been evaluated in apples, in which it favors the maintenance of firmness, and acidity, and reduces the development of some physiopathologies without affecting the sensory quality of the fruit (Calvo and Caldan, 2012).

6.5 CONTROLLED ATMOSPHERE APPLICATION

The atmosphere to be used for a given product is based on recommendations derived from research carried out by many fruit research centers around the world; it is necessary to contrast these recommendations in each specific area given the influence of environmental factors on the behavior of the species and/or varieties (Graell and Ortiz, 2003).

It is recommended that the desired product temperature and atmosphere composition be reached within a maximum of 7–10 days after the food is placed in storage before ripening begins, that is, when the ethylene in the fruit is less than 1 ppm (Bishop, 1990).

CAs are currently used commercially almost exclusively for apple, pear, kiwi, and cabbage storage. In Mexico, they are mainly used to store apples and pears in the states of Chihuahua and Coahuila (Yahia, 1995b). Table 6.1 shows gas concentrations and temperatures for CA storage of various fruits and vegetables.

TABLE 6.1 Recommended Gas Concentrations for the Equilibrium Atmosphere and Temperature Range for the Preservation of Fruit and Vegetable Products.

Product	Temperature (°C)	%O_2	%CO_2
Apricot	0–5	2–3	2–3
Avocado	5–13	2–5	3–10
Broccoli	0–5	1–2	5–10
Pumpkin	0–5	3–5	5–7
Cherry	0–5	3–10	10–12
Brussels sprouts	0–5	1–2	5–7
Cauliflower	0–5	2–5	2–5
Mushroom	0–5	Air	10–15
Asparagus	0–5	Air	5–10
Spinach	0–5	Air	10–20
Strawberry	0–5	10	15–20
Pomegranate	4–6	5–10	0–2
Kiwi	0–5	2	5
Lettuce	0–2	2–5	0
Lemon	12–14	5–10	0–2
Tangerine	4–5	10–12	0–2
Mango	10–15	5	5
Apple	0–5	2–3	1–2
Cantaloupe melon	3–7	3–5	10–15
Peach	0–5	1–2	5
Orange	2–4	10–12	0–2
Papaya	10–15	5	10
Pear	0–5	2–3	0–1
Cucumber	8–12	3–5	0
Semi-ripe tomato	8–12	3–5	0
Pineapple	10–15	5	10
Olive	8–12	2–5	2–10

Source: Artés (1987); Kader (1990), Marcellin (1992), IIR (1995), Yahia et al. (2019).

6.6 DISADVANTAGES OF USING CONTROLLED ATMOSPHERES

Improper use of the CA and MA may cause some problems and result in product loss (Yahia, 1995b). Some of the potential problems are discussed in the following subsections.

6.6.1 ANAEROBIOSIS

Exposure of fresh products to very low O_2 and/or very high CO_2 concentrations that are outside the tolerance limits for a particular product may cause respiration to change from aerobic (with oxygen) to anaerobic (without oxygen). Anaerobiosis refers to the fermentation of a fruit or vegetable, which causes bad odors and flavors (Day, 1993; Farber, 1991).

6.6.2 DEVELOPMENT OF BAD FLAVORS AND ODORS

Long-term storage of fruits in CA may decrease their ability to develop their characteristic flavor and aroma (Yahia, 1995b). Bad tastes and odors may also occur at very low O_2 concentrations as a result of anaerobic respiration and very high CO_2 concentrations due to fermentative metabolism (Kader, 2004).

6.6.3 PHYSIOLOGICAL DISORDERS

CA can stimulate or aggravate some physiological disorders of fruit and vegetable products, for example, internal browning in pears and apples, brown spots in lettuce, chilling injury of some products (Kader, 2004), and freezing damage to texture.

6.6.4 INCREASED SUSCEPTIBILITY TO FUNGAL ATTACK

Inadequate gas levels can facilitate the penetration of fungi and other microorganisms through plant tissues (Yahia, 1995b).

6.6.5 BACTERIOLOGICAL ISSUES

Very low O_2 and/or very high CO_2 concentrations can stimulate the growth of certain pathogenic bacteria such as Clostridium and Campylobacter, which is a potential risk in products stored in modified atmospheres (Farber, 1991).

6.6.6 *ABNORMAL RIPENING OF FRUITS AND VEGETABLES*

Inadequate atmospheres can cause abnormal or irregular ripening of products (Kader et al., 1989). This can occur in banana, mango, pear, and tomato with O_2 levels below 2% and CO_2 levels above 5% (Kader, 2004).

6.6.7 *HIGH INITIAL COST*

A relatively high initial investment is required for equipment acquisition and chamber construction with adequate conditions for CA storage, as well as hiring trained personnel to operate the system. Furthermore, gas concentrations are different for each particular product (Day, 1993; Farber, 1991).

6.7 ONION PRESERVATION

In onions, the organ of consumption is generally the bulb, which, unlike other vegetable species, has some characteristics that favor conservation for a relatively long time. Under photoperiod conditions suitable for bulb formation, the plant initiates reserve accumulation at the base of the leaves, forms cataphylls without laminae, and ends with the leaf primordia initiation, before entering a dormancy period. During this period the bulbs have a reduced metabolism and do not have the capacity to emit leaves and roots, which facilitates conservation. Depending on the variety, growth restarts after a variable number of days. Inner cataphylls are thick and firm, while outer cataphylls are formed by dead cells, which form a barrier to gas exchange and a protective element against mechanical damage and fungal attack. Bulbs have low respiration rates, which can be considered as an indicator of low metabolic activity, a characteristic also favorable for conservation (Furlani and Rivero, 2010; UCDavis, 2022).

The main purpose of onion storage in CA is to preserve bulbs with commercial qualities for a prolonged period of time.

Postharvest quality of onion is strongly influenced by cultivar, cultural practices, climatic conditions during the growing and development period, as well as storage conditions. Well-preserved onions are generally long-day onions with prolonged dormancy periods, while short-day cultivars generally have storability. Fertilization, mainly nitrogen fertilization, also affects onion storage, as well as late harvests (Poldma et al., 2012).

6.8 ONION QUALITY CRITERIA

Onion quality indices can be objective or subjective and should serve to accurately express the quality and postharvest life of the product in different situations.

Some of the main characteristics to be taken into account to define onion quality are color, absence of sprouting and roots, firmness, soluble solids, Dry matter, size, weight, pungency, and phytopathological damage.

a. **Color.** Onion commercial value is strongly dictated by its aesthetic appearance, related in an important way to skin color, due to the high sensitivity of the consumer to this quality factor (Kiura et al., 2021).

   Onion color is defined according to the coloration of the epidermis or skin, this parameter is verified visually or by means of a colorimeter. Onions consumed in Mexico are white, purple, and yellow in color. In white onions, the bulb must have a white skin covering at least 80% of its surface at the time of harvesting, while purple and yellow onions must be 100% covered by these colors at the time of harvesting (Norma Mexicana NMX-FF-021-SCFI-2003).

   Bulb color can be affected by handling. It has been observed that on onions of white varieties or those with few protective cataphylls, a greenish coloration can appear on the external scales, which is caused by exposing the bulbs to direct sunlight for a few days or when they are kept in brightly lit places. This disorder detracts from the commercial quality and may impart a strange taste to the product.

   Onion color has been related to its antioxidant capacity. It was found that red onions have a higher antioxidant capacity than yellow and white onions, although yellow onions are richer in phenolic compounds (Gokce et al., 2010).

b. **Root sprouting and emission.** The term sprouting in onion refers to the new leaves that begin to develop from the bulb after having been at rest for a period of time (Mexican Standard NMX-FF-021-SCFI-2003).

   Sprouting rate during onion storage is closely related to different agronomic parameters such as soil moisture, nitrogen fertilization, bulb ripening at harvest time, water stress, and variety (Sorensen and Grevsen, 2002; Miedema, 1994).

Root sprouting and emission from bulbs not only accelerates spoilage but also determines the end of its shelf life since the consumer rejects the presence of sprouts and/or roots emerging from the product. Root emission affects the visual quality of the onion and may furthermore predispose to further deterioration due to root rot. It is caused by excessive humidity during storage.

Once they have completed their development, bulbs enter a state of "rest" or dormancy, characterized by a very reduced physiological activity that does not respond to environmental conditions, that is, they do not sprout even under optimal humidity and temperature conditions. Different studies have shown that sprouting inhibitory compounds such as abscisic acid predominate in this stage over promoters such as gibberellins, auxins, and others. This balance changes with storage time leading to a state in which it sprouts or emits roots if exposed to favorable environmental conditions. There is no clear differentiation between one state and another, but rather a slow transition. As time goes by, promoters predominate, and sprouting irremediably takes place. Refrigerated storage and CAs slow down sprouting and rooting (Lopez, 2003; Sharma et al., 2016). Sprouting in onions is considered to be the main contributor to postharvest losses (20–40%) of total storage losses (50–90%), and as a result, sprouted onions are discarded and placed in containers, so no attention has been paid to onions with this condition, however, it has been found that these onions contain high phenolic content and high antioxidant activity (Majid et al., 2016).

Another physiological disorder of onion is sun damage, which is the result of prolonged exposure of the onion to solar radiation. The problem is associated with high temperatures and/or ultraviolet rays. It manifests itself in the form of blisters and rupture of the peridermis, especially when the temperature of the product is close to 50°C (Furlani and Rivero, 2010).

c. **Firmness.** Bulb consistency or firmness is a very important quality characteristic. Bulbs should be firm and not yield to strong pressure. Soft bulbs are the result of excessive dehydration, sun or mechanical damage, or due to the presence of a physiological disorder known as translucent cataphyll. Also, the neck of the bulb should be narrow and dry. The presence of wide necks indicates poor curing, initiation of bud break, flower scape production, or root growth (Furlani and Rivero, 2010).

d. **Soluble solids.** Soluble solids values are considered a good approximation of the onion sugar content. This parameter is related to the shelf life and sweetness of this vegetable. There is a direct relationship between the content of soluble solids and the percentage of dry matter in onion and high amounts of these components indicate a better aptitude for conservation (Fondo Europeo Agrícola de Desarrollo Rural, 2008).

e. **Pungency.** This parameter refers to the level of onion pungency and is established by determining the pyruvic acid content of the bulbs. The secondary organosulfur compounds of onion are responsible for the pungent flavor and aroma. Thiosulfinates are the predominant flavor compounds in raw onion and are formed when unflavored *S*-alk(en) yl-l cysteine sulfoxides (ACSO) are lysed by alliinase after tissue breakdown. The initial products of this reaction are sulfenic acid, pyruvate, and ammonia. Depending on the alkenyl substituent present, sulfenic acids either condense in pairs to form thiosulfinate or organize to form thioprapanal sulfoxides or lachrymatory factors (Lee et al., 2010).

There are several scales to measure the level of pungency, a simplified scale is used by the sweet onion industry in Georgia in the United States of America, where pyruvic acid levels vary from 0 to 18 μmol/g of fresh onion (μmol/g). When values are between 0 and 3 μmol/g, they are classified as low pungency, when they are between 3 and 7 medium pungency and if they are higher than 7, the onions are classified as high pungency. Another scale is the one used by Vidalia Labs International, Georgia, USA, where the scale ranges from 0 to 10 μmol/g of fresh onion. When pyruvic acid values are less than 3, they are classified as very mild, between 3 and 4 they are classified as mild, between 4 and 5.5 they are classified as slightly pungent, between 5.5 and 6 as pungent and greater than 6 as very pungent. Thus, onions with a pyruvic acid level below 3.5 are called extra sweet onions, between 3.6 and 5.5 sweet onions, and with values above 5.6 spicy onions. The pungency level is a parameter directly related to the variety and other agronomic parameters such as fertilization, irrigation, climatic conditions, and soil type (Raigón, 2006).

f. **Size onion.** Size is determined by the weight or by the equatorial diameter of the vegetable. According to Boyhan et al. (2004), the small onion should measure <6.4 cm; medium, from ≥ 6.4 to <7.6 cm; large from ≥ 7.6 to ≤ 8.9 cm; and jumbo, >8.9 cm. Meanwhile, in Mexico, ball onions must comply with the specifications presented in Table 6.2 (NMX-FF-021-SCFI-2003).

TABLE 6.2 Size Specifications of Mexican Ball Onions.

Size	Equatorial diameter range (cm)
Small	Less than 3.5 cm
Medium small	3.6–5.0 cm
Medium	5.1–7.0 cm
Large	7.1–9.0 cm
Extra	Greater than 9.0

Source: NMX-FF-021-SCFI-2003.

g. **Dry matter.** Dry matter is one of the most important factors in onion quality, about 80% of the *dry matter of the bulb is made up of nonstructural carbohydrates such as sucrose, glucose, fructose*, and low molecular weight fructans. Sugar metabolism is closely linked to onion dormancy and sprouting (Lee et al., 2010). Bulbs with high amounts of dry matter are more suitable for prolonged storage and also tend to contain a higher concentration of fructans, which represent the main carbohydrate reserve in most bulbs, and are converted to fructose as a source of energy for sprouting (Chope et al., 2012).

h. **Phytopathological damage.** Numerous pathogens can cause diseases that shorten shelf life and affect product quality. Most common are *Aspergillus niger*, *Botrytis* spp., *Erwinia carotovora*, *Fusarium* spp., *Penicillium* spp., and *Rhyzopus* spp.

In some cases, disease development can be initiated as a result of nematode attacks that produce deformations and cracks in some bulbs, which are not always eliminated in the selection process. Overall, proper temperature management should be considered the main weapon of control. Low temperatures coupled with low relative humidity are generally more effective than any other treatment, including chemicals, in controlling diseases.

6.8.1 FUNGAL DAMAGE

A. niger (black mold). It is the main microorganism that attacks onions during storage, especially at high temperatures and relative humidity. It affects the outer parchment of the onion, producing black powdery spots (fungal spores), which significantly detract from the commercial quality of the product. Moreover, it can attack the fleshy cataphylls, showing more or less sunken rotting grooves of dry consistency, depending on the environmental humidity. It can move from diseased to healthy bulbs. The onion surface must be kept dry during and after harvesting (storage) (Fernandez et al., 1999).

When the entire bulb surface turns black, the onion shrivels, wrinkles becomes secondarily infected with bacteria, and rots. After harvest, black mold is avoided by transporting and storing the bulbs below 15°C or at very low humidity. Infection is minimized by avoiding hitting and cutting the bulbs.

Basal rot (Fusarium spp.*).* Generally, infection occurs in the field and manifests itself in storage. Dry or semi-wet rot appears at the bulb base and can affect the entire bulb. Damaged tissue becomes aqueous. Infection can spread from diseased to healthy onions. It is one of the main causes of post-harvest losses. Onion cultivation should be avoided in fields where onion disease attacks have occurred in previous years (Fernández et al., 1999).

Blue mold. Watery rot on the collar and external scales, followed by the appearance of blue–green (occasionally yellow–green) spores, is caused by the fungus *Penicillium*. Bruising and other mechanical damage, sun scald, and freeze damage should be minimized (UCDavis, 2022).

6.8.2 BACTERIAL DISEASES

These are characterized by soft rotting and often have an unpleasant odor. Bacteria enter through wounds produced before or after harvest. The main genera are Erwinia and Pseudomonas. Low temperatures and careful management to avoid injury are the best preventive control of these diseases.

Bacterial Rot/Soft Rot. Characterized by watery, foul-smelling, slimy liquid areas, this rot is caused by the bacterium *E. carotovora* subsp. Carotovora (UCDavis, 2022). *Erwinia* can infect plants in the field. Leaves wilt and eventually dry up. If a longitudinal cut of an infected plant is made, it will be observed that the middle part of the new bulb appears completely loose and very sticky. Coats become soft and watery in appearance. Later on, a light yellow to light brown gelatinous substance will form. Affected bulbs are soft and watery and, when squeezed, a sticky fluid or pulp comes out of the bulbs (Bejo, 2011).

The bacteria that cause these problems in onion crops are only found in soil, irrigation water, and in the remains of previous crops. Infection almost always starts through wounds caused by insect damage, hail, heavy rain, and strong wind. On the contrary, if it rains gently for a prolonged period, the leaves may become soggy, either because they are held together or because the leaf axils fill with water and begin to rot. Bacteria mostly enter the bulb through wounds in the neck, but this is not always the case. The higher the temperature, the faster the infection progresses. Bacteria are inactivated at temperatures below 3°C (Bejo, 2011).

6.9 FACTORS AFFECTING THE CONSERVATION OF ONIONS

a. *Variety.* Generally, short-day varieties are less suitable for prolonged storage than long-day varieties, as the latter have a higher dry matter content and therefore have lower water losses and are less susceptible to sprouting and postharvest diseases. However, significant differences in dry bulb storability have been reported among cultivars of the same photoperiod type, which may differ up to 3–3.5 months of storage period. Cultivar selection and pre-harvest conditions, such as plant establishment methods (transplanting rather than direct seeding) and harvest stage (early harvest when 20–50% of the top of the plant has fallen off) can reduce the incidence of sprouting after long-term storage. Other important crop characteristics that can affect bulb storage capacity include the number and thickness of coat layers and the pungency of inner layers (Petropoulos et al., 2016; Grevsen et al., 2004).

b. *Curing.* Immediately after harvest, bulbs must undergo a drying ("curing") process to harden their outer scales, reduce skin cracks, and allow the necks to narrow, thus inhibiting pathogen infections. The curing method (field or forced air curing) and conditions (temperature and relative humidity), as well as growing conditions and harvest stage, can be of great importance to achieve maximum quality of dried bulbs and minimum losses due to water losses and pathogen infections (Petropoulos et al., 2016).

c. *Temperature.* It is one of the factors that have the greatest influence on the metabolic activity of fruit and vegetable products. If the temperature increases, the rate of respiration will also increase, generating a greater amount of heat, which decreases the postharvest life of the products.

Temperature acts directly on physiological processes related to sprouting inhibition and to a lesser degree on bulb root formation. In turn, high temperature favors the growth of disease-causing microbes.

The ideal temperature for a good conservation of onion is 0–5°C or 28–30°C. Freezing temperature is − 0.4 to − 0.8°C. Low temperatures cause softening and blackening of tissues. When stored at low temperatures, 1–2°C is recommended, followed by a period of reconditioning before sale to avoid water condensation. Moreover, the maximum temperature tolerated by the bulbs is 38–41°C, but at this temperature, there is a risk of microorganism growth (Brice et al., 1997).

High temperatures cause damage, and softening, leaving the appearance of cooked tissues, as well as promoting the development of pathogens (fungi and bacteria) at a higher rate. Sprouting is later at 0°C than at storage temperatures of 5–10°C. Temperatures between 16 and 22°C promote sprouting, the optimum being 15°C, which is inhibited at 30°C. A similar response to temperature changes is seen in root formation, except that the optimum temperature is found in lower ranges (9–11°C) (Brice et al., 1997).

Onion can be affected by freezing, its freezing point is approximately -1°C depending on the variety and solids content. Bulbs may be affected, showing a yellow-grayish coloration and a watery consistency. This disorder progresses from the surface to the bulb center and from the periphery to the central part of the vegetable. Some measures to alleviate damage in affected bulbs consist of unfreezing the product slowly to about 4°C and marketing it immediately (Furlani and Rivero, 2010).

Another problem that can occur in onions is that of translucent cataphylls, a problem that occurs more frequently after 3 or 4 months of cold storage, being its incidence more common in large bulbs. It usually starts in the most curved portion of the cataphylls and may involve the base of the outermost leaves. Cataphylls become partially transparent, in contrast to the opaque appearance of healthy tissue, and conducting vessels become more noticeable. In severe cases, the bulb softens and is invaded by various pathogens. Its incidence is significantly reduced by cooling the product as soon as possible after curing (Furlani and Rivero, 2010).

d. *Relative humidity (RH).* High RH in the environment (above 80%) is one of the most important enemies for onion conservation since it stimulates the formation of new roots and sprouting in the bulbs and promotes the development of pathogens (fungi and bacteria) in a wide range of temperatures.

   High relative humidity, with water condensation on the peel, causes fermentation in the external cataphylls and consequently, the presence of stains, deteriorating the presentation and external quality of the bulbs.

   Low relative humidity (less than 60%) increases weight loss due to bulb dehydration, resulting in wrinkling, which is undesirable for marketing. An optimum RH for storage is 65–75%. Overall, the loss of as little as 5% of a fruit's weight results in an unattractive appearance, decreasing its market value (Brice et al., 1997).

6.10 ONION STORAGE IN CONTROLLED ATMOSPHERES

Onions to be stored in CAs must have the correct maturity, be very well cured, dry, and healthy, so they must be rigorously selected, and they must not be hit during the whole postharvest process so that they are free of any damage. It should be remembered that storage will only preserve the quality with which the onion enters.

A good quality onion should be undamaged, flawless, clean-skinned, free of pathogens, and have the expected color for the variety (Muhie, 2022).

Controlled atmospheres have been found to give better results when stored early in the season just after bulb curing. Factors such as irrigation, harvest date, environmental conditions, variety, use of growth regulators, fertilization, and type of curing have an effect on onion storage in CA.

The following are some general recommendations for onion storage in CA (Sumner, 2000).

- The chamber must be airtight and moisture resistant.

  Construction must provide the following:

An absolute vapor barrier to prevent moisture migration in the CA chamber.

Adequate wall, ceiling, and floor insulation.

High tightness (chamber must contain a pressure medium, pressure change 0.8–0.4 or 0.6–0.3 in of water for at least 30 min). A leaking chamber will require excessive addition of nitrogen and carbon dioxide. Leaks around the door and improper sealing are major contributors to the loss of chamber tightness.

The CA chamber must be protected from wind and direct contact with sunlight and rain to avoid sudden temperature changes that can cause leakage.

Each chamber should have a vacuum–pressure release valve and should be equipped with an airbag to prevent atmosphere loss when small fluctuations in pressure occur.

A large air pocket may be necessary for chambers in onion storage due to the large increase in air volume during the evaporator defrost cycle.

Chambers should be interconnected (with PVC piping and valves) to allow more efficient atmosphere establishment and pressure normalization. This type of interconnection is especially important in small chambers.

- Storage temperature.

Some portions of onions may freeze at temperatures of about -0.6°C (30.8°F).

Each chamber must have at least two evaporator units with a condensing unit (sized to meet the heat gain of the chamber) equipped with reheat. Evaporator defrost cycles should be staggered with PVC pipe interconnections between chambers to serve as atmosphere expansion/contraction regulators. A backup cooling system may be advisable.

- Air circulation.

Storage air circulation should be forced to 100–200 m^3/ton of onion. Forced air circulation is needed to maintain uniform temperature and relative humidity throughout the CA chamber. Fans should be placed at the rear of the room (at the opposite end of the entrance door). The stacking of bins should allow air circulation through them. The fan speed should be variable and controlled to achieve relative humidity conditions as uniform as possible throughout the chamber. Fan operation should start when the first onions are placed inside the chamber and continue to run until the last onions are removed from storage.

- Relative humidity.

  Onions may lose a certain amount of moisture during CA storage. For example, the Vidalia onion could lose 1–1.5% weight per month. The dehumidification system must be capable of removing this amount of water without significantly affecting the storage temperature. The operation of the dehumidification units is controlled by the hygrostat. Improper control of relative humidity, air circulation, and temperature can cause complete loss of onions during storage.

- Low oxygen concentration in CA chamber.

  A low oxygen level, for example, 3%, reduces respiration, and sugar loss and prolongs the storage life of onions. In this environment, humans can lose consciousness in less than 30 seconds, they cannot breathe and their heartbeat stops within minutes. Therefore, do not enter a CA chamber with low oxygen concentration without breathing apparatus.

- High CO_2 concentration in CA chamber.

  A high carbon dioxide atmosphere, for example, 5%, inhibits respiration and pungency development and prolongs storage and shelf life after onion storage. The CO_2 produced by the onion is probably not enough to develop and maintain a high level of this gas.

- CA chamber monitoring and control system.

  It is essential to control the temperature, humidity, and atmosphere in the CA chambers without entering them.

This control includes the measurement of at least twice a day of the following:

a. Temperature in the warmest and coldest areas of the chamber with an alarm for high and low temperatures.

b. Oxygen and carbon dioxide concentrations. Adjustments can be made manually or automatically with a computerized system.
c. Relative humidity. Relative humidity sensors located at the front and rear of the chamber are used to monitor RH differences on which the airflow rate is based, when this difference is too large the fan speed should be increased. This fan speed adjustment can also be done manually or automatically with a computerized system. A record of all measurements and adjustments should be kept. Sensors should operate for at least 6 months without recalibration.

- Heating the onion before removing it from the CA chamber.

  Onions should be "warmed" before leaving the CA chamber to avoid moisture condensation on the bulbs. Fans should operate during the entire heating period. Additional heat may be needed for this process, but the difference between air temperature and bulb surface temperature should not exceed the dew point (approximately 15°F or -9°C). The chamber must be ventilated before opening the door.

- Chamber size should reflect the sales rate of onions.

  Overall, each CA chamber should contain the volume of onions to be sold in 1 week. If weekly sales are less than the capacity of a chamber, onions should be conditioned in another area outside the chamber.

  It should be noted that the removal of a portion of the onions from a CA chamber will reduce the shelf life and quality of the remaining onions left in the chamber if storage conditions are not restored. Furthermore, regeneration of the CA atmosphere can be prohibitively expensive if there are very large empty spaces in the chamber (partially filled chambers).

  Onions should be sold before starting the heating procedure. Shelf life loss due to sprouting can increase rapidly after about 3 weeks. Onions should have good air circulation through marketing to prevent deterioration by root development and surface fungus growth.

a. Onion storage conditions in CAs.
Commercial storage of onion bulbs in CA is limited in part due to variable success and inconsistent effects on bulb quality. Response to CA varies among cultivars, therefore, experiments should be

conducted under local conditions to determine the appropriate levels and composition of gases for each specific variety or local cultivar.

It has been found that the onion preservation in CA has significantly improved its postharvest quality with respect to bulbs stored both refrigerated and at room temperature in several varieties. Conditions such as 1% oxygen and 5% CO_2 at a temperature of 1–2°C and a RH of 58–75% reduced storage losses, and showed higher dry matter and total soluble solids content. With this technology, onions with high pungency maintained good marketable quality for more than 8 months (Poldman et al., 2012). According to Tanaka et al. (1996), a mixture of 1% O_2 and 1% CO_2 at 1°C and 65–70% humidity inhibited sprouting and preserved the vegetable for up to 1 year. Chávez-Mendoza et al. (2016) and Ríos-González et al. (2018) on the contrary, worked with storage conditions of 1% O_2 + 1% CO_2 and 3% O_2 + 5% CO_2, 2.5°C and 60–75% humidity, reporting minimal quality changes in onion kept in CA for 7 months. With both CA treatments, sprouting was inhibited by 100%, as well as microbial growth and there were no significant changes in color, but a slight increase in pungency was observed.

Chope et al. (2007) reported that an atmosphere consisting of 5 kPa CO_2 and 3 kPa O_2 has been recommended for soft onions of the Granex variety, low oxygen concentration inhibited bulb sprouting and reduced weight losses, however, this may have a detrimental effect on sweet onion quality through increased pungency, although the effect is not as great as in onions stored in normal atmosphere or air, furthermore it may limit the subsequent bulb shelf life with accelerated sprouting 3 weeks after removal from CA storage.

Uddin et al. (2003) stored Hysam variety bulbs in regular atmosphere and CA with two treatments 2% O_2 + 2% CO_2 and 2% O^2 and 8% CO_2, 0.5 ± 0.5°C, 80% RH for 9 weeks, finding that pungency and flavor precursors increased by 9–11% in onion stored in regular atmosphere and were significantly reduced by the CA effect.

Smittle (1989) reported that the combination of CA storage (5% CO_2, 3% O_2) and refrigerated storage (1°C) resulted in 99% of bulbs being considered marketable after 7 months of storage; however, there was a 9% weight loss. This author also recommended a CA of 3% O_2, 5% CO_2, 92% N_2, 1 °C, 70–80% RH, and a ventilation rate of 5.m3.h-1.m3 of onions for Vidalia sweet onions, which are known for their sweetness and low pungency.

Finally, Table 6.3 shows various CA compositions recommended for onion storage (Thompson, 1996).

TABLE 6.3 Recommended CA Composition for Storage of Onion Bulbs.

Carbon dioxide (%)	Oxygen (%)	Temperature (°C)
0	1–2	–
0	1–2	0–5
0–5	1–2	0–5
0–5	0–1	0–5
5	3	1
5	5	4–5
10	3	4–5

Source: Thompson (1996).

6.11 CONCLUSION

AC storage technology is a very old technique that has evolved and expanded its application in the conservation of more products of both animal and vegetable origin. Its use allows preserving the quality of these for long periods. However, its effect on each particular fruit must be previously researched to obtain the desired results.

Onion is a widely cultivated and consumed crop in the world, it is a seasonal product and although it is not very perishable like other crops, it does suffer from storage losses, so AC storage technology is a good alternative for its conservation, also taking advantage of the fact that it is not climacteric, which favors its conservation for longer periods.

The quality of the onion during storage is influenced by a large number of factors, both pre and postharvest, which must be considered for a correct application of CA. Up to now, atmospheres with 1% O_2 + 1% CO_2 and 3% O_2 + 5% CO_2, in combination with low-temperature storage (0.5–2.5°C) and relative humidity of 58–75%, are the most researched and the most employed, which have allowed to maintain the quality of the onion from 7 to 12 months. The most important impact of AC technology is in the pungency levels which tended to increase slightly during prolonged storage.

KEYWORDS

- **controlled atmospheres**
- **conservation**
- **quality index**
- **onion storage**

REFERENCES

Ali, S.; Khan, A. S.; Malik, A. U.; Anwar, R.; Anjum, M. A.; Nawaz, A.; Naz, S. Combined Application of Ascorbic and Oxalic Acids Delays Postharvest Browning of Litchi Fruits Under Controlled Atmosphere Conditions. *Food Chem.* **2021,** *350*, 129277. https://doi.org/10.1016/j.foodchem.2021.129277

Arshad, M. O.; Chauhan, Y.; Singh, P.; Srivastav, P.; Gupta, M.; Patwa, N. Advancements in Controlled Atmosphere Storage Technology—A Review. In *Proceedings of Second International Conference in Mechanical and Energy Technology*; Yadav, S., Haleem, A., Arora, P. K., Kumar, A., Eds.; Springer, 2023; pp 399–410. https://link.springer.com/chapter/10.1007/978-981-19-0108-9_42

Artés, F. Refrigeración y comercialización hortofrutícolas en la Región de Murcia [Fruit and Vegetable Refrigeration and Marketing in the Region of Murcia]. II Edición. Ed. CEBAS-CSIC, 1987; 150 p.

Bejo Z. B. V. Enfermedades y plagas importantes en cebollas [Important Diseases and Pests in Onions], 2011; 17 pp. https://baixardoc.com/preview/20130126104951-bejo-enfermedades-y-plagas-cebollas-5cba3001bfd7a

Benkeblia, N.; Varoquaux, P. Effects of Gamma Irradiation, Temperature and Storage Time on the Status of the Glucose, Fructose and Sucrose in Onion Bulbs *Allium cepa* L. *Int. Agrophys.* **2003,** *17*, 1–5.

Bishop, D. Controlled Atmosphere Storage. In: *Cold and Chilled Storage Tecnology*; Dellino, C. V. J., Ed.; Blackie and Son Ltd.: Glasgow, 1990.

Bohyan, G. E.; Purvis, A. C.; Hurst, W. C.; Torrance, R. L.; Paulk, J. T. Harvest Date Effect on Yield and Controlled-Atmosphere Storability of Short-Day Onions. *HortScience* **2004,** *39* (7), 1623–1629. https://doi.org/10.21273/HORTSCI.39.7.1623

Brackmann, A.; Anese, R. D. O.; Thewes, F. R.; Fronza, D.; Hamann, J. J. Storability of 'Tupy'and 'Guarani' Blackberries in Controlled Atmosphere. *Bragantia* **2016,** *75*, 240–246. https://doi.org/10.1590/1678-4499.321

Brice, J. L.; Currah, L.; Malins, A.; Bancroft, R. *Onion Storage in the Tropics: A Practical Guide to Methods of Storage and Their Selection*; Natural Resourxes Institute. The University of Greenwich, 1997; 116 p.

Calvo, G.; Candan, A. P. Tendencias para la conservación de frutas de pepita [Trends for the Preservation of Pome Fruits]. *Revista Iberoamericana de Tecnología Postcosecha* **2012,** *13* (2), 153–159.

Chávez-Mendoza, C.; Vega-García, M. O.; Guevara-Aguilar, A.; Sánchez, E.; Alvarado-González, M.; Flores-Córdova, M. A. Effect of Prolonged Storage in Controlled Atmospheres on the Conservation of the Onion (*Allium Cepa* L.) Quality. *Emirates J. Food Agric.* **2016,** *28* (12), 842–852. DOI: 10.9755/ejfa.2016-06-643.

Chirinos, R.; Campos, D.; Martínez, S.; Llanos, S.; Betalleluz-Pallardel, I.; García-Ríos, D.; Pedreschi, R. The Effect of Hydrothermal Treatment on Metabolite Composition of Hass Avocados Stored in a Controlled Atmosphere. *Plants* **2021,** *10* (11), 2427. https://doi.org/10.3390/plants10112427

Chope, G. A.; Cools, K.; Hammond, J. P.; Thompson, A. J.; Terry, L. A. Physiological, Biochemical and Transcriptional Analysis of Onion Bulbs During Storage. *Ann. Bot.* **2012,** *109* (4), 819–831. https://doi.org/10.1093/aob/mcr318

Chope, G. A.; Terry, L. A.; White, P. J. The Effect of the Transition Between Controlled Atmosphere and Regular Atmosphere Storage on Bulbs of Onion Cultivars SS1, Carlos and Renate. *Postharv. Biol. Technol.* **2007,** *44* (3), 228–239. https://doi.org/10.1016/j.postharvbio.2006.12.018

Crisosto, C. H.; Garner, D.; Crisosto, G. High Carbon Dioxide Atmospheres Affect Stored 'Thompson Seedless' Table Grapes. *HortScience* **2002,** *37* (7), 1074–1078. https://doi.org/10.21273/HORTSCI.37.7.1074

Day, B. P. F. *Fruit and Vegetables, Principles and Applications of Modified Atmosphere Packaging*, Ed. R. T. Parry; Blackie: Glaskow, 1993; pp 114–133.

Dilmaçünal, T.; Bayar, B.; Demirhan, Ö. Impact of Modified Atmosphere Packaging and Controlled Atmosphere Applications on 'Seval F1' Tomato Fruit Quality and Marketability. *J. Inst. Sci. Technol.* **2022,** *12* (2), 527–538. https://doi.org/10.21597/jist.1037827

Farber, J. M. Microbiological Aspects of Modified Atmosphere Packaging—A Review. *J. Food Prot.* **1991,** *54*, 58–70. https://doi.org/10.4315/0362-028X-54.1.58

Fernández, L. J.; Liverotti, O.; Wocca, F. Manejo postcosecha de cebolla, Control de calidad, Inspección de Frutas y Hortalizas [Onion Postharvest Handling, Quality Control, Fruit and Vegetable Inspection].Corporación del Mercado Central de Buenos Aires [Corporation of the Central Market of Buenos Aires], 1999; 30 pp. https://www.yumpu.com/es/document/read/14747439/cebolla-mercado-central-de-buenos-aires

Fondo Europeo Agrícola de Desarrollo Rural. Cebolla Fuentes de Ebro: Estudios sobre la calidad del material vegetal en Aragón. Centro de Transferencia Agroalimentaria. Informaciones técnicas No. 194, 2008. https://bibliotecavirtual.aragon.es/es/catalogo_imagenes/grupo.do?path=3705355

Fuentealba, C.; Vidal, J.; Zulueta, C.; Ponce, E.; Uarrota, V.; Defilippi, B. G.; Pedreschi, R. Controlled Atmosphere Storage Alleviates Hass Avocado Black Spot Disorder. *Horticulturae* **2022,** *8* (5), 369. https://doi.org/10.3390/horticulturae8050369

Furlani, M. A.; Rivero, M. L. Manual del Cultivo de la cebolla. Centro Regional Cuyo INTA, 2010. 10 p. https://inta.gob.ar/documentos/manual-del-cultivo-de-la-cebolla

Garba, U. A.; Mohammed, D.; Ja´afaru, A. Study and Analysis of a Refrigerated and Controlled Atmosphere Onion Storage System. *JORIND* **2014,** *12*, 246–252.

Gökçe, A. F.; Kaya, C.; Serçe, S.; Özgen, M. Effect of Scale Color on the Antioxidant Capacity of Onions. *Sci. Hortic.* **2010,** *123* (4), 431–435. https://doi.org/10.1016/j.scienta.2009.11.007

González, M. V.; Miravé, J. P.; Saluzzo, J. A. Relaciones entre el tamaño y la conservación de bulbos de cebolla para diferentes cultivares y condiciones de almacenamiento. *Revista FAVE- Ciencias Agrarias* **2005,** *4*, 21–30.

Gorreapti, K.; Thangasamy, A.; Bhagat, Y.; Murkute, A. A. Curing of Onion: A Review. *Indian Hortic. J.* **2017,** *7* (1), 8–14.

Graell, J.; López, M. L.; Larrigaudiere, C. Manzanas y peras. Avances técnicos en atmosfera controlada [Apples and Pears: Technical Advances in Controlled Atmosphere]. *Horticultura Internacional Agosto* **1998,** 67–69. https://www.mapa.gob.es/ministerio/pags/biblioteca/revistas/pdf_hortint/hortint_1998_21_66_69.pdf

Graell, J.; Ortiz, A. Recomendaciones para almacenamiento en atmosfera controlada [Recommendations for Controlled Atmosphere Storage]. *Revista Horticultura* **2003,** *172*, 38–43. https://xdoc.mx/preview/recomendaciones-para-almacenamiento-en-atmosfera-controlada-60348e2d61c42

Grevsen, K.; Sorensen, J. N. Sprouting and Yield in Bulb Onions (*Allium cepa* L.) as Influenced by Cultivar, Plant Establishment Methods, Maturity at Harvest and Storage

Conditions. *J. Hortic. Sci. Biotech.* **2004,** *79* (6), 877–884. https://doi.org/10.1080/14620316.2004.11511860

International Institute of Refrigeration (IIR). *Guide to Refrigerated Transport*; Ed. Int. Inst. Refrigerationm 1995; 150 pp.

Kader, A. A. A Summary of CA Requirements and Recommendations for Fruit Other Than Pome Fruits. In: *Proceedings of the Fifth International Controlled Atmosphere Research Conference*, Wenatchee, Washington, USA, Vol. 2; 14–16 June, 1989; pp 303–328.

Kader, A. A. Controlled Atmosphere Storage. In: *Agricultural Handbook Number 66*; Gross, K. C., Wang, C. Y., Saltveit, M. E., Eds.; The Commercial Storage of Fruits, Vegetables and Florist and Nursery Stocks; USDA, 2004.

Kader, A. A. Modified Atmospheres During Transport and Storage of Fresh Fruits and Vegetables. *I Int. Cong. Food Technol. Develop. Murcia.* **1990,** *1*, 149–163.

Kader, A. A. *Postharvest Technology of Horticultural Crops*, 3rd ed.; UC ANR Publications, Davis, 2002; p 40.

Kader. A. A. Fruit, Maturity, Ripening and Quality Relationships. *Perishables Handling Newslett.* **1994,** *80*, 2.

Kiura, I. N.; Bernard M. Gichimu, B. M.; Rotich, F. Visual and Keeping Quality of Stored Bulb Onions as Affected by Harvest and Postharvest Treatments. *Int. J. Agron.* **2021,** *2021*. https://doi.org/10.1155/2021/9969571

Lee, E. J.; Jeon, J. K.; Suh, J. K. Pyruvic Acid and Sugar Contents during Storage Duration in Onion (*Allium cepa* L.). *J. Bio-Environ. Control.* **2010,** *19* (4), 377–381.

López, C. A. F. Manual Para la Preparación y Venta de Frutas y Hortalizas. Del campo al mercado [Manual for the Preparation and Sale of Fruits and Vegetables: From the Field to the Market]; FAO Agricultural Services Bulletin 15, 2003. https://www.fao.org/3/y4893s/y4893s.pdf

López, V. J. Á.; Valverde Juárez, F. J.; Mejía Torres, S. L.; López Angulo, G.; Vega García, M. O. Efecto del almacenamiento en atmósfera controlada sobre la calidad postcosecha y nutricional del tomate. *Revista Chapingo. Serie Horticultura* **2011,** *17* (2), 115–128.

Magaña-Benítez, W.; Balbín, A. M.; Corrales, J.; Saucedo, G. V. C.; Sauri, D. E. Frutas de pitahaya (*Hylocereus undatus*) frigoconservadas a 4°C en atmósferas controladas. *Revista Iberoamericana de Tecnología Postcosecha* **2010,** *11* (2), 143–147.

Majid, I.; Dhatt, A. S.; Sharma, S.; Nayik, G. A.; Nanda, V. Effect of Sprouting on Physicochemical, Antioxidant and Flavonoid Profile of Onion Varieties. *Int. J. Food Sci. Technol.* **2016,** *51* (2), 317–324. https://doi.org/10.1111/ijfs.12963

Marcellin, P. Les maladies physiologiques du froid. In: *Les végétaux et le froid*; Côme, D., Ed.; Hermann: París, 1992; pp 53–105.

Martínez-Damián M. T.; M. Cantwell de Trejo. Cambios de calidad en espinaca almacenada en atmósferas controladas [Quality Changes in Spinach Stored in Controlled Atmospheres]. *Revista Chapingo Serie Horticultura* **2002,** *8* (1), 49–62.

Martins, D. R.; de Resende, E. D. Quality of Golden Papaya Stored Under Controlled Atmosphere Conditions. *Food Sci. Technol. Int.* **2013,** *19* (5), 473–481. https://doi.org/10.1177/10820132124622

Miedema, P.; Bulb Dormancy in Onion. I: The Effects of Temperature and Cultivar on Sprouting and Rooting. *J. Hortic. Sci.* **1994,** *69*, 29–39. https://doi.org/10.1080/14620316.1994.11515245

Muhie, S. H. Preharvest Production Practices, and Postharvest Treatment and Handling Methods for Best Quality Onion Bulbs. *J. Hortic. Sci. Biotechnol.* **2022,** *97* (5), 552–559. https://doi.org/10.1080/14620316.2022.2041493

Nega, G.; Mohammed, A.; Menamo, T. Effect of Curing and Top Removal Time on Quality and Shelf Life of Onions (*Allium cepa L.*). *GJSFR* **2015,** *15*, 26–36.

NMX-FF-021-SCFI-2003. Productos alimenticios no industrializados para Consumo humano–bulbos–cebolla (*Allium cepa* L.) [Non-Industrialized Food Products for Human Consumption–Bulbs–Onion (*Allium cepa* L.)-] Especificaciones (cancela a la NMX-FF-021-1986) CDU: 635.25.

Parry, R. T. Introduction. In: *Principles and Applications of Modified Atmosphere Packaging of Food*; Parry, R. T., Ed.; Blackie Academic & Professional: Glasgow, 1993; pp 1–18.

Petropoulos, S. A.; Ntatsi, G.; Ferreira, I. C. F. R. Long-Term Storage of Onion and the Factors That Affect Its Quality: A Critical Review. *Food Rev. Int.* **2017,** *33* (1), 62–83. https://doi.org/10.1080/87559129.2015.1137312

Põldma, P.; Moor, U.; Merivee, A.; Tõnutare, T. Effect of Controlled Atmosphere Storage on Storage Life of Onion and Garlic Cultivars. In: *IV International Conference Postharvest Unlimited*, **2011,** *945*; pp 63–69. https://www.actahort.org/books/945/945_7.htm

Raigón, M. D. El nivel de pungencia en las cebollas [The Level of Pungency in Onions]. *Revista Horticultura Internacional* **2006,** *51*, 48–51.

Ríos-González, I. J.; Guevara-Aguilar, A.; Sida-Arreola, J. P.; Sánchez, E.; Ronquillo-Aboite, J. O.; Alvarado-González, M.; Chávez-Mendoza, C. Effect of Controlled Atmospheres on the Quality of White Onion (*Allium cepa* L.). *J. Food Sci. Technol.* **2018,** *55* (9), 3564–3574.

Sharma, K.; Assefa, A. D.; Kim, S.; Ko, E. Y.; Park, S. W. Change in Chemical Composition of Onion (*Allium cepa* L0. cv. Sunpower) During Post-Storage Under Ambient Conditions. *New Zealand J. Crop Hortic. Sci.* **2014,** *42* (2), 87–98. https://doi.org/10.1080/01140671.2013.860039

Sharma, K.; Rok Lee, Y.; Park, S. W.; Nile, S. H. Importance of Growth Hormones and Temperature for Physiological Regulation of Dormancy and Sprouting in Onions. *Food Rev. Int.* **2016,** *32* (3), 233–255.

Smittle, D. A. Controlled Atmosphere Storage of Vidalia Onions. In: *International Controlled Atmosphere Conference 5th Proceedings*, Wenatchee, Washington, USA, Vol. 2, Other Commodities and Storage Recommendations, 1989; pp 171–177.

Smittle, D. A. Evaluation methods for 'Granex' onions. *J. Am. Soc. Hortic. Sci.* **1988,** *113*, 877–880.

Sorensen, J. N.; Grevsen, K. Nitrogen and Water Stress Affects Sprouting in Bulb Onions Stored Over Winter. *Acta Hortic.* **2002,** *571*, 98–86.

Sumner, P. E. Vidalia onion CA Storage Requirements. Cooperative Extension Service. The University of Georgia College of Agricultural and Environmental Sciences. USA, 2000. http://www.caes.uga.edu/commodities/fruits/veg/pubs/documents/VIDALIAONIONCASTORAGEREQUIREMENTS.pdf.

Thompson, A. K. *Controlled Atmosphere Storage of Fruits and Vegetables*, 2nd ed.; CAB International Edition, MPG Books Group, 2010; pp 1–2.

Thompson, A. K. *Fruit and Vegetable Storage Hypobaric, Hyperbaric and Controlled Atmosphere*; Springer Briefs in Food, Health, and Nutrition, Springer: Huddersfield, 2016; p 8.

Thompson, A. K. *Postharvest Technology of Fruits and Vegetables*; Blackwell Science Ltd.: London, 1996.

UCDavis. Cebollas de Guarda: Recomendaciones para Mantener la Calidad Postcosecha. Postharvest Center. University of California, 2022. https://postharvest.ucdavis.edu/Commodity_Resources/Fact_Sheets/Datastores/Vegetables_Spanish/?uid=10&ds=803

Uddin, M. M.; MacTavish, H. S. Controlled Atmosphere and Regular Storage-Induced Changes in *S*-alk (en) yl-l-cysteine Sulfoxides and Alliinase Activity in Onion Bulbs (*Allium cepa* L. cv. Hysam). *Postharv. Biol. Technol.* **2003,** *28* (2), 239–245. https://doi.org/10.1016/S0925-5214(02)00195-3

Yahia, E. Uso de atmósferas modificadas y controladas como medio de conservación de frutas y hortalizas. Memorias del simposio internacional de manejo postcosecha y perspectivas de comercialización de frutas y hortalizas [Use of Modified and Controlled Atmospheres as a Means of Preserving Fruit and Vegetables. Proceedings of the International Symposium on Post-Harvest Handling and Marketing Prospects for Fruits and vegetables], Universidad Nacional de Colombia [National university of Colombia], 27–30 June, 1995a.

Yahia, E. La tecnología de las atmosferas modificadas y controladas parte I. [The Technology of Modified and Controlled Atmospheres Part I]. *Horticultura Internacional* **1995b,** *7*, 37–39.

Yahia, E. La tecnología de las atmosferas modificadas y controladas parte II [The Technology of Modified and Controlled Atmospheres Part II]. *Horticultura Internacional* **1995c,** *8*, 20–25.

Yahia, E. M.; Fadanelli, L.; Mattè, P.; Brecht, J. K. Controlled Atmosphere Storage. In: *Postharvest Technology of Perishable Horticultural Commodities*; Woodhead Publishing, 2019; pp 439-479.

Yamashita, K.; Sasaki, W.; Fujisaki, K.; Haga, H. Effects of Controlled Atmosphere Storage on Storage Life and Shelf Life of the Japanese Onion Cultivar Super-Kitamomiji. In: *Proceedings of the 10th International Controlled & Modified Atmosphere Research Conference*; Turkey, 2009; p 53.

SECTION 2
FOOD BIOTECHNOLOGY

CHAPTER 7

Larrea tridentata Extracts and the Capacity of Its Biocompounds in Medicine and Agriculture

JOLANTA E. MARSZALEK[1], DANIELA M. SÁNCHEZ-PÉREZ[2], ERIKA FLORES LOYOLA[1], AGUSTINA RAMÍREZ-MORENO[1], and LUCIO RODRÍGUEZ-SIFUENTES[1]

[1]*Facultad de Ciencias Biológicas, Universidad Autónoma de Coahuila, Coahuila, México*

[2]*Instituto Tecnológico de Torreón (ITT)-DEPI, Carretera Torreón-San Pedro, Torreón, Coahuila, México*

ABSTRACT

The use of plants is of great interest in traditional medicine and when looking for new pharmacological materials for drug development. Creosote bush, Larrea tridentata (LT), is a perennial shrub growing in southern parts of the USA and northern semideserts of Mexico. The extracts of that plant contain a variety of metabolites, among which lignans are the most prevalent. Nordihydroguaiaretic acid (NDGA) is an extensively studied lignan with antibacterial, antiviral, anthelmintic, antifungal, and anticancer properties. However, other lignans and phenolic compounds found in LT extracts demonstrate similar activities. Even though extended exposure to LT extracts is believed to be toxic to the organism, it is valid to exploit the use of its isolated biocompounds. This review aims to give a good overview of the research into possible applications of these metabolites in medicine and agriculture.

Engineering Principles for Food Processing Technology and Product Realization. Leonardo Sepúlveda Torre, Porteen Kannan, A. K. Haghi, PhD (Eds.)

7.1 INTRODUCTION

Research always directs its interests in improving human life; recently, novel treatments for diverse human diseases have been sought. The reasons can be found in increased bacterial resistance to antibiotics (Andersson et al., 2017), developed resistance to antifungal treatments (Perlin et al., 2017), and a high number of difficult-to-treat cancers (Holohan et al., 2013; Tsimberidou et al., 2015). Consequently, alternatives that would have a strong activity with no toxicity and preferably low cost are wanted. In that aspect, more researchers explore natural sources, such as compounds from plants, bacteria, marine plants, and others, that could fulfill that need (Firenzuoli and Gori, 2007). Diverse studies demonstrate the presence of phytocompounds in plants that have activity against several human illnesses (Wen et al., 2012; Thangapazham et al., 2016, Hosseini and Ghorbani, 2015; Aboody and Mickymaray , 2009). In different regions of the World, many of these plants have been used in traditional medicine for hundreds of years. Some plants native to North America possess many of these qualities. They are found in arid zones, and because of their environmental conditioning, these plants contain many phytocompounds with exciting properties. One such plant is the creosote bush, Larrea tridentata (Zygophyllaceae) (LT), also locally referred to as greasewood and chaparral in the USA, and gobernadora in Mexico (Fig. 7.1), which is a common shrub in North American deserts. It is abundant in the semidesert areas of San Luis Potosí, Coahuila, Chihuahua, Durango, Sonora, Zacatecas, Baja California, and Baja California Sur in Mexico, as well as in the Southwest United States in Arizona, California, Nevada, Texas, and New Mexico (Rzedowski and Huerta, 1994). These plant extracts have been commonly used in broad applications in traditional Mexican medicine (Arteaga et al., 2005). In this chapter, we examine information about L. tridentata and its biocompounds that found applications in the field of medicine and agriculture.

7.2 PLANT COMPOSITION AND ANTIOXIDANT CAPACITY

The leaves of L. tridentata contain around 5% moisture and within the solids: reducing sugars (1.9%), nonreducing sugars (12%), fat (3.4%), crude fiber (12.3%), and ash (10.7%) (Treviño-Cueto et al., 2007; Mendez et al., 2012). As a result, the plant is an outstanding source of natural compounds, with approximately 50% of the dry weight of leaves being extractable compounds (Arteaga et al., 2005).

FIGURE 7.1 Image of the plant *L. tridentata* taken in the area of Torreon, Coahuila, México.

It has been reported that up to 62% of the plants' phytocompounds belong to the tannins (Treviño-Cueto et al., 2007). Besides tannins, saponins, sterols, monoterpenes, sesquiterpenes, lignans, flavonoids, and terpenes were detected in the aqueous and organic extracts of *L. tridentata* (Mendez et al., 2012; Gnabre et al., 2015). These varied phytocompounds are accompanied by essential oils (0.1% of the dry weight) and 67 nonvolatile compounds (Arteaga et al., 2005).

Among bioactive compounds present in this plant, nordihydroguaiaretic acid (NDGA), kaempferol, and quercetin are found at considerably high concentrations (Hyder et al., 2002). Phytoestrogens, including flavonoids comprising isoflavonoids and flavanols, derivates, lignans, and coumestans, are secondary plant metabolites. All of these have attracted sizable attention due to their protective action against several health disorders. They prevent and treat menopausal symptoms, skin aging, osteoporosis, cancer, and cardiovascular, neurodegenerative, immune, and metabolic diseases (Sirotkin and Halim Harrath, 2014).

LT's principal compounds are phenolics, which are present in high concentrations in leaves (36.2 mg/g), green steams (40.8 mg/g), and roots (28.6 mg/g) (Fig. 7.2A). Condensed tannins also are present in all tissues with high concentrations, specifically in flowers (1.7 mg/g), seeds (1.1 mg/g), and roots less than 5 mm in diameter (1.1 mg/g) (Fig. 7.2B). The most studied lignan, NDGA, is found in flowers, leaves, green stems, and small woody stems (<5 mm in diameter), with the highest concentrations in leaves (38.3 mg/g) and green steams (32.5 mg/g) (Fig. 7.2C) (Hyder et al., 2002). In the aerial parts of LT, about 25 main phytocompounds are found. Their structure was mostly triterpene glycoside (Jitsuno and Mimaki, 2010). In 2011 two new lignans glycosides were described, larreatricin A and larreatricin B (Yokosuka et al., 2011). Recently, thymol (5.38 mg/mL) and carvacrol (4.18 mg/mL) were detected in an extract exhibiting strong antibacterial properties (Morales-Ubaldo et al., 2022). A list of some of the described phytocompounds is presented in Table 7.1, with an indication of what properties they exhibited.

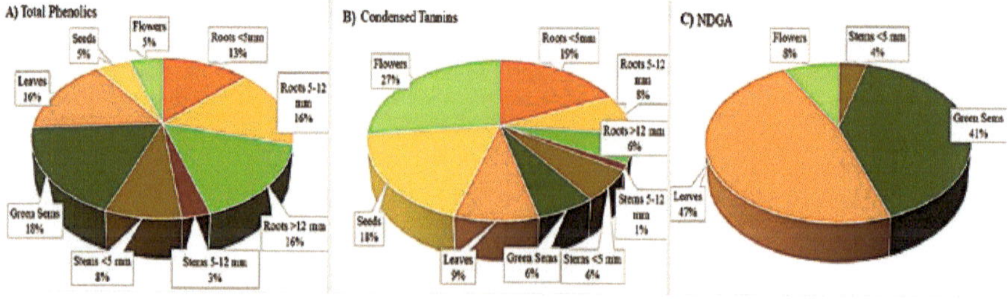

FIGURE 7.2 Percent distribution of (A) total phenolics, (B) condensed tannins, and (C) NDGA in different plant parts of L. tridentata adapted from Hyder et al. (2002).

Source: Adapted from Hyder et al., 2002

TABLE 7.1 NMR-Identified Compounds from L. tridentate and Their Cytotoxic, Antiprotozoal, and Antioxidant Activity.

Compound	Antiprotozoal Activity	Antibacterial Activity	Cytotoxic activity against (µg/mL)					References
			HL-60 cells IC50	ACC-375 cells	L-6 cells	MCF-7 cells	SW-480 cells	
(+)-Dihydroisorhamnetin			>50.0					Abou-Gazar et al. (2004)
(7r,7′r)-4,4′-Dihydroxy-8,8’-dehydro-7,7′-epoxylignan	✓				>90			Schmidt et al. (2012)
(7s,7′s,8s,8′s)-3,3′,4,4′-Tetrahydroxy-7,7′-epoxylignan	✓				154			Schmidt et al. (2012)
(7s,7′s,8s,8′s)-3,3′,4′-Trihydroxy-4-methoxy-7,7′-epoxylignan	✓				188			Schmidt et al. (2012)
(7s,8s,7′s,8′s)-3,3′,4′-Thrihydroxy-4-methoxy-7′,7′-epoxy-lignan			17.3					Abou-Gazar et al. (2004)
(s)-4′,5-Dihydroxy-7-methoxyflavone				>80		60	>80	Lambert et al. (2005)
3,3′-di-*o*-Demethylisoguaiacin	✓				132			Schmidt et al. (2012)
3,4-Dehydrolarreatricin			27.6					Abou-Gazar et al. (2004)
3,4-Dihidroxi-3′,4′-dimetoxi-6,7′-ciclolignano		✓				51		Morales-Ubaldo et al. (2022)
3,4′-Dihidroxy-3′,4-domethoxy-6,7′-cyclolignan				50		25	50	Lambert et al. (2005)
3′-Demethoxy-6-*o*-demethylisoguaicin		✓	13.6					Favela-Hernández et al. (2012); Abou-Gazar et al. (2004)
3′-Demethoxy-6-demethylisoguaiacin 6-*O*-*β*-D-glucopyranoside			>20					Yokosuka et al. (2021)
3′-Demethoxy-6-demethylisoguaiacin 6,40-di-*O*-*β*-D-glucopyranoside			>20					Morales-Ubaldo et al. (2022); Yokosuka et al. (2021)

TABLE 7.1 *(Continued)*

Compound	Antiprotozoal Activity	Antibacterial Activity	Cytotoxic activity against (μg/mL)					References
			HL-60 cells IC50	ACC-375 cells	L-6 cells	MCF-7 cells	SW-480 cells	
6,30-Didemethylisoguaiacin 6,30,40-tri-*O*-*β*-D-glucopyranoside			>20					Yokosuka et al. (2021)
3″-hydroxy-4-*epi*-larreatricin			13.6					Abou-Gazar et al. (2004)
3'-*o*-Methyltaxifolin	✓				27			Schmidt et al. (2012)
4,4′-Dihydroxy-7,7′-dioxolign-8(8′)-ene			>50					Abou-Gazar et al. (2004)
4,4′-Dihidroxi-3-metoxi-6,7′-ciclolignano		✓				56		Morales-Ubaldo et al. (2022)
1,2,4-Trihydro-1-(3,4-dihydroxyphenyl)-2,3-dimethyl-3,4,6,7-naphthalenetetrol.			16					Yokosuka et al. (2021)
5,7,4′-Trihydroxy-3,8,3′-trimethoxyflavone			>50.0					Abou-Gazar et al. (2004)
5,4′-Dihydroxy-3,7,8,3′-tetramethoxyflavone		✓						Favela-Hernández et al. (2012)
5,4′-Dihydroxy-7-methoxyflavone		✓						Favela-Hernández et al. (2012)
5,4′-Dihydroxy-3,7,8-trimethoxyflavone		✓						Favela-Hernández et al. (2012)
2-[1-(3,4-Dihydroxyphenyl)-2-methyl-3-oxobutyl]-4,5-dihydroxybenzaldehyde			17					Yokosuka et al. (2021)
(2*R*,3*R*)-2,3-Dihydro-2-(3,4-dihydroxyphenyl)-3- methyl-5-(*E*)-propenylbenzofuran.			4.1					Yokosuka et al. (2021)
(2*R*,3*R*)-2,3-Dihydro-2-(3-hydroxy-4-methoxyphenyl)-3- methyl-5-(*E*)-propenylbenzofuran			15					Yokosuka et al. (2021)

TABLE 7.1 *(Continued)*

Compound	Antiprotozoal Activity	Antibacterial Activity	Cytotoxic activity against (μg/mL)					References
			HL-60 cells IC50	ACC-375 cells	L-6 cells	MCF-7 cells	SW-480 cells	
Butane-type di-*o*-methylated lignan				10		12	31	Lambert et al. (2005)
Butane-type di-*o*-methylated lignan				7		10	30	Lambert et al. (2005)
Butane-type tri-*o*-methyl lignan				7		5.3	55	Lambert et al. (2005)
Carvacrol		✓						Morales-Ubaldo et al. (2022)
Cyclolignan-3′-demethoxyisoguaiacin				22		9.6	55	Lambert et al. (2005)
Herbacetin-3,8,4′-trimethyl ether			>50.0					Abou-Gazar et al. (2004)
Herbacetin-3,7′-dimethylether	✓				8			Schmidt et al. (2012)
Kaempferol-3,7-dimethylether	✓				15			Schmidt et al. (2012)
Kaempferol-3-methyl1 ether (isokempferide)			28.3					Abou-Gazar et al. (2004)
Lignoids-4-*epi*-larreatricin		✓	>50					Abou-Gazar et al. (2004); Favela-Hernández et al. (2012)
meso-(rel 7s,8s,7′s,8′s)-3,4,3′,4′-Tetrahydroxy-7,7′-epoxylignan			>50.0					Abou-Gazar et al. (2004)
meso-Dihydroguaiaretic acid		✓						Schmidt et al. (2012)
meso-Nordihydroguaiaretic acid	✓	✓			33			Schmidt et al. (2012)
Naringenin	✓				>90			Schmidt et al. (2012)
NDGA	✓	✓	2.6					Bashyal et al. (2017); Abou-Gazar et al. (2004); Favela-Hernández et al. (2012)
Thymol		✓						Morales-Ubaldo et al. (2022)

7.2.1 LIGNANS

Lignans, a class of secondary plant metabolites, are produced by oxidative dimerization of two phenylpropanoid units. They are optically active phenolic dimers possessing a 2,3-dibenzylbutane structure (Zitterman, 2003; Saleem et al., 2005). Although their molecular backbone consists only of two phenylpropane (C6–C3) units, lignans show an enormous structural diversity. In rye wheat, for example, these compounds appear mainly in the glycosidic form. In many plants, the lignans are the building blocks of the plant cell wall (Zitterman, 2003). Thus, they are widely distributed in the plant kingdom and have been found in species belonging to more than 70 families. Lignans are found in roots, rhizomes, stems, leaves, seeds, and fruits. Unfortunately, with some exceptions, the plants studied do not provide commercially practical quantities of lignans. Flaxseed has been reported to have the highest concentrations of dietary lignans in the form of secoisolariciresinol diglucoside, and these are available for purchase as a supplement (Source Naturals Lignan Extract Caps, n.d.).

Regarding biological interactions, lignans are phytoestrogens with estrogenic or antiestrogenic activity (Barakat et al., 2018). In addition, they respond to various biological issues in the form of anticancer, antioxidant, antihypertensive, antiviral, and even insecticidal properties (Simpson and Amos, 2017). These varied activities increased the interest in lignans and their synthetic derivatives and proposed their applications as therapeutics; for example, in cancer treatment.

7.2.1.1 NORDIHYDROGUAIARETIC ACID

Nordihydroguaiaretic acid [1,4-bis-(3,4-dihydroxyphenyl)-2,3-dimethylbutan] (NDGA) is a simple lignan that is linear with phenylpropane groups connected via a b–b′ linkage. NDGA could be called the dominant lignan of LT because it is present in all parts of the plant, representing approximately 5–10% of the leaves' dry weight and 80% of all phenolics. This phytocompound is of great interest due to its biological properties as a potent antioxidant (Floriano-Sánchez et al., 2006).

7.2.2 ANTIOXIDANT ACTIVITY OF LIGNANS

The efficiency of lignans as antioxidants is related to their structure and solubility. Antioxidants are defined as any molecule that inhibits oxidation, that is, loss of electrons and increase in the oxidative state of another molecule. Oxidation forms free radicals that are unstable atoms and molecules possessing an insufficient number of electrons. Antioxidants usually terminate many reactions by removing free radical intermediates and inhibiting other oxidation reactions. Thus, antioxidants often serve as reducing agents (e.g., thiols, ascorbic acid, and polyphenols). Antioxidants work through single or combined mechanisms: free radical scavenging, complexing of pro-oxidant, reducing activity, quenching of singlet oxygen, and scavenging lipid peroxyl radicals. Preventive actions are represented by the antioxidants that serve as inhibitors of free radical oxidation reactions (Kasote, 2013; Mehta and Gowder, 2015).

During the normal function of the human body, free radicals, reactive oxygen species (ROS), and nitrogen species are continuously generated. They are important for energy supply, chemical signaling, detoxification, and immune functions (Dimitrios, 2006). However, overproduction of these species due to exposure to external oxidant substances or a failure in the body's defense mechanisms such as endogenous enzymes (superoxide dismutase, glutathione peroxidase, and catalase), leads to damage of valuable biomolecules such as DNA, lipids, and proteins (Mehta and Gowder, 2015; Aruoma, 1998; Dimitrios, 2006).

It has been established that these damages are associated with an increased risk of various degenerative diseases such as cardiovascular disease, cancer, and others. On the contrary, phenolic compounds are excellent natural antioxidants. They are known to counteract the excess ROS and erase their pathological effects. The antioxidant capacity of phenolic compounds is mainly due to their redox properties, which allow them to act as reducing agents, hydrogen donors, singlet oxygen quenchers, or metal chelators (Mehta and Gowder, 2015).

In LT, the antioxidants are represented by lignans, flavonoids, condensed tannins, and other phenolics. As aforementioned, the main constituent of the lignans in LT is NDGA. Its polyphenol-bearing o-dihydroxy structure, with four phenolic hydroxyl groups, promotes radical scavenging ability (Floriano-Sánchez et al., 2006; Choi and Jung, 2016). The antioxidant activity of NDGA has been studied extensively over the years. Interestingly, NDGA's antioxidant ability was explored as a food protection agent in the 1950s (Bickoff and Williams, 1945; Lundberg et al., 1944; Stull et al., 1948). Recently, the characteristics of NDGA have been well-reviewed (Manda et al., 2020); thus, here we summarize shortly only its action. The potent scavenging activity of NDGA against multiple types of ROS comes from donating one electron and one proton from each of its hydroxyl groups contained in the catechol rings, which converts NDGA structure into an oxidized catechol-quinone. The quinone is, firstly, a semi-quinone radical, which is further converted into an ortho-quinone and superoxide anion. Ortho-quinone may be changed back to the reactive semi-quinone by cellular NADPH-dependent reductases, but this redox cycle is highly toxic, especially at high NDGA concentrations. As a consequence, NDGA at a high dose might lead to cytotoxic effects in the organism, as discussed later in the chapter.

Free radical scavenging activity and stabilization of newly formed transient radicals are facilitated in compounds that do not possess steric hindrance (Kitts et al., 1999; Foti et al., 1996) but have an ortho-hydroxyl group (Lu and Liu, 1992). In the case of lignans found in LT, due to tetrahydrofuran moiety, the number and the position of the phenyl moieties in epoxylignans may contribute to robust antioxidant effects.

7.3 MEDICINAL APPLICATION OF *L. TRIDENTATA*

Creosote bush has been used in traditional medicine as a treatment for around 50 different illnesses, as reported by Arteaga et al. (2005). These include skin problems (acne, dandruff, psoriasis); blood problems (altered blood pressure, anemia, blood purifier, bruises, and hemorrhoids); digestive system issues (bowel cramps and inflammation, stomach pain and diarrhea, diseases of the liver, kidney, and gallbladder stones, kidneys pain and cystitis, ulcer, indigestion, urinary tract infections, and diuretic); bacterial and varial infections (cold, influenza, bronchitis, chickenpox, tuberculosis); women issues (menstrual pains, inflammation after delivery, cramping, contraceptive, and sterility) as well as allergic treatment, diabetes, headache and toothache, and snakebite pain. In addition, tonics and extracts from LT are believed to have anti-inflammatory, antiamoebic, antibiotic, antifungal, antineoplastic, antiseptic, and antiviral properties (Arteaga et al., 2005; Heron and Yarnell, 2001).

7.3.1 CYTOTOXICITY OF LIGNANS, ANTICANCER APPLICATION

Cytotoxicity study using human umbilical vein endothelial cells (HUVEC) and isolated compounds from LT extracts showed to inhibit HUVEC cell viability with half maximal inhibitory concentration (IC50) values of 86 and 59 μM for NDGA and 3′-O-methyl-NDGA, respectively. These two compounds were correspondingly 2.4- and 1.6-fold less toxic to human cells than to Naegleria fowleri (the parasite studied in the research) (Bashyal et al., 2017).

For eight lignans and cyclolignans, including NDGA, isolated from the flowering tops of LT (Table 7.1), the median inhibition concentration, IC50, was evaluated against three cancerous cell lines (Lambert et al., 2005). The results showed IC50 values of 7–80 μM, with the linear butane-type lignans being the most potent and colon cancer cells being the least sensitive cell type. The relative capacity of linear butane-type lignans against human breast cancer appears to be aligned with the number of O-methyl groups in the NDGA molecule. The potency of O-methylation is seen in the increased lipophilicity, which allows the compounds to easier cross the cell's plasma membrane, or in the molecule's ability to bind to a lipophilic target, such as the estrogen receptor expressed in MCF-7 cells (Lambert et al., 2005)40.

Two meso-nordihydroguaiaretic acid glycoside derivatives, larrealignans A and B, tetra- and tri-desmosides lignans, respectively, did not exhibit cytotoxicity against HL-60 human leukemia cells even at a concentration of 20 μM. Other compounds isolated from LT showed moderate cytotoxicity (two aglycone compounds had IC50 values of 7.0 and 4.9 mM) (Yokosuka et al., 2011). The same group isolated 17 lignan compounds from LT leaves and determined 10 as described for the first time. The cytotoxicity of these compounds was evaluated against promyeoloblasts HL-60 cells, and the majority of these lignans presented IC50 values in the range of 2.7–17 μM. Compound (2R,3R)-2,3-dihydro-2-(3,4-dihydroxyphenyl)-3- methyl-5-(E)-propenylbenzofuran showed the highest cytotoxicity and induced apoptotic cell death (Yokosuka et al., 2021).

NDGA exhibits anticancer activity in numerous in vivo and in vitro studies (Rowe et al., 2008; Leon et al., 2016; Park et al., 2005; Wang et al., 2011). The ability of NDGA to hinder the proliferation of breast cancer cells is directly related to the inhibition of the function of two receptor tyrosine kinase (RTKs), the insulin-like growth factor receptor (IGF-1R) and the c-erbB2/HER2/neu (HER/neu) receptor (Youngren et al., 2005). Other aspects of NDGA's anticancer activity were established as an inhibitor of lipoxygenase isoenzymes and showed the strongest growth inhibition with an IC50 of 1.9 ± 0.5 μg (Hausott et al., 2003).

To improve the anticancer action of NDGA, a new derivate NDGA-P21 was synthesized by methylation of -OH groups and linearization of the molecule. With these changes, the new compound exhibited a higher suppression of glioma (GBM) cells by arresting the cell cycle of GBM cells in G0/G1 phase, blocking cell proliferation sequentially, and suppressing the stemness of glioma stem-like cells (GSLCs) (Zhao et al., 2017).

NDGA has been tested as a potential treatment for skin cancer and restored reduced glutathione levels and activities of antioxidant enzymes. As a result, elevated myeloperoxidase activities, xanthine oxidase, and skin edema formation in TPA-treated mice (treated with 12-O-tetradecanoylphorbol-13-acetate) were also lowered by NDGA, indicating a restrained inflammatory response. At the same time, the histological results demonstrated an inhibitory effect of NDGA on cellular inflammatory responses (Raisuddin et al., 2011).

Other reports (Lambert et al., 2002) established that intraperitoneal administration of NDGA is lethal to the mouse at LD50 of 75 mg/kg, showing an increase in serum alanine aminotransferase levels, which indicates liver damage. At the same, freshly isolated mouse hepatocytes were more sensitive to NDGA than human melanoma cells. Furthermore, glucuronidation was identified as a potential detoxification mechanism for NDGA. Both mono and diglucuronide conjugates of NDGA were formed after intravenous dosing. The monoglucuronide was also formed after the cultivation of NDGA with human hepatic microsomes. This suggested that glucuronide conjugation is essential in the metabolism of NDGA by humans (Lambert et al., 2002).

Additionally, NDGA has been tested for animal medicine as an effective anti-angiogenic agent in ovalbumin-sensitized guinea pigs (Bergren and Valentine, 2016).

Luo et al. (1988) determined the reduction in cell viability of a methanol LT extract on P388, a murine leukemia line with an EC50 of 0.57 μg/mL (Luo et al. 1988). Gil Salido et al. (2016) studied the antiproliferative effect of various extracts from Mexican plants against various cancer lines. In murine cell lines, LT had one of the highest hindering effects and showed a dominant selective index, which is a measure of the selectivity of the extract for eliminating cancer cells line rather than normal cells. The cancer cells were eliminated 3.17 times more often than normal cells suggesting that LT would be an excellent candidate for cancer treatment. In the human cell lines, the extracts of LT showed IC50 of below 200 μg/mL (range, 64.28–163.73 μg/mL) (Gil Salido et al., 2016).

7.3.2 ANTIMICROBIAL ACTIVITY

It has been proven that the extracts from LT show antibacterial activity against a wide range of bacterial strains (Morales-Ubaldo et al., 2021, 2022).

In addition, solutions of purified compounds such as NDGA, methylated-NDGA (Reyes-Melo et al., 2017), 3′-demethoxy-6-O-demethyl-isoguaiacin (Favela-Hernández et al., 2015), and others (Favela-Hernández et al., 2012), also have antibacterial effects.

When testing for antibacterial activity, the most straightforward and well-established method is the agar-substrate diffusion method (Das et al., 2010). Martins et al. (2013) screened various LT extracts (methanol, dichloromethane, and ethyl acetate as solvents) against Gram-positive and Gram-negative bacteria. In general, the crude extracts hindered Gram-positive bacteria's growth more effectively; however, all of the extracts had antibacterial properties and showed MIC between 62.5–125 μg/mL. The ethyl acetate fractions from LT leaves were the most efficient in blocking the growth of Staphylococcus aureus (Martins et al., 2013). When 95% ethanolic extract (1:3 plant to solvent ratio) was used against methicillin-resistant S. aureus the MIC was determined to be 60 μg/mL, and retardation activity saw a 10−6-fold reduction in bacterial replication as compared to untreated samples (Snowden et al., 2014).

Similarly, when crude aqueous and ethanolic extracts were tested against food-borne pathogen bacteria Enterobacter aerogenes, Escherichia coli, Salmonella typhi, and S. aureus, they showed high growth hindering activity (80% growth suppression). In this case, however, the authors attributed the LT extracts' activity to the presence of tannins (Mendez et al., 2012). Several mechanisms for tannin inhibition of bacterial growth can be found in the literature, but they can be classified into the interaction of tannins with bacterial and substrate proteins, interaction with bacterial cell wall plasma membrane, and chelation of metal ions (Štumpf et al., 2020).

It also has been shown that ethanolic extracts (80%, 3 g/20 mL) of LT were effective against several pathogens: Listeria monocytogenes (10 mg/mL), Clostridium perfringens (12 mg/mL), Shigella dysenteriae (14 mg/mL), Yersinia enterocolitica (19 mg/mL), Proteus vulgaris (10 mg/mL), the actinomycetes, and some of the molds tested (Verástegui et al., 1996).

Looking more closely at specific lignans and flavonoids found in LT extracts, Favela-Hernández et al. (2012) isolated seven compounds (NDGA; 4-epi-larreatricin; 3′-demethoxy-6-O-demethylisoguaiacin; 5,4′-dihydroxy-3,7,8,3′-tctramethoxyflavone; 5,4′-dihydroxy-3,7,8-trimethoxyflavone; 5,4′-dihydroxy-7-methoxyflavone; and 5,8,4′-trihydroxy-3,7-dimethoxyflavone). They showed that except for the last compound (5,8,4′-trihydroxy-3,7-dimethoxyflavone), all compounds tested had some degree of activity against five of the 12 bacteria tested. The main compound responsible for the antimicrobial and antimycobacterial activity was the lignan 3′-demethoxy-6-O-demethylisoguaiacin which displayed activity against resistant S. aureus (MIC 25 mg/mL), Enterococcus faecalis (MIC 12.5 mg/mL), E. coli (MIC 50 mg/mL), Enterobacter cloacae (MIC 12.5 mg/mL), and multidrug-resistance strains of Mycobacterium tuberculosis (MIC 12.5–50 mg/mL). In addition, the other lignans (NDGA and 4-epi-larreatricin) showed activity against S. aureus MR (MIC 50 and >50 mg/mL) and E. cloacae (MIC 50 and 12.5 mg/mL), respectively. As the lignans responded to antimycobacterial strains, the flavonoids had limited activity (Favela-Hernández et al., 2012). Lignan meso-dihydroguaiaretic acid isolated from LT targets the α-subunit of coenzyme A transferase of M. tuberculosis H37Rv present in both geraniol and 1-and 2-methylnaphthalene degradation pathways (Clemente-Soto et al., 2014). On the contrary, NDGA inhibits the growth of M. tuberculosis at low concentrations (8 μg/mL) due to destabilization of the bacterial cell wall and blocking the synthesis of fatty acids and mycolic acids (Guzmán-Beltrán et al., 2016; Tsuchiya et al., 1996).

Morales-Ubado et al. (2022) proved that hydroalcoholic crude extracts of LT, as well as their various fractions, exhibit antibacterial activity against various bacteria: S. aureus6538, L. monocytogenes19113, E. coli35218, Pseudomonas aeruginosa9027, as well as against multidrug-resistants S. aureus01, S. aureus02, Bacillus cereus, E. coli01, E. coli02, Klebsiella pneumoniae, and Pasteurella multocida. The crude extract was more sensitive to inhibiting Gram-positive bacteria. In addition, the authors showed the presence of thymol and carvacrol in the extract, which are proven antibacterial compounds. Their mechanism of action is related to the inhibition of biofilm formation and cell membrane damage. The presence of other compounds (coumarin, caffeic acid derivatives, rutin, quercetin glucoside, kaempferol, and nor 3′-demethoxyisoguaiacin) was also shown, and all of these have been reported to have antibacterial activity, especially nor 3′-demethoxyisoguaiacin displays bactericidal activity against multidrug-resistant bacteria associated with bovine mastitis (Morales-Ubado et al., 2022). In a similar study, antibacterial activity against phytopathogenic multidrug-resistant bacteria: Clavibacter michiganensis subsp. michiganensis, Pseudomona syringae, and Xanthomonas campestris (strains causing plant diseases: bacterial canker, bacterial speck, and bacterial spot, respectively) were shown in ethanolic crude extract and its subfractions (Morales-Ubaldo et al., 2021). Gram-negative bacteria, including E. coli and P. aeruginosa, show resistance to 40% ethanoic LT extract treatment (MIC >1000 μg/mL). However, Gram-positive bacteria are more sensitive to the treatments. Further analysis suggested that the LT extracts act similar to β-lactam antibiotics; thus, they inhibit the synthesis of the peptidoglycan layer of bacterial cell walls (Turner et al., 2021).

7.3.3 ANTHELMINTIC

As mentioned above, tea from LT has been used in traditional medicine against many different illnesses and within them as a treatment for intestinal parasites (Arteaga et al., 2005; Mabry et al., 1978). Treatments against parasites were tested in an in vitro study of LT methanolic extracts on sheathed and exsheathed larvae of Homonchus contortus. The anthelmintic activity at low concentrations exhibited a low mortality rate (16–34%). At the highest concentration tested (200 mg/mL), LT extracts displayed some cytotoxicity for the sheathed larvae (30%), but a higher mortality rate was recorded in the exsheathed larvae (about 70%). The extract of LT damaged the cuticle of the larvae, with the degree of mortality dependent on the dose of the hydromethanolic extracts but not on the length of exposure time, which proves that these extracts possess anti-H. contortus properties, predominantly during the exsheathed stage of this nematode (García et al., 2018).

Compounds isolated from LT were studied against three parasites: Entamoeba histolytica, Giardia lamblia, and N. fowleri. The ethyl acetate fraction showed antiparasitic activity at 50 μg/mL. In this fraction, nine compounds were identified; lignans: NDGA, 3′-O-methylnordihydroguairetic acid, nor-3′-demethoxyisoguaiacin, two analogs of nor-3′-demethoxyisoguaiacin, ((7R, 7′R)-7, 7′-bis(4′,3,4-trihydroxyphenyl)-(8R, 8′S)-8,8′-dimethyltetrahydrofuran), and a flavonoid (3-Methoxy-6,7,4′-trihydroxyflavonol). The compounds displayed a dose–response antiparasitic activity against all three pathogens by reducing the culture density by 50% compared to untreated trophozoite cultures. Looking at structure–activity relationships, it has been suggested that the more flexible straight chain structure offers more results in higher anthelmintic activity. On the contrary, the difference in hydrophilicity may be a physical property of two compounds preventing diffusion into the parasite trophozoites (Bashyal et al., 2017).

7.3.4 ANTIPROTOZOA ACTIVITY

Protozoa-caused diseases are grave health problems that have a profound global impact on the human and animal populations (Nash, 2014). Protozoan parasites that cause the majority of organ-damaging and often fatal diseases are Giardia intestinalis, Trypanosoma cruzi, Plasmodium falciparum, L. mexicana, and Trichomonas vaginalis (Lee et al., 2019). Unfortunately, the treatment and medications available for these parasites are often toxic, induce resistance, and present many adverse effects in the patients (Lee et al., 2019; Gelb and Hol, 2002). In traditional Mexican medicine, LT tea has been used as a treatment for parasites such as amoebiasis (Arteaga et al., 2005; Heron and Yarnell, 2001); however, only a few proper investigations have been carried out to evaluate these extracts as antiprotozoal agents. Nevertheless, as early as 1978 (Segura, 1978), there was a report of ethanoic LT extract containing NDGA (10−6 to 10−8 M) inhibitory activity against Entamoeba.

Among biocompounds in LT, meso-nordhydroguariaretic acid (m-NDGA) shows the highest antiprotozoal activity (Schmidt et al., 2012). Regardless of its composition, a crude dichloromethane extract of LT acted as a natural antiprotozoal agent against Trypanosoma brucei rhodesiense, T. cruzi, Leishmania donovani, and P. falciparum. The major detected lignan, m-NDGA, was responsible for the anti-inflammatory effects and was found to have the IC50 values: of 4.5, 33.1, 12, and 7.7 μM, respectively, against the listed above parasites (Schmidt et al., 2012). Bashyal et al. (2017) evaluated nine lignans separated from ethanolic LT extract against E. histolytica, G. lamblia, and N. fowleri. Six compounds exhibited a good antiparasitic response against the pathogens by reducing the culture density by 50% (EC50 83–236 μM for E. histolytica, 36–188 μM for G. lamblia, and 37–155 μM for N. fowleri) compared to untreated trophozoite cultures. The authors draw a connection between the structures of the compounds and their antiparasitic activity due to flexible chains and additional methoxy groups in the structure. Two hydrophobic compounds did not display antiparasitic activity, most likely due to difficulty diffusing into the parasite. The only isolated hydrophilic flavonol had activity against G. lamblia (EC50 = 153 μM) and N. fowleri (EC50 = 235 μM) and was believed to diffuse actively through organism membranes. NDGA and m-NDGA were also indicated to have the highest potential for future study as novel drugs against N. fowleri since they are probably inhibitors of cysteine protease activity in that parasite (Bashyal et al., 2017).

7.3.5 ANTIVIRAL ACTIVITY

Viruses as intracellular parasites are composed of either DNA or RNA genetic information enclosed in a lipid covering. They are nonautonomous and require a host body or living cell to perform their living function, such as replication (Behl et al., 2021). The treatment against viruses focuses on controlling the development of high infections via immunization (vaccination) or suppression of varial replication after exposure to the viral vector. Typically, antiviral drugs target a virus-encoded factor that can prevent the virus's genetic information's adhesion, inclusion, or replication. More novel approaches target the host cells' factors the virus needs but are not mandatory for host cell functions (Kumar et al., 2020). Biocompounds active against viruses are found within three main groups: alkaloids, terpenes, and polyphenols (Behl et al., 2021).

Within biocompounds described in LT lignans, specifically, NDGA and mNDGA, show antiviral activity (Cui et al., 2020). It is speculated that NDGA inhibits viral infection by targeting genome replication and viral assembly in the dengue virus (Soto-Acosta et al., 2014). In vitro studies of NDGA displayed its action against the hepatitis C virus (Syed et al., 2011), West Nile virus, and Zika virus (Merino-Ramos et al., 1996). NDGA probably reduces genome replication in these viruses and disturbs lipid metabolism by interfering with the sterol regulatory element binding proteins pathway.

Compounds from methylated derivate NDGA, such as 3-O-methyl-NDGA, and other lignans with various degrees of O-methylation, were first shown to possess activity against HIV in the mid-1990s (Gnabre et al., 1996). These compounds were effective at inhibiting HIV replication in cell cultures. The creosote bush lignans have structures that promote binding to DNA sections loaded in cytosine/guanine connections and Specificity Protein One (Sp1). If bound to Sp1, the blockade by the plant molecules, as tetra-O-methylnordihydroquaiaretic acid, prevents Sp1 factors from connecting to their related site, and the virus replication is revoked (Gnabre et al., 2015).

In the primary research, anti-HIV lignans from LT were isolated by multistep solvent fractionation. Each fraction was evaluated by secreted alkaline phosphatase assay against Tat transactivation of HIV-1. Obtained compounds representing O-methyl and O-acetyl-substituted NDGA were recognized as a new class of anti-HIV agents with crucial clinical relevance (Gnabre et al., 1996).

7.3.6 TOXICITY

As with other phytocompounds, when consuming plant extracts, there is always a possibility of toxicity or even poisoning by long-term exposure (Harutyunyan et al., 2019; Wojcikowski et al., 2004; Licata et al., 2013).

In the late 1990s, reports surfaced on renal and hepatotoxicity issues due to chronic use of LT tea and NDGA supplementation (Arteaga et al., 2005), triggering several studies on the safety of LT compounds to properly assess the level of toxicity, especially when it relates to NDGA exposure.

Sheikh et al. (1997) described 18 patients with LT-associated toxicity, 72% of whom revealed liver injury, with different clinical presentations, containing cholestatic hepatitis, mild hepatitis, cirrhosis, and even acute liver failure (Sheikh et al., 1997). As mentioned in the previous section, the redox cycle of NDGA at specific concentrations leads to cytotoxic effects in an organism; thus, the evaluation of its toxicity has been studied by various research groups. One of these reported that LT toxicity is caused by NDGA activity which inhibits cyclooxygenase and cytochrome P-450-dependent monooxygenase activity in epidermal and hepatic microsomal preparations (Agarwal et al., 1991).

NDGA is an inhibitor of intracellular vesicular transport at concentrations between 50 and 100 μM it interrupts the secretory vesicular route and the endocytic pathway in human dendritic cells (Drecktrah et al., 1998). At lower concentrations (below 12 μM), it has a reversible cholestatic effect in hepatocyte couplets (Arteaga et al., 2005), which strongly suggests that it needs to be used in a specific range of effective concentrations. In their review, Arteaga et al. (2005) propose a range between 10–100 μM as safe concentrations since higher concentrations cause irreversible system damage.

Aarland et al. (2015) demonstrated the low toxicity of an LT-methanolic extract in an in vivo model of Artemia salina, with a lethal dose of 50% (DL50) of 1592 ppm when toxicity is considered at 200 ppm (Aarland et al., 2015). Noncancerous Hs27 fibroblast cells were treated with LT extracts (ethanol, ethanol: water, and water) at different concentrations (3.75–120 μg/mL) to evaluate their toxicity, and no considerable cell death was observed at concentrations below 120 μg/mL compared to the negative control (Skouta et al., 2018). Thus, regarding a whole extract of LT, if a small dosage is administered to patients, no visible damage is detected. Additionally, medicating with LT should be done cautiously in persons with a history of liver disease (Heron and Yarnell, 2001).

7.4 THE USE OF *L. TRIDENTATA* IN AGRICULTURE

Other studies looked for utilization of LT extracts in other than medical applications due to the previously mentioned unique set of properties. For example, purified polyphenols from LT were mixed into a bioactive and biodegradable coating deposited on avocados cv Hass to prolong their shelf life. As a result, the bioactive film was able to significantly reduce loss of weight (40% and 58% for room and refrigerated conditions, respectively), preserve firmness (9 times under refrigerated conditions), maintain brightness (50% more than the control), and inhibit the development of internal grey pulp (100%) (Aguirre-Joya et al., 2017). The tannin-rich aqueous extract of LT has been used as a carbon source in solid-state fermentation inoculated with Aspergillus niger. In the fermentation, gallic and ellagic acids were produced at a fair rate within 96 h (Ventura et al., 2008).

7.4.1 ANTIFUNGAL

Due to the composition of high polyphenolic content, extracts from LT have been explored as potential antifungal treatments. The plant phytoanticipins, as generally secondary metabolites, generate molecular interaction of the organelles and tissues of the fungi through targeting biomembranes, proteins, or nucleic acids (Engelmeier and Hadacek, 2006). Alcohols are enzyme inhibitors due to their lipophilic characteristic, especially phenolic alcohols' free OH groups (Asael et al., 2018). Isothiocyanates and phenolic compounds inactivate fungal proteins by binding to the amino group or sulfhydryl group of amino acids (Tiznado and Troncoso, 2008). The alkaloids get introduced into the DNA and cause mutations in the microorganism. Quinones, flavones, flavonoids, tannins, and flavonols form nucleophilic complexes with proteins catalyzing inactivation (Cowan, 1999; Selitrennikoff, 2001).

Tequida et al. (2002) reported the fungicide activity of several plant extracts, including LT, against various fungi. They saw higher antifungal activity in ethanolic vs. methanolic LT extract for all tested fungi (Aspergillus niger; 77 vs 41, Aspergillus flavus; 76 vs 72, Penicillium chrysogenum; 63 vs 52, Penicillium expansum; 72 vs 58, Fusarium poae; 100 for both, and Fusarium moniliforme.; 100 vs 88) (Tequida-Meneses et al., 2002). The polyphenolic extract prepared from dry LT material in 70% acetone was determined to be a strong fungicide that inhibited 100% inhibition in a variety of moderately sensible and very sensible fungi (Pythium sp., Colletotrichum truncatum, Colletotrichum coccodes, Alternaria alternata, Fusarium verticillioides, Fusarium solani, Fusarium sambucinum, and Rhizoctonia solani) (Tequida et al., 2002).

Previously, it has been believed that tannins as poly and monomeric phenols present in LT extracts exhibit a potent antifungal activity because they can form complexes with polysaccharides and proteins of external layers of the fungal cells interfering with the functions of cell walls and membranes, eventually causing the death of microorganisms (Aguilar et al., 2007).

Extracts from LT showed good activity against A. flavus and Aspergillus parasiticus. Within the extract, two lignans, NDGA and methyl-NDGA, were identified as the phytoanticipins. The 100% inhibition for both fungi treated was achieved at 300 and 500 μg/mL for methyl-NDGA and NDGA, respectively (Vargas-Arispuro et al., 2005).

At the same time, aqueous extracts of LT are effective against Botrytis cinerea, C. coccodes, and Fusarium oxysporum (Lira-Saldivar et al., 2006).

Polyphenols from LT ethanol extracts can inhibit 100% Phytophthora cinnamomi at the concentration of 1000 ppm but also showed high growth inhibitions in other solvents (water, lanolin, and cocoa), and the minimum inhibitory concentration value of 6.96 mg/L (Castillo Reyes et al., 2015).

Leaf dichloromethane (DCM) and methanolic extracts can inhibit 100% F.oxysporum f. sp. radicis-lycopersici at 3000 and 4000 ppm, respectively, both in vitro and in vivo. The LT extract with DCM exhibited inhibition of 98% at 750 ppm; 94% for methanolic at 2000 ppm, and 97% for ethanolic at 500 ppm, which indicates that the compounds extracted with EtOH had greater fungicidal power over the fungus in vitro. When directly applied to tomato plants, it reduced the severity and incidence of disease caused by phytopathogenic fungus without significantly affecting the plant biology, such as chlorophyll synthesis, plant height, and the number of leaves (Peñuelas-Rubio et al., 2017).

7.4.2 ANTIPROTOZOA ACTIVITY

LT can be used as an unconventional and functional fooder for livestock, even though it might not be preferential for the animals in taste respect. Protozoa (Eimeria spp. oocysts) count was measured in rumen fluid and feces of sheep fed with 5 and 10% LT substituted fodder. 10% creosote bush as a source of fodder in sheep diets did not increase daily weight gain (very similar nutritive value to control) but reduced the count of Eimeria spp. oocysts in rumen fluid (down by 55%) and feces (down by 40%) (Hernández-Baez et al., 2019).

7.5 OTHER APPLICATIONS

Recently, aqueous extracts from fresh leaves of LT were used to synthesize silver nanoparticles (Méndez-Andrade et al., 2022). The authors used the nanoparticles to control the growth of C. michiganensis subsp. michiganensis in tomato plants and determined that their presence had a positive influence on the antioxidant protection of the plants.

7.6 CONCLUSIONS

Creosote bush (L. tridentata) has been used in traditional medicine and research for many years. The abundance of various extractable polyphenols creates an opportunity for many applications of that plant in medicine and agriculture. These applications are related to LT's antiseptic, antiamoebic, antibiotic, antifungal, and antiviral properties. Recently described lignans with potent antibacterial activity against multidrug-resistant bacteria give hope for the development of a new line of medical treatment against these microbes. However, there is still much to research in this field. It is unfortunate that, thus far, there is only limited research on the antiviral application of LT extracts and phytocompounds.

Similarly, there is a high expectation of using these biomolecules as novel treatments in fighting various cancer diseases. Thus, expanding and deepening the applied research of LT materials as antiviral and anticancer agents is suggested. On the contrary, the application of polyphenols as antifungal treatment in agriculture could replace the use of toxic chemicals on plants.

KEYWORDS

- ***Larrea tridentate***
- **lignans**
- **phytocompounds**

REFERENCES

Aarland, R. C.; Peralta-Gómez, S.; Sanchéz, C. M.; Parra-Bustamante, F.; Villa-Hernández, J. M.; de León-Sánchez, F. D.; Pérez-Flores, L. J.; Rivera-Cabrera, F.; Mendoza-Espinoza, J. A. A Pharmacological and Phytochemical Study of Medicinal Plants Used in Mexican Folk Medicine. *Indian J. Tradit. Knowl.* **2015,** *14* (4), 550–557.

Aboody, M. S. Al; Mickymaray, S. Anti-Fungal Efficacy and Mechanisms of Flavonoids. *Antibiotics* **2020,** *9* (2). https://doi.org/10.3390/antibiotics9020045.

Abou-Gazar, H.; Bedir, E.; Takamatsu, S.; Ferreira, D.; Khan, I. A. Antioxidant Lignans from *Larrea tridentata. Phytochemistry* **2004,** *65* (17), 2499–2505. https://doi.org/10.1016/j.phytochem.2004.07.009.

Agarwal, R.; Wang, Z.; Bik, D.; Mukhtar, H. Nordihydroguaiaretic Acid, an Inhibitor of Lipoxygenase, Also Inhibits Cytochrome P-450-Mediated Monooxygenase Activity in Rat Epidermal and Hepatic Microsomes. *Drug Metab. Dispos.* **1991,** *19* (3), 620–624.

Aguilar, C. N.; Rodríguez, R.; Gutiérrez-Sánchez, G.; Augur, C.; Favela-Torres, E.; Prado-Barragan, L. A.; Ramírez-Coronel, A.; Contreras-Esquivel, J. C. Microbial Tannases: Advances and Perspectives. *Appl. Microbiol. Biotechnol.* **2007,** *76* (1), 47–59. https://doi.org/10.1007/s00253-007-1000-2.

Aguirre-Joya, J. A.; Ventura-Sobrevilla, J.; Martínez-Vazquez, G.; Ruelas-Chacón, X.; Rojas, R.; Rodríguez-Herrera, R.; Aguilar, C. N. Effects of a Natural Bioactive Coating on the Quality and Shelf Life Prolongation at Different Storage Conditions of Avocado (Persea Americana Mill.) Cv. Hass. *Food Packag. Shelf Life* **2017,** *14* (September), 102–107. https://doi.org/10.1016/j.fpsl.2017.09.003.

Andersson, D. I.; Hughes, D. Selection and Transmission of Antibiotic-Resistant Bacteria. *Microbiol. Spectr.* **2017,** *5* (4), 1–17. https://doi.org/10.1128/microbiolspec.mtbp-0013-2016.

Arteaga, S.; Andrade-Cetto, A.; Cárdenas, R. *Larrea tridentata* (Creosote Bush), an Abundant Plant of Mexican and US-American Deserts and Its Metabolite Nordihydroguaiaretic Acid. *J. Ethnopharmacol.* **2005,** *98* (3), 231–239. https://doi.org/10.1016/j.jep.2005.02.002.

Aruoma, O. I. Free Radicals, Oxidative Stress, and Antioxidants in Human Health and Disease. *J. Am. Oil Chem. Soc.* **1998,** *75* (2), 199–212. https://doi.org/10.1007/s11746-998-0032-9.

Asael, R. G. H.; Guevara-Gonzalez, R. G.; De Jesus, R. G. S.; Angelica, F. P. A. Antifungal Activity of Mexican Endemic Plants on Agricultural Phytopathogens: A Review. *2018 14th Int. Eng. Congr. CONIIN 2018* **2018,** No. October. https://doi.org/10.1109/CONIIN.2018.8489793.

Barakat, R.; Park, C. J.; Perez, P. A.; Chiu, K.; Ko, C. Female Antiestrogens. In: *Encyclopedia of Reproduction*; Elsevier, 2018; pp 740–747. https://doi.org/10.1016/B978-0-12-801238-3.64414-8.

Bashyal, B.; Li, L.; Bains, T.; Debnath, A.; LaBarbera, D. V. *Larrea tridentata*: A Novel Source for Anti-Parasitic Agents Active Against *Entamoeba histolytica*, *Giardia Lamblia* and *Naegleria fowleri*. *PLoS Negl. Trop. Dis.* **2017,** *11* (8), e0005832. https://doi.org/10.1371/journal.pntd.0005832.

Behl, T.; Rocchetti, G.; Chadha, S.; Zengin, G.; Bungau, S.; Kumar, A.; Mehta, V.; Uddin, M. S.; Khullar, G.; Setia, D. et al. Phytochemicals from Plant Foods as Potential Source of Antiviral Agents: An Overview. *Pharmaceuticals* **2021,** *14* (4), 1–46. https://doi.org/10.3390/ph14040381.

Bergren, D. R.; Valentine, J. L. Anti-Anaphylactic Action of Nordihydroguaiaretic Acid in Antigen Sensitized Guinea Pigs. In: *Respiratory Physiology and Neurobiology*; Elsevier B.V., 2016; pp 26–31. https://doi.org/10.1016/j.resp.2016.09.003.

Bickoff, E.; Williams, K. T.; Sparks, M. Stabilization of Carotene with Nordihydrouaiaretic Acid and Other Antioxidants. *Oil Soap* **1945,** *5* (19), 128–131.

Castillo Reyes, F.; Hernandez Castillo, F. D.; Clemente Constantino, J. A.; Gallegos Morales, G.; Rodriguez Herrera, R.; Aguilar, C. N. In Vitro Antifungal Activity of Polyphenols-Rich Plant Extracts against Phytophthora *cinnamomi* Rands. *African J. Agric. Res.* **2015,** *10* (50), 4554–4560. https://doi.org/10.5897/ajar2013.8072.

Choi, Y. S.; Jung, M. Y. Kinetic Study on the Singlet Oxygen Quenching Activity of Nordihydroguaiaretic Acid (NDGA) Using Methylene Blue Sensitized Photooxidation of α-Terpinene. *Food Sci. Biotechnol.* **2016,** *25* (5), 1333–1336. https://doi.org/10.1007/s10068-016-0209-1.

Clemente-Soto, A. F.; Balderas-Rentería, I.; Rivera, G.; Segura-Cabrera, A.; Garza-González, E.; Del Rayo Camacho-Corona, M. Potential Mechanism of Action of meso-Dihydroguaiaretic Acid on *Mycobacterium tuberculosis* H37Rv. *Molecules* **2014,** *19* (12), 20170–20182. https://doi.org/10.3390/molecules191220170.

Cowan, M. M. Plant Products as Antimicrobial Agents. *Clin. Microbiol. Rev.* **1999,** *12* (4), 564–582. https://doi.org/10.1128/CMR.12.4.564.

Cui, Q.; Du, R.; Liu, M.; Rong, L. Lignans and Their Derivatives from Plants as Antivirals. *Molecules* **2020,** *25* (1), 1–17. https://doi.org/10.3390/molecules25010183.

Das, K.; Tiwari, R. K. S.; Shrivastava, D. K. Techniques for Evaluation of Medicinal Plant Products as Antimicrobial Agent: Current Methods and Future Trends. *J. Med. Plants Res.* **2010,** *4* (2), 104–111. https://doi.org/10.5897/JMPR09.030.

Dimitrios, B. Sources of Natural Phenolic Antioxidants. *Trends Food Sci. Technol.* **2006,** *17* (9), 505–512. https://doi.org/10.1016/j.tifs.2006.04.004.

Drecktrah, D.; De Figueiredo, P.; Mason, R. M.; Brown, W. J. Retrograde Trafficking of Both Golgi Complex and TGN Markers to the ER Induced by Nordihydroguaiaretic Acid and Cyclofenil Diphenol. *J. Cell Sci.* **1998,** *111* (7), 951–965. https://doi.org/10.1242/jcs.111.7.951.

Engelmeier, D.; Hadacek, F. Chapter 17 Antifungal Natural Products: Assays and Applications; 2006; pp 423–467. https://doi.org/10.1016/S1572-557X(06)03017-0.

Favela-Hernández, J. M. J.; Clemente-Soto, A. F.; Balderas-Rentería, I.; Garza-González, E.; Camacho-Corona, M. D. R. Potential Mechanism of Action of 3′-Demethoxy-6-O-Demethylisoguaiacin on Methicillin Resistant *Staphylococcus aureus*. *Molecules* **2015,** *20* (7), 12450–12458. https://doi.org/10.3390/molecules200712450.

Favela-Hernández, J. M. J.; García, A.; Garza-González, E.; Rivas-Galindo, V. M.; Camacho-Corona, M. R. Antibacterial and Antimycobacterial Lignans and Flavonoids from *Larrea tridentata*. *Phyther. Res.* **2012,** *26* (12), 1957–1960. https://doi.org/10.1002/ptr.4660.

Firenzuoli, F.; Gori, L. Herbal Medicine Today: Clinical and Research Issues. *Evid. Based Complement. Altern. Med.* **2007,** *4* (SUPPL. 1), 37–40. https://doi.org/10.1093/ecam/nem096.

Floriano-Sánchez, E.; Villanueva, C.; Medina-Campos, O. N.; Rocha, D.; Sanchez-González, D. J.; Cardenas-Rodríguez, N.; Pedraza-Chaverrí, J. Nordihydroguaiaretic Acid Is a Potent in Vitro Scavenger of Peroxynitrite, Singlet Oxygen, Hydroxyl Radical, Superoxide Anion and Hypochlorous Acid and Prevents in Vivo Ozone-Induced Tyrosine Nitration in Lungs. *Free Radic. Res.* **2006,** *40* (5), 523–533. https://doi.org/10.1080/10715760500419365.

Foti, M.; Piattelli, M.; Baratta, M. T.; Ruberto, G. Flavonids, Coumarins, and Cinnamic Acids as Antioxidants in a Micellar System. Structure–Activity Relationship. *J. Agric. Food Chem.* **1996,** *44* (2), 497–501. https://doi.org/10.1021/jf950378u.

García, J. E.; Gómez, L.; Mendoza-de-Gives, P.; Rivera-Corona, J. L.; Millán-Orozco, J.; Ascacio, J. A.; Medina, M. A.; Mellado, M. Anthelmintic Efficacy of Hydro-Methanolic Extracts of *Larrea tridentata* against Larvae of *Haemonchus contortus*. *Trop. Anim. Health Prod.* **2018,** *50* (5), 1099–1105. https://doi.org/10.1007/s11250-018-1535-5.

Gelb, M. H.; Hol, W. G. J. Tropical Protozoan Parasites. *Science* **2002,** *297*, 343–344.

Gil Salido, A. A.; Iloki Assanga, S. B.; Lewis Luján, Lidianys Maria Fernández Ángulo, D.; Lara Espinoza, C. L.; Acosta Silva, A. L.; Rubio Pino, J. L. Composition of Secondary Metabolites in Mexican Plant Extracts and Their Antiproliferative Activity towards Cancer Cell Lines. *Int. J. Sci.* **2016,** *2* (03), 63–77. https://doi.org/10.18483/ijsci.971.

Gnabre, J. N.; Ito, Y.; Ma, Y.; Huang, R. C. Isolation of Anti-HIV-1 Lignans from *Larrea tridentata* by Counter-Current Chromatography. *J. Chromatogr. A* **1996,** *719* (2), 353–364. https://doi.org/10.1016/0021-9673(95)00727-X.

Gnabre, J.; Bates, R.; Huang, R. C. Creosote Bush Lignans for Human Disease Treatment and Prevention: Perspectives on Combination Therapy. *J. Tradit. Complement. Med.* **2015,** *5* (3), 119–126. https://doi.org/10.1016/j.jtcme.2014.11.024.

Guzmán-Beltrán, S.; Rubio-Badillo, M. Á.; Juárez, E.; Hernández-Sánchez, F.; Torres, M. Nordihydroguaiaretic Acid (NDGA) and α-Mangostin Inhibit the Growth of *Mycobacterium tuberculosis* by Inducing Autophagy. *Int. Immunopharmacol.* **2016,** *31*, 149–157. https://doi.org/10.1016/j.intimp.2015.12.027.

Harutyunyan, K.; Balayan, K.; Tadevosyan, G.; Hayrapetyan, M.; Musayelyan, R.; Grigoryan, R.; Khondkaryan, L.; Sarkisyan, N.; Babayan, N. Genotoxic Potential of Selected Medicinal Plant Extracts in Human Whole Blood Cultures. *J. HerbMed. Pharmacol.* **2019,** *8* (2), 160–162. https://doi.org/10.15171/jhp.2019.25.

Hausott, B.; Greger, H.; Marian, B. Naturally Occurring Lignans Efficiently Induce Apoptosis in Colorectal Tumor Cells. *J. Cancer Res. Clin. Oncol.* **2003,** *129* (10), 569–576. https://doi.org/10.1007/s00432-003-0461-7.

Hernández-Baez, I.; García-López, J. C.; Espinosa-Reyes, G.; Lee-Rangel, H. A.; Faz-Colunga, D. Á.; Pinos-Rodríguez, J. M. Creosote Bush (*Larrea tridentata*) Biomass as Fodder for Sheep. *Rev. Chapingo Ser. Zo. Áridas* **2019,** *18* (2), 1–9. https://doi.org/10.5154/r.rchsza.2019.09.023.

Heron, S.; Yarnell, E. The Safety of Low-Dose *Larrea tridentata* (DC) Coville (Creosote Bush or Chaparral): A Retrospective Clinical Study. *J. Altern. Complement. Med.* **2001,** *7* (2), 175–185. https://doi.org/10.1089/107555301750164262.

Holohan, C.; Van Schaeybroeck, S.; Longley, D. B.; Johnston, P. G. Cancer Drug Resistance: An Evolving Paradigm. *Nat. Rev. Cancer* **2013,** *13* (10), 714–726. https://doi.org/10.1038/nrc3599.

Hosseini, A.; Ghorbani, A. Cancer Therapy with Phytochemicals: Evidence from Clinical Studies. *Avicenna J. Phytomed.* **2015,** *5* (2), 84–97. https://doi.org/10.22038/ajp.2015.3872.

Hyder, P. W.; Fredrickson, E. L.; Estell, R. E.; Tellez, M.; Gibbens, R. P. Distribution and Concentration of Total Phenolics, Condensed Tannins, and Nordihydroguaiaretic Acid (NDGA) in Creosotebush (*Larrea tridentata*). *Biochem. Syst. Ecol.* **2002,** *30* (10), 905–912. https://doi.org/10.1016/S0305-1978(02)00050-9.

Jitsuno, M.; Mimaki, Y. Triterpene Glycosides from the Aerial Parts of *Larrea tridentata*. *Phytochemistry* **2010,** *71* (17–18), 2157–2167. https://doi.org/10.1016/j.phytochem.2010.09.012.

Kasote, D. M. Flaxseed Phenolics as Natural Antioxidants. *Int. Food Res. J.* **2013,** *20* (1), 27–34.

Kitts, D. D.; Yuan, Y. V.; Wijewickreme, A. N.; Thompson, L. U. Antioxidant Activity of the Flaxseed Lignan Secoisolariciresinol Diglycoside and Its Mammalian Lignan Metabolites Enterodiol and Enterolactone. *Mol. Cell. Biochem.* **1999,** *202* (1–2), 91–100. https://doi.org/10.1023/a:1007022329660.

Kumar, N.; Sharma, S.; Kumar, R.; Tripathi, B. N.; Barua, S.; Ly, H.; Rouse, B. T. Host-Directed Antiviral Therapy. *Clin. Microbiol. Rev.* **2020,** *33* (3), e00168-19 1-36. https://doi.org/10.1128/CMR.00168-19.

Lambert, J. D.; Sang, S.; Dougherty, A.; Caldwell, C. G.; Meyers, R. O.; Dorr, R. T.; Timmermann, B. N. Cytotoxic Lignans from *Larrea tridentata*. *Phytochemistry* **2005,** *66* (7), 811–815. https://doi.org/10.1016/j.phytochem.2005.02.007.

Lambert, J. D.; Zhao, D.; Meyers, R. O.; Kuester, R. K.; Timmermann, B. N.; Dorr, R. T. Nordihydroguaiaretic Acid: Hepatotoxicity and Detoxification in the Mouse. *Toxicon* **2002,** *40* (12), 1701–1708. https://doi.org/10.1016/S0041-0101(02)00203-9.

Lee; Kim; Hayat; Shin. Recent Advances in the Discovery of Novel Antiprotozoal Agents. *Molecules* **2019,** *24* (21), 3886. https://doi.org/10.3390/molecules24213886.

Leon, D.; Parada, D.; Vargas-Uribe, M.; Perez, A. A.; Ojeda, L.; Zambrano, A.; Reyes, A. M.; Salas, M. Effect of Nordihydroguaiaretic Acid on Cell Viability and Glucose Transport in Human Leukemic Cell Lines. *FEBS Open Bio.* **2016,** *6* (10), 1000–1007. https://doi.org/10.1002/2211-5463.12106.

Licata, A.; Macaluso, F. S.; Craxì, A. Herbal Hepatotoxicity: A Hidden Epidemic. *Intern. Emerg. Med.* **2013,** *8* (1), 13–22. https://doi.org/10.1007/s11739-012-0777-x.

Lira-Saldivar, R. H.; Hernández-Suárez, M.; Hernández-Suárez, F. D. Activity of *Larrea tridentata* (D .C .) Coville L . Extracts and Chitosan against Fungi That Affect Horticultural Crops. *Chapingo Ser. Hortic.* **2006,** *12* (2), 211–216.

Lu, H.; Liu, G.-T. Anti-Oxidant Activity of Dibenzocyclooctene Lignans Isolated from Schisandraceae. *Planta Med.* **1992,** *58* (04), 311–313. https://doi.org/10.1055/s-2006-961473.

Lundberg, W. O.; Halvorson, H. O.; Burr, G. O. The Antioxidant Properties of Nordihydroguaiaretic Acid. *Oil Soap* **1944,** *21* (2), 33–35. https://doi.org/10.1007/BF02593156.

Luo, Z.; Meksuriyen, D.; Erdelmeier, C. A. J.; Fong, H. H. S.; Cordell, G. A. Larreantin, a Novel, Cytotoxic Naphthoquinone from *Larrea tridentata*. *J. Org. Chem.* **1988,** *53* (10), 2183–2185. https://doi.org/10.1021/jo00245a011.

Mabry, T. J.; Hunziker, J. H.; Jr., D. R. D. *Creosote Bush: Biology and Chemistry of Larrea in New World Deserts*; Dowden, Hutchinson & Ross, Inc., USA, 1978.

Manda, G.; Rojo, A. I.; Martínez-Klimova, E.; Pedraza-Chaverri, J.; Cuadrado, A. Nordihydroguaiaretic Acid: From Herbal Medicine to Clinical Development for Cancer and Chronic Diseases. *Front. Pharmacol.* **2020,** *11* (Feb), 1–21. https://doi.org/10.3389/fphar.2020.00151.

Martins, S.; Amorim, E. L. C.; Sobrinho, T. J. S. P.; Saraiva, A. M.; Pisciottano, M. N. C.; Aguilar, C. N.; Teixeira, J. A.; Mussatto, S. I. Antibacterial Activity of Crude Methanolic Extract and Fractions Obtained from *Larrea tridentata* Leaves. *Ind. Crops Prod.* **2013,** *41* (1), 306–311. https://doi.org/10.1016/j.indcrop.2012.04.037.

Mehta, S. K.; Gowder, S. J. T. Members of Antioxidant Machinery and Their Functions. In *Basic Principles and Clinical Significance of Oxidative Stress*; Thatha Gowder, J. S., Ed.; InTech, 2015; p 13. https://doi.org/10.5772/61884.

Mendez, M.; Rodríguez, R.; Ruiz, J.; Morales-Adame, D.; Castillo, F.; Hernández-Castillo, F. D.; Aguilar, C. N. Antibacterial Activity of Plant Extracts Obtained with Alternative Organics Solvents against Food-Borne Pathogen Bacteria. *Ind. Crops Prod.* **2012,** *37* (1), 445–450. https://doi.org/10.1016/j.indcrop.2011.07.017.

Méndez-Andrade, R.; Vallejo-Perez, M. R.; Loera-Alvarado, E.; los Santos-Villarreal, G.; García-Cerda, L. A.; Vera-Reyes, I. Efficacy of Biosynthesized Silver Nanoparticles from *Larrea tridentata* against *Clavibacter michiganensis*. *J. Phytopathol.* **2022,** *170* (2), 91–99. https://doi.org/10.1111/jph.13058.

Merino-Ramos, T.; Jiménez De Oya, N.; Saiz, J. C.; Martín-Acebes, M. A. Antiviral Activity of Nordihydroguaiaretic Acid and Its Derivative Tetra-O-Methyl Nordihydroguaiaretic Acid against West Nile Virus and Zika Virus. *Antimicrob. Agents Chemother.* **2017,** *61* (8). https://doi.org/10.1128/AAC.00376-17.

Morales-Ubaldo, A. L.; Gonzalez-Cortazar, M.; Zaragoza-Bastida, A.; Meza-Nieto, M. A.; Valladares-Carranza, B.; A. Alsayegh, A.; El-Saber Batiha, G.; Rivero-Perez, N. Nor 3′-Demethoxyisoguaiacin from *Larrea tridentata* Is a Potential Alternative against Multidrug-Resistant Bacteria Associated with Bovine Mastitis. *Molecules* **2022,** *27* (11), 3620. https://doi.org/10.3390/molecules27113620.

Morales-Ubaldo, A. L.; Rivero-Perez, N.; Avila-Ramos, F.; Aquino-Torres, E.; Prieto-Méndez, J.; Hetta, H. F.; El-Saber Batiha, G.; Zaragoza-Bastida, A. Bactericidal Activity of *Larrea tridentata* Hydroalcoholic Extract against Phytopathogenic Bacteria. *Agronomy* **2021,** *11* (5), 957. https://doi.org/10.3390/agronomy11050957.

Nash, T. E. Parasitic Diseases That Cause Seizures. *Epilepsy Curr.* **2014,** *14* (2_suppl), 29–34. https://doi.org/10.5698/1535-7511-14.s2.29.

Osorio, E.; Flores, M.; Hernández, D.; Ventura, J.; Rodríguez, R.; Aguilar, C. N. Biological Efficiency of Polyphenolic Extracts from Pecan Nuts Shell (Carya Illinoensis), Pomegranate Husk (Punica Granatum) and Creosote Bush Leaves (*Larrea tridentata* Cov.) against Plant Pathogenic Fungi. *Ind. Crops Prod.* **2010,** *31* (1), 153–157. https://doi.org/10.1016/j.indcrop.2009.09.017.

Park, R.; Chang, C. C.; Liang, Y. C.; Chung, Y.; Henry, R. A.; Lin, E.; Mold, D. E.; Huang, R. C. C. Systemic Treatment with Tetra-O-Methyl Nordihydroguaiaretic Acid Suppresses the Growth of Human Xenograft Tumors. *Clin. Cancer Res.* **2005,** *11* (12), 4601–4609. https://doi.org/10.1158/1078-0432.CCR-04-2188.

Peñuelas-Rubio, O.; Arellano-Gil, M.; Verdugo-Fuentes, A. A.; Chaparro-Encinas, L. A.; Hernández-Rodríguez, S. E.; Martínez-Carrillo, J. L.; Vargas-Arispuro, I. D. C. Extractos de *Larrea tridentata* Como Una Estrategia Ecológica Contra *Fusarium oxysporum* Radicis-Lycopersici En Plantas de Tomate Bajo Condiciones de Invernadero. *Rev. Mex. Fitopatol. Mex. J. Phytopathol.* **2017,** *35* (3), 360–376. https://doi.org/10.18781/r.mex.fit.1703-3.

Perlin, D. S.; Rautemaa-Richardson, R.; Alastruey-Izquierdo, A. The Global Problem of Antifungal Resistance: Prevalence, Mechanisms, and Management. *Lancet Infect. Dis.* **2017,** *17* (12), e383–e392. https://doi.org/10.1016/S1473-3099(17)30316-X.

Raisuddin, S.; Rahman, S.; Ansari, R. A.; Rehman, H.; Parvez, S. Nordihydroguaiaretic Acid from Creosote Bush (*Larrea tridentata*) Mitigates 12- O -Tetradecanoylphorbol-13-Acetate-Induced Inflammatory and Oxidative Stress Responses of Tumor Promotion

Cascade in Mouse Skin. *Evid. Based Complement. Altern. Med.* **2011,** *2011*. https://doi.org/10.1093/ecam/nep076.

Reyes-Melo, K.; García, A.; Romo-Mancillas, A.; Garza-González, E.; Rivas-Galindo, V. M.; Miranda, L. D.; Vargas-Villarreal, J.; Favela-Hernández, J. M. J.; Camacho-Corona, M. del R. Meso-Dihydroguaiaretic Acid Derivatives with Antibacterial and Antimycobacterial Activity. *Bioorganic Med. Chem.* **2017,** *25* (20), 5247–5259. https://doi.org/10.1016/j.bmc.2017.07.047.

Rowe, D. L.; Ozbay, T.; Bender, L. M.; Nahta, R. Nordihydroguaiaretic Acid, a Cytotoxic Insulin-like Growth Factor-I Receptor/HER2 Inhibitor in Trastuzumab-Resistant Breast Cancer. *Mol. Cancer Ther.* **2008,** *7* (7), 1900–1908. https://doi.org/10.1158/1535-7163.MCT-08-0012.

Rzedowski, J.; Huerta, L. *Vegetación de México*, 1st ed.; Editorial Limusa, S. A. Noriega Editores, 1994.

Saleem, M.; Kim, H. J.; Ali, M. S.; Lee, Y. S. An Update on Bioactive Plant Lignans. *Nat. Prod. Rep.* **2005,** *22* (6), 696. https://doi.org/10.1039/b514045p.

Schmidt, T. J.; Rzeppa, S.; Kaiser, M.; Brun, R. *Larrea tridentata*—Absolute Configuration of Its Epoxylignans and Investigations on Its Antiprotozoal Activity. *Phytochem. Lett.* **2012,** *5* (3), 632–638. https://doi.org/10.1016/j.phytol.2012.06.011.

Segura, J. J. Effects of Nordihydroguairetic Acid and Ethanol on the Growth of Entamoeba Invadens. *Arch. Invest. Med. (Mex).* **1978,** *1*, 157–162.

Selitrennikoff, C. P. Antifungal Proteins. *Appl. Environ. Microbiol.* **2001,** *67* (7), 2883–2894. https://doi.org/10.1128/AEM.67.7.2883-2894.2001.

Sheikh, N. M. Chaparral-Associated Hepatotoxicity. *Arch. Intern. Med.* **1997,** *157* (8), 913. https://doi.org/10.1001/archinte.1997.00440290099011.

Simpson, D.; Amos, S. Other Plant Metabolites. In *Pharmacognosy*; Elsevier, 2017; pp 267–280. https://doi.org/10.1016/B978-0-12-802104-0.00012-3.

Sirotkin, A. V.; Halim Harrath, A. Phytoestrogens and Their Effects. *Eur. J. Pharmacol.* **2014,** *741*, 230–236. https://doi.org/10.1016/j.ejphar.2014.07.057.

Skouta, R.; Morán-Santibañez, K.; Valenzuela, C. A.; Vasquez, A. H.; Fenelon, K. Assessing the Antioxidant Properties of *Larrea tridentata* Extract as a Potential Molecular Therapy against Oxidative Stress. *Molecules* **2018,** *23* (7), 1–17. https://doi.org/10.3390/molecules23071826.

Snowden, R.; Harrington, H.; Morrill, K.; Jeane, L. D.; Garrity, J.; Orian, M.; Lopez, E.; Rezaie, S.; Hassberger, K.; Familoni, D.; et al. A Comparison of the Anti-*Staphylococcus aureus* Activity of Extracts from Commonly Used Medicinal Plants. *J. Altern. Complement. Med.* **2014,** *20* (5), 375–382. https://doi.org/10.1089/acm.2013.0036.

Soto-Acosta, R.; Bautista-Carbajal, P.; Syed, G. H.; Siddiqui, A.; Del Angel, R. M. Nordihydroguaiaretic Acid (NDGA) Inhibits Replication and Viral Morphogenesis of Dengue Virus. *Antiviral Res.* **2014,** *109* (1), 132–140. https://doi.org/10.1016/j.antiviral.2014.07.002.

Source Naturals Lignan Extract Caps. https://www.swansonvitamins.com/p/source-naturals-lignan-extract-70-mg-60-caps (accessed 18 Sept 2022).

Stull, J. W.; Herreid, E. O.; Tracy, P. H. A Study of the Use of the Antioxidant Nordihydroguaiaretic Acid in Dairy Products. II: Its Antioxygenic Properties in Unsweetened Frozen Cream. *J. Dairy Sci.* **1948,** *31* (12), 1024–1028. https://doi.org/10.3168/jds.S0022-0302(48)92292-9.

Štumpf, S.; Hostnik, G.; Primožič, M.; Leitgeb, M.; Salminen, J.-P.; Bren, U. The Effect of Growth Medium Strength on Minimum Inhibitory Concentrations of Tannins and

Tannin Extracts against *E. coli*. *Molecules* **2020,** *25* (12), 2947. https://doi.org/10.3390/molecules25122947.

Syed, G. H.; Siddiqui, A. Effects of Hypolipidemic Agent Nordihydroguaiaretic Acid on Lipid Droplets and Hepatitis C Virus. *Hepatology* **2011,** *54* (6), 1936–1946. https://doi.org/10.1002/hep.24619.

Tequida-Meneses, M.; Cortez-Rocha, M.; Rosas-Burgos, E. C.; López-Sandoval, S.; Corrales-Maldonado, C. Efecto de Extractos Alcohólicos de Plantas Silvestres Sobre La Inhibición de Crecimiento de *Aspergillus flavus*, *Aspergillus niger*, *Penicillium chrysogenum*, *Penicillium expansum*, *Fusarium moniliforme* y *Fusarium poae*. *Rev. Iberoam. Micol.* **2002,** *19*, 84–88.

Thangapazham, R. L.; Sharad, S.; Maheshwari, R. K. Phytochemicals in Wound Healing. *Adv. Wound Care* **2016,** *5* (5), 230–241. https://doi.org/10.1089/wound.2013.0505.

Tiznado, E.; Troncoso, R. Control of Fungal Diseases with Isothiocyanates. *Stewart Postharvest Rev.* **2008,** *2* (1), 1–14. https://doi.org/10.2212/spr.2006.1.4.

Treviño-Cueto, B.; Luis, M.; Contreras-Esquivel, J. C.; Rodríguez, R.; Aguilera, A.; Aguilar, C. N. Gallic Acid and Tannase Accumulation during Fungal Solid State Culture of a Tannin-Rich Desert Plant (*Larrea tridentata* Cov.). *Bioresour. Technol.* **2007,** *98* (3), 721–724. https://doi.org/10.1016/j.biortech.2006.02.015.

Tsimberidou, A. M. Targeted Therapy in Cancer. *Cancer Chemother. Pharmacol.* **2015,** *76* (6), 1113–1132. https://doi.org/10.1007/s00280-015-2861-1.

Tsuchiya, H.; Sato, M.; Miyazaki, T.; Fujiwara, S.; Tanigaki, S.; Ohyama, M.; Tanaka, T.; Iinuma, M. Comparative Study on the Antibacterial Activity of Phytochemical Flavanones against Methicillin-Resistant *Staphylococcus aureus*. *J. Ethnopharmacol.* **1996,** *50* (1), 27–34. https://doi.org/10.1016/0378-8741(96)85514-0.

Turner, T.; Ruiz, G.; Gerstel, J.; Langland, J. Characterization of the Antibacterial Activity from Ethanolic Extracts of the Botanical, *Larrea tridentata*. *BMC Complement. Med. Ther.* **2021,** *21* (1), 177. https://doi.org/10.1186/s12906-021-03344-9.

Vargas-Arispuro, I.; Reyes-Báez, R.; Rivera-Castañeda, G.; Martínez-Téllez, M. A.; Rivero-Espejel, I. Antifungal Lignans from the Creosotebush (*Larrea tridentata*). *Ind. Crops Prod.* **2005,** *22* (2), 101–107. https://doi.org/10.1016/j.indcrop.2004.06.003.

Ventura, J.; Belmares, R.; Aguilera-Carbo, A.; Gutiérrez-Sanchez, G.; Rodríguez-Herrera, R.; Aguilar, C. N. Fungal Biodegradation of Tannins from Creosote Bush (*Larrea tridentata*) and Tar Bush (Fluorensia Cernua) for Gallic and Ellagic Acid Production. *Food Technol. Biotechnol.* **2008,** *46* (2), 213–217.

Verástegui, M. A.; Sánchez, C. A.; Heredia, N. L.; García-Alvarado, J. S. Antimicrobial Activity of Extracts of Three Major Plants from the Chihuahuan Desert. *J. Ethnopharmacol.* **1996,** 175–177. https://doi.org/10.1016/0378-8741(96)84802-1.

Wang, B.; Yu, S.-c.; Jiang, J.-y.; Porter, G. W.; Zhao, L.-t.; Wang, Z.; Tan, H.; Cui, Y.-h.; Qian, C.; Ping, Y.-f. et al. An Inhibitor of Arachidonate 5-Lipoxygenase, Nordy, Induces Differentiation and Inhibits Self-Renewal of Glioma Stem-Like Cells. *Stem Cell Rev. Reports* **2011,** *7* (2), 458–470. https://doi.org/10.1007/s12015-010-9175-9.

Wen, C. C.; Chen, H. M.; Yang, N. S. *Developing Phytocompounds from Medicinal Plants as Immunomodulators*, Vol. 62; 2012. https://doi.org/10.1016/B978-0-12-394591-4.00004-0.

Wojcikowski, K.; Johnson, D. W.; Gobé, G. Medicinal Herbal Extracts—Renal Friend or Foe? Part One: The Toxicities of Medicinal Herbs. *Nephrology* **2004,** *9* (5), 313–318. https://doi.org/10.1111/j.1440-1797.2004.00310.x.

Yokosuka, A.; Iguchi, T.; Jitsuno, M.; Mimaki, Y. Structure and Cytotoxicity of Novel Lignans and Lignan Glycosides from the Aerial Parts of *Larrea tridentata*. *Molecules* **2021,** *26* (20), 6186. https://doi.org/10.3390/molecules26206186.

Yokosuka, A.; Matsuo, Y.; Jitsuno, M.; Adachi, K.; Mimaki, Y. Larrealignans A and B, Novel Lignan Glycosides from the Aerial Parts of *Larrea tridentata*. *Chem. Pharm. Bull.* **2011,** *59* (12), 1467–1470. https://doi.org/10.1248/cpb.59.1467.

Youngren, J. F.; Gable, K.; Penaranda, C.; Maddux, B. A.; Zavodovskaya, M.; Lobo, M.; Campbell, M.; Kerner, J.; Goldfine, I. D. Nordihydroguaiaretic Acid (NDGA) Inhibits the IGF-1 and c-ErbB2/HER2/Neu Receptors and Suppresses Growth in Breast Cancer Cells. *Breast Cancer Res. Treat.* **2005,** *94* (1), 37–46. https://doi.org/10.1007/s10549-005-6939-z.

Zhao, Q. W.; Lin, Y.; Xu, C. R.; Yao, Y. L.; Cui, Y. H.; Zhang, X.; Bian, X. W. NDGA-P21, a Novel Derivative of Nordihydroguaiaretic Acid, Inhibits Glioma Cell Proliferation and Stemness. *Laboratory Investigation*. Nature Publishing Group 2017; pp 1180–1187. https://doi.org/10.1038/labinvest.2017.46.

Zitterman, A. DIETARY FIBER | Bran. In *Encyclopedia of Food Sciences and Nutrition*; Elsevier, 2003; pp 1844–1850. https://doi.org/10.1016/B0-12-227055-X/00346-1.

CHAPTER 8

Perspectives for Innovation of New Food Preservatives from Euphorbiaceae

REBECA VILLANUEVA CHÁVEZ, LUIS ENRIQUE COBOS-PUC, ELDA PATRICIA SEGURA-CENICEROS, ANNA ILINÁ, and SONIA YESENIA SILVA BELMARES

Department of Food Research, School of Chemical Sciences, Autonomous University of Coahuila, Saltillo, Coahuila, Mexico

ABSTRACT

The microbial deterioration of food and its intake represents a public health problem due to bacterial resistance. According to this, the losses of food products cause great economic losses in the food industry. Therefore, the food industry incorporates preservatives to prolong the shelf life of foods. Preservatives frequently added to food production are sorbates and benzoates. However, they present undesirable effects on health that can be mild or severe. The most common are bloating and changes in the gut microbiota, but they can cause tachycardia, hives, and heart disease. Therefore, it is required to innovate preservatives with fewer adverse health effects. Plants represent a great alternative to obtaining natural preservatives that, when incorporated into food, improve the quality of life. Since ancient times, plants have been traditionally used as food preservatives because they contain antimicrobial compounds such as essential oils. However, due to its volatility, the effect may be lost. Euphorbiaceae represents a potential source of innovative new preservatives since its species contain compounds with remarkable antibacterial effects. These plants synthesize different compounds as a defense mechanism to survive against pathogens. Therefore, the generation of

Engineering Principles for Food Processing Technology and Product Realization. Leonardo Sepúlveda Torre, Porteen Kannan, A. K. Haghi, PhD (Eds.)

knowledge about the chemical structure of antimicrobial compounds could lay the groundwork for innovating new preservatives.

8.1 INTRODUCTION

Food spoilage occurs frequently causing economic losses for the food industry. The deterioration begins with a natural oxidative process triggered by a series of chemical and enzymatic reactions that result in the senescence of the product (Anwar et al., 2019). Therefore, the food gradually loses its properties over time, reflected in physiological and organoleptic changes (De Oliveira et al., 2021). For this reason, preservation methods have been used for years to preserve foods of animal or plant origin to feed ever-growing populations. Conservation allows the integrity of the food to be maintained throughout the production process until the final product is marketed (Floros et al., 2010). However, the most important thing is its conservation during handling, transport, and storage (Thyberg and Tonjes, 2008). Consequently, different preservation methods have been used but some modify the nutritional composition of foods during the process. Additionally, synthetic additives have been conventionally added into processed foods for years (Joardder and Masud, 2019). Some preservatives maintain the integrity of nutrients due to antioxidant properties mean prevention of oxidative deterioration, while others inhibit microbial growth. However, they have side effects that harm health because they exceed the recommended daily intake (Trasande et al., 2018). They are generally recognized as safe (GRAS) substances, but some cause serious adverse effects such as cancer, multiple sclerosis, and heart disease (Mazdeh et al., 2014; Baran et al., 2021). Therefore, natural preservatives in processed foods could reduce health risks from the intake of synthetic additives (Sharif et al., 2017). Plant terpenoids and phenolic compounds inhibit or retard the growth of microorganisms (Gonelimali et al., 2018; Negi, 2012).

Essential oils are an example of terpenoids used since ancient times as food preservatives. Thymol, eugenol, and carvacrol have an effect against bacteria transmitted by contaminated food and are used to preserve meat, dairy, and vegetable products (Pandey et al., 2017). The main mechanism of the antibacterial effect is damage to the bacterial membrane (Faustino et al., 2019). Consequently, plants represent a source to discover new food preservatives that replace conventional additives with even higher activities in the short term (Pisoschi et al., 2019). The plants used since ancient times to preserve food are aromatic due to terpenoids, but the Euphorbiaceae contain them and have an antimicrobial effect. Euphorbiaceae grow widely in Africa and America (Islam et al., 2019). Therefore, they easily adapt to different habitats and produce a variety of bioactive compounds (Mwine and van Damme, 2011). One of its most outstanding effects is the antimicrobial due to monoterpenes, diterpenes, and triterpenes (Félix-Silva et al., 2018). However, it has antioxidant activity due to the content of flavonoids, alkaloids, and tannins (Greay and Hammer, 2015; Cavalcante et al., 2020). In consequence, they can be incorporated as preservatives in foods to improve shelf life and prevent their senescence due to environmental factors and microorganisms (Pandey et al., 2017). Some species of Euphorbiaceae that could be used as preservatives are described below since they have an antioxidant or antimicrobial effect.

Euphorbia heterophylla has an antibacterial effect attributed to its aqueous extract (Ughachukwu et al., 2014). *Manihot multifida* (L.) Crantz's antioxidant and antimicrobial effect is attributed to its extracts (Fabri et al., 2015). *Euphorbia ebracteolata* Hayata has an antimicrobial effect attributed to terpenoids, acetophenones, and flavonoids (Ting et al., 2020). *Euphorbia neriifolia* L. contains β-amyrin recognized as responsible for the antibacterial effect. However, it is also related to flavonoids, terpenoids, and alkaloids (Sultana et al., 2022). *Euphorbia tirucalli* contains alkaloids, flavonoids, tannins, and terpenoids. The stem has polyphenols with antioxidant activity (Nguyen et al., 2021) and peroxidases used to synthesize aromatic chemicals and remove peroxides from food products and industrial waste (Shukla et al., 2017). According to this, Euphorbiaceae plants are of industrial interest (Keneni and Marchetti, 2017); *Jatropha curcas* and *Jatropha dioica* are examples employed to obtain biodiesel from their oils (Zahan and Kano, 2018). However, phorbol esters from *J. curcas* extracted from a by-product of the biodiesel fuel industry affect phytopathogenic fungi (Saetae and Suntornsuk, 2012). In addition, *J. dioica* has an antimicrobial effect attributed to β-sitosterol (Silva-Belmares et al., 2014). *Hevea brasiliensis* contains antioxidant compounds that can be used in the food industry to inhibit the formation of free radicals to preserve the existing properties of food matrices (Oleinik et al., 2022). Therefore, extraction methods have been implemented to recover bioactive extracts with environmentally friendly solvents (Oleinik et al., 2022)].

Manihot esculenta is used to formulate industrial products (Okezie et al., 1983), but it generates waste rich in bioactive compounds. Therefore, the waste has the potential to produce products with added value (Cui et al., 2017). On the contrary, starch has been incorporated into biocomposite materials to preserve food since it includes antimicrobial compounds such as essential oils (De Souza et al., 2014). Since these materials have advantages over conventional ones formulated based on synthetic plastics, the *M. esculenta* films are edible (Bharti et al., 2020). Therefore, bioactive compounds and plant polymers from Euphorbiaceae could be used to formulate food preservatives, previously evaluating their toxicity (Saetae and Suntornsuk, 2012).

8.2 FOOD PRESERVATION AND ADVERSE EFFECTS OF SYNTHETIC PRESERVATIVES

Preservation methods have been used for years to preserve hunted foods or vegetables gathered from the field to feed ever-growing populations. Therefore, the integrity of the food must be maintained throughout the production process, from its cultivation at the time of planting food to the commercialization of the final product (Floros et al., 2010). Therefore, changes in a society focused on nutritional value, taste, texture, and quality of food have emerged throughout Consequently, food preservation focuses on reducing food waste and maintaining accessibility, including in geographic areas where food is not produced. However, the most important thing is its conservation during handling, transport, and storage (Thyberg and Tonjes, 2008; Kumari et al., 2019). Consequently, food preservation focuses on reducing food waste and maintaining accessibility, including in geographic areas where food is not produced. However, the most important thing is its conservation during handling, transport, and storage (Thyberg and Tonjes, 2008). Therefore, throughout history has been used conservation methods such as cooking (50,000 BC), drying (12,000 BC), fermentation (10,000 BC), sautéing, pickling (3600 BC), smoking (3500 BC), refrigeration (1784), and canning (1809). Although these methods are effective, some of them modify the nutritional composition of the food during the process due to temperature. For this reason, since 1950 synthetic additives have been incorporated into conventionally processed foods (Joardder and Masud, 2019).

Some preservatives focus on maintaining the nutrient integrity of the food through the use of antioxidant compounds since they prevent oxidative deterioration of the food. However, others are used to inhibit the growth of bacteria, yeasts, and molds. Butylated hydroxyanisole (BHA) is a widely used food preservative for its antioxidant effect. Additionally, it is used in the pharmaceutical and cosmetic industries to prevent lipid oxidation. On the contrary, A synthetic antimicrobial additive is a substance added to fruits, vegetables, and prepared foods to increase their shelf life and maintain quality and safety. Because it retards the growth of microbial contamination, fermentative processes and decomposition. Benzoates; sorbates; sodium or potassium nitrite and nitrate; sulfites; and glutamate are the most used. However, it has been seen that they have secondary effects due to their excessive use, which is why they harm the health of children and adults (Trasande et al., 2018). Since they frequently exceed the recommended daily intake (Chaleshtori et al., 2018). They are generally GRAS substances, but some salts, such as benzoate and sorbate produce allergies, intolerance, cancer, multiple sclerosis, nausea, heart disease, and hives at high concentrations (Mazdeh et al., 2014). BHA has mutagenic and carcinogenic potential (Baran et al., 2021). Sodium nitrate is frequently used to preserve cured meat products but has a toxic effect on humans and animals. It also causes methemoglobinemia due to the N-nitrosamines that are produced during metabolic conversion after ingestion (Özen et al., 2014).

Accordingly, the effect produced by the intake of preservatives that are ingested regularly through the consumption of processed foods that exist in the market is alarming. On the contrary, it is advisable to include scientific information on adverse effects in the labeling to reduce excessive intake due to misinformation towards the consumer. Therefore, to reduce health risks due to the intake of synthetic additives proposed addition of natural preservatives on food such as organic acids, vinegar, sugar, and spices (Sharif et al., 2017), since these are more tolerable for humans.

8.3 NATURAL ALTERNATIVES FOR FOOD PRESERVATION

Currently, leaves, stems, roots, or the entire plant are known to contain essential oils, phenolic compounds, terpenes, alkaloids, aldehydes, ketones, organic acids, saponins, thiosulfates, and sterols that inhibit or retard the growth of bacteria, fungi, and yeasts (Gonelimali et al., 2018; Lyu et al., 2019). Terpenoids are a group of compounds derived from the mevalonic acid pathway characterized by their antibacterial effects and include terpenes, carotenoids, and steroidal compounds. Therefore, for years, humans have explored new sources of natural preservatives in nature, focusing mainly on plants that contain essential oils since their compounds have the main advantage of having no harmful effects on health. Thymol, eugenol, and carvacrol are obtained from leaves, stems, or roots through different extraction techniques to improve their yield (Bulama et al., 2015). These compounds have an effect against *Listeria monocytogenes*, *Salmonella typhimurium*, *Escherichia coli* O157:H7, *Shigella dysenteriae*, *Bacillus cereus*, and *Staphylococcus aureus*. Therefore, used to preserve fish, dairy products, vegetables, fruits, and cooked rice (Burt et al., 2004). Carvacrol, thymol, and eugenol alter bacterial membrane functions by inhibiting ATPase activity, followed by the excretion of ATP and other cellular components. Carvacrol and thymol dissolve the phospholipid bilayer through alignment between the fatty acid chains (Moghaddam and Mehdizadeh, 2017).

The main mechanism of the antibacterial effect is the damage to the bacterial membrane (Faustino et al., 2019). The cell permeability of the microbial membrane is altered by the interaction of hydroxyl groups with membrane proteins, causing a change in its structure and function due to a loss of internal biomolecules such as ribose and sodium glutamate. Therefore, electron transport, nutrient absorption, protein, and nucleic acid synthesis are affected, and the enzymatic activity of the pathogen is affected (Rahman and Sarker, 2020). Another mechanism of antibacterial activity of phenolic compounds is alkyl substitution towards the aromatic nucleus (Bajpai et al., 2008). It has been known for years that terpenoids, phenolic compounds, and plant enzymes have an antibacterial effect (Bubonja-Šonje et al., 2020). Bean protease is an example, given that it has more antioxidant and antimicrobial effects than sorbate and benzoate salts (El-Saadony et al., 2020). Consequently, plants represent a source for discovering new food preservatives that replace conventional additives with even higher activities in the short term. In addition, they are safer, increase the shelf life of foods, and have more effect than conventional preservatives (Pisoschi et al., 2019)]. Table 8.1 describes some of the more common bioactive compounds of plant origin used as food preservatives.

TABLE 8.1 Antimicrobial Compounds Isolated from Plants Bioactive Antimicrobial Effects.

Plant	Bioactive compound	Microorganism	Minimal inhibitory concentration (MIC) (mg/mL)	References
Syzygium aromaticum	Eugenol	*L. monocytogenes*	0.5	Cui et al. (2018)

TABLE 8.1 *(Continued)*

Plant	Bioactive compound	Microorganism	Minimal inhibitory concentration (MIC) (mg/mL)	References
Origanum vulgare	Carvacrol and thymol	*S. aureus*, *Staphylococcus epidermidis*, *Sthaphylococcus sciuri*, and *E. coli*	0.25–0.5	Wijesundara and Rupasinghe (2018)
Kaffir lime	d-Limonene, β-pinene, and sabinene	*S. aureus*, *Elizabethkingia meningoseptica*, and *L. monocytogenes*	1.0–8.0	Sreepian et al. (2019)
Cuminum cyminum	1,8-cineole, limomene and pinene	*S. dysenteriae*, *L. monocytogenes*, *Salmonella typhi* y *Streptococcus pyogenes*	0.3–5	Sharafati Chaleshtori et al. (2106)
Elletaria cardamomum	α-Terpinyl acetate, 1.8-cineole, linalool acetate, sabinene, and linalool	*Candida albicans*, *Streptococcus mutans*, *S. aureus*, *L. monocytogenes*, *B. cereus*, and *S. typhimurium*	5–10	Abdullah et al. (2017)
Myristica fragrans Houtt	Safrole, β-phellandrene, 3-p-menthene and l-terpinen-4-ol	*S. mutans*	2	Ali et al. (2021)
Zingiber Officinale	Mono y di terpenos del aceite esceencial (cineole, eugenol y geraniol).	*L. monocytogenes*, *S. typhimurium*, and *S. aureus*	8.7	Beristain-Bauza et al. (2019)
Hibiscus sabdariffa L.	Cyanidine-3-sambubiose, cyanidine-3-glucoside, flavonoids, tannins, and saponins	*S. typhi* and *E. coli*	0.20	Sutiany et al. (2019)

As described, the plants used from ancient times to conserve food are known as aromatic due to the terpenes they contain in essential oils that give them an antibacterial effect. However, other groups of plants contain terpenoids with antimicrobial potential, such as Euphorbiaceae.

8.4 ANTIMICROBIAL POTENTIAL OF EUPHORBIACEAE

Euphorbiaceae plants grow in tropical and subtropical areas of Africa and America (Islam et al., 2019). Euphorbiaceae includes around 300 genera and 8000 species known for their medicinal components (Govea-Salas et al., 2017). Euphorbiaceae has a ubiquitous geographic distribution resulting in the adaptation of this group of plants to multiple habitats. Therefore, they produce an enormous variety of bioactive compounds (Mwinea and van Damme, 2011). The effect of each Euphorbiaceae plant depends on its chemical composition since it produces a wide variety of metabolites for their growth, reproduction, and survival (Azmir et al., 2013). These plants produce both primary and secondary metabolites. Primary metabolites include amino acids, carbohydrates, lipids, and nucleic acids. These compounds form cell structures and regulate their vital functions (Pagare et al., 2015), so they are necessary for plant life, while secondary metabolites improve survival against environmental stress (Ashraf et al., 2018).

Therefore, its species are used in the traditional medicine of different countries. One of the most outstanding effects of Euforbiaceae is antimicrobial. Euphorbiaceae produces antimicrobial compounds, highlighting terpenoids such as monoterpenes, diterpenes, and triterpenes (Felix-Silva et al., 2018). These compounds are detected in different concentrations in the oily phase of their extracts, and their concentration depends on each species (Elshamy et al., 2019).

In addition, it synthesizes flavonoids, alkaloids, and tannins (Cavalcante et al., 2020). On the contrary, compounds of Euforbiaceae present antioxidant activity (Greay and Hammer, 2015). Therefore, they can be incorporated into foods as additives to improve shelf life and prevent early senescence due to environmental factors and the presence of microorganisms (Pandey et al., 2017). Additionally, some of its compounds have cytotoxic, antispasmodic, antitumor, analgesic, and gastroprotective effects (Cavalcante et al., 2020; Sabandar et al., 2013). Some Euhorbiaceae species used as antimicrobials in different countries are described below. Therefore, they could be used as food preservatives.

8.4.1 *EUPHORBIA HETEROPHYLLA*

The plant is used in India and Nigeria for its many properties, but its most prominent use is to treat gastrointestinal, respiratory, and skin infections and to treat malaria (Okeniyi et al., 2012). Ethanol extracts have been used since ancient times and were extracted in gin. However, the extraction and efficiency of its components depend on the geographical location and soil conditions (Moghaddam and Mehdizadeh, 2017). The aqueous extracts also have an inhibitory effect on Klebsiella *pneumoniae*, *E. coli*, *Pseudomonas aeruginosa*, and *S. aureus*. However, *S. aureus* is more sensitive to extracts (Ughachukwu et al., 2014).

8.4.2 EUPHORBIA EBRACTEOLATA HAYATA

The dried root of the plant is used in China, Korea, and Japan to treat bloating and cough. The chemical components are terpenoids, acetophenones, and flavonoids. The extracts possess antiviral and antimicrobial effects (Ting et al., 2020).

8.4.3 EUPHORBIA NERIIFOLIA L.

The plant contains compounds with antimicrobial properties. The extracts have an effect against various pathogens. Consequently, used to treat emerging infectious diseases, focusing on bacterial, viral, fungal, and parasitic infections. This plant contains Euhol, monohydroxy triterpene, nerifoliol, taraxerol, β-amyrin, glut-5-(10)-en-1-one, neriifolone, and cycloartenol. In addition, quercetin, rutin, friedelin taraxerol, epitaraxerol, taraxeryl acetate, 3β-friedelanol, 3β-acetoxy friedelane, 3β-simiarenol, afzelin, 24-methylene cycloarenol, ingenol triacetate, and β-amyrin. β-Amyrin is one of those responsible for the antibacterial effect but to is associated with flavonoids, monoterpenoids, diterpenoids, triterpenoids, and alkaloids (Sultana et al., 2022).

8.4.4 MANIHOT MULTIFIDA (L.) CRANTZ

The plant is used in folk medicine to treat infected wounds. It has antioxidant and antimicrobial potential against *C. albicans*. The methanolic extract of fruits, and aqueous and hexane extract of leaves have activity with MIC of 39 μg/mL. In addition, the methanolic extract has the potential to purify DPPH (1,1-diphenyl-2-picrylhydrazyl) with an IC_{50} of 46.9 μg/mL, while the hexane extract has an IC_{50} of 59.2 μg/mL. However, the ethanol extract has activity against brine shrimp at an LC_{50} of 15.6 μg/mL (Fabri et al., 2015).

8.4.5 *MALLOTUS OPPOSITIFOLIUS*

El género Mallotus presenta alrededor de 150 especies, incluyendo árboles, arbustos y algunas plantas trepadoras (Abban et al., 2013). In Africa, the leaves of *M. oppositifolius* are used as an antiparasitic since they cause the expulsion of tapeworms and the elimination of ameba. Consequently, it is used to treat diarrhea and dysentery. Furthermore, it is used to treat urinary and venereal infections (Agyare et al., 2014). The antimicrobial effect is attributed to phloroglucinols, benzopyrans, cardenolides, flavonoids, steroids, and triterpenoids (Abban et al., 2013). However, the leaves of *M. oppositifolius* produce compounds derived from acylphloroglucinol with effect against *E. coli*, *S. aureus*, *S. typhi*, *P. aeruginosa* (Tchangoue et al., 2020).

8.4.6 *CNIDOSCOLUS QUERCIFOLIUS*

The plant, endemic to northeastern Brazil, is used in popular medicine as an antibiotic. Leaf extract has low activity against *S. aureus* and coagulase-negative *Staphylococcus* with MIC $\geq$ 2000 μg/mL. The plant contains linamarin cyanoglycoside, cinnamic acid, and a mixture of steroids and terpenoids. However, Linamarin is responsible for the effect on *E. coli*, *S. aureus*, and *P. aeruginosa* with MIC of 1000, 2000, and 2000 μg /mL, respectively (Félix-Silva et al., 2018).

8.4.7 *CROTON BETACEUS*

The Croton genus contains an enormous diversity of antimicrobial compounds against pathogens. Therefore, their species have a high potential to develop alternative pharmacological treatments to treat infections or to use them as food preservatives (Wright, 2017). This plant contains promising antibacterial compounds since its extract with hexane has an effect against *Enterococcus faecalis*. This effect is attributed to its fractions rich in flavonoid-type phenolic compounds. In addition, the most effective fractions contain tannins and free steroids (Das Rocha et al., 2021).

8.4.8 CROTON LUNDIANUS

Croton lundianus (Didr.) Müll. Arg. is native to Brazil, but not endemic. The hexane extract also affects *E. faecalis* attributed to flavonoid-type phenolic compounds, tannins, and free steroids (Das Rocha et al., 2021).

8.4.9 CROTON DICHOGAMOUS

This plant is a shrub used in East Africa to treat infectious diseases. Extracts in acetone, ethanol (50%), and distilled water at 250 mg/mL are effective against *B. cereus*, *S. aureus*, and *C. albicans*, but not on *P. aeruginosa* and *E. coli*. The bacteria most sensitive to its extracts is *B. cereus*, with an MIC of 10.42 mg/mL (Matara et al., 2021).

8.4.10 JATROPHA CURCAS

The name comes from the Greek "jatros" meaning doctor and "trophe" meaning food. This group of plants is distributed from the west to tropical areas of Latin America due to the export and migration of ancient civilizations. *J. curcas* has been used in traditional medicine in Asia and Africa to treat gastrointestinal infections such as diarrhea, typhoid fever, and cholera (Dada et al., 2014). The effect of the plant is due to diterpenes with antimicrobial and anti-proliferative activity (Devappa et al., 2011). Antimicrobial diterpenes such as rhamnofolan, daphnano, latirano, tigliano, and dinorditerpenes have been detected.

8.4.11 JATROPHA NEOPAUCIFLORA

The Jatropha genus has about 175 species and is native to tropical America. However, it is distributed in the tropics and subtropics of Asia and Africa. This genus contains compounds with antimicrobial and antioxidant effects. In addition, lignans, flavonoids, coumarins, alkaloids, eudesmenoic acids, but mainly terpenes have been detected (Cavalcante et al., 2020) In Mexico, different species of the Europhobiaceae family grow, which includes around 48 genera. Most grow in semi-arid areas such as the state of Coahuila (Devappa et al., 2011). *Jatropha neopauciflora* grows in northern Mexico and has antimicrobial potential.

8.4.12 *JATROPHA DIOICA*

The plant grows like a bush from 50–150 cm it is known as Sangre de Drago because it contains a colorless latex that turns red on contact with air. The plant grows in dry climates and rocky soils on the mountains. Consequently, it is easily found in semi-desert areas. Therefore, it grows in the northeast of Mexico and infusion is used in Mexican traditional medicine for skin infections, vaginitis, urethritis, and conjunctivitis (Govea-Salas et al., 2017; Sabandar et al., 2013). Hexane extracts have an effect against bacteria that cause gastrointestinal diseases and dermatophyte fungi such as *S. aureus*, *E. coli*, *B. cereus*, *Salmonella thypimurium*, *C. albicans*, *Candida parapsilosis*, *Cryptococcus neoformans*, and *Sporotrix schenckii* (Silva-Belmares et al., 2014).

Information on the chemical composition of the plant is scarce. However, Jatrophona, riolozatrione, and β-sitosterol have been detected in extracts. Additionally, the plant contains polyphenols, tannins, flavonoids, ellagitannins, and terpenes as the natural defense of the plant. These compounds have biological activities such as antimicrobials and antioxidants (Castro-Ríos et al., 2020). Accordingly, *J. dioica* has the potential to design a food preservative based on the standardization of its extract based on the content of β-sitosterol or riolozatrione. To contribute to the decrease in the consumption of synthetic preservatives (Pérez-Pérez et al., 2020)]. Table 8.2 shows some antimicrobial compounds detected in extracts of different species of Euphorbiaceae, their MIC, and the spectrum of action.

8.5 INDUSTRIAL APPLICATIONS OF EUPHORBIACEAE

The constant change in society and its demands over the years led to the development of new technologies to replace the usual ones due to the exponential demographic increase since it causes the overexploitation of ecosystems and gives rise to problems such as climate change (Keneni and Marchetti, 2017).

Therefore, in recent years, the use of natural resources has been resumed to manufacture products for constant consumption, such as food. The demand for conventional energy is scarce in some countries. Therefore, research is focused on producing biofuels from natural sources such as Euphorbiaceae because they synthesize terpenoids that can used to obtain biodiesel and others to obtain preservatives for the food industry due to their antimicrobial properties (Zahan and Kano, 2018). Therefore, obtaining biodiesel has been proposed to correct the environmental impact of conventional energy sources

TABLE 8.2 The Spectrum of Action and MIC of Antimicrobial Compounds Detected in Plant Extracts of Euphorbiaceae.

Plant	Antimicrobial compound	Extraction type	MIC (mg/mL)	Microorganism	References
J. curcas L.	Δ-Cadinene, α-epi-cadinol, pulegone, thymol	Hydrodistillation	0.3–0.6	*P. aeruginosa*	Babahmad et al. (2018)
Jatropha integerrima	Cineole, p-cymene, α-pinene	Hydrodistillation	625–1250	*B. cereus, S. aureus* and *P. aeruginosa*	Eshilokuna et al. (2007)
Jatropha podagrica	Japodagrine, Japodagrone	Methanol	5–20	*Bacillus subtilis*	Minh et al. (2019)
Jatropha multifida	Multifidanol, Jatrophen	Ethanol	12.5	*P. aeruginosa* y *S. aureus*	Senou et al. (2022)
Jatropha gossypiifolia L.	Jatrofeno, jatrofenone, jatrofatrione, curcusone, japogadrol, multifolone	*Ethanol*	0.025–0.05	*E. coli*, Enterococcus *faecium*, and *S. aureus*	Wu et al. (2019)
J. neopauciflora	β-Pinene	*Methanol/hexane*	2	*S. aureus*	Hernández-Hernández et al. (2016)
Croton pseudopulchellus	Lupeol	*Metanol/hexane*	16–32	*S. aureus, P. aeruginosa* y *E. coli*	Tatsimo et al. (2020)
Croton ceanothifolius Baill	α-Pinene, (E)-caryophyllene, β-elemene, sesquiterpenes	Hydrodistillation	12–16	*S. aureus* y *E. coli*	De Araújo et al. (2020)
Croton adipatus	α-Thujene, β-myrcene, d-limonene	Hydrodistillation	286	*B. subtilis*	Cucho-Medrano et al. (2021)
Croton thurifer	β-Elemene, germacrene-D, α-thujene	Hydrodistillation	296	*S. aureus*	Cucho-Medrano et al. (2021)
Croton collinus	B-Caryophyllene, γ-terpinene, α-thujene, d-limonene	Hydrodistillation	72	*B. subtilis*	Cucho-Medrano et al. (2021)

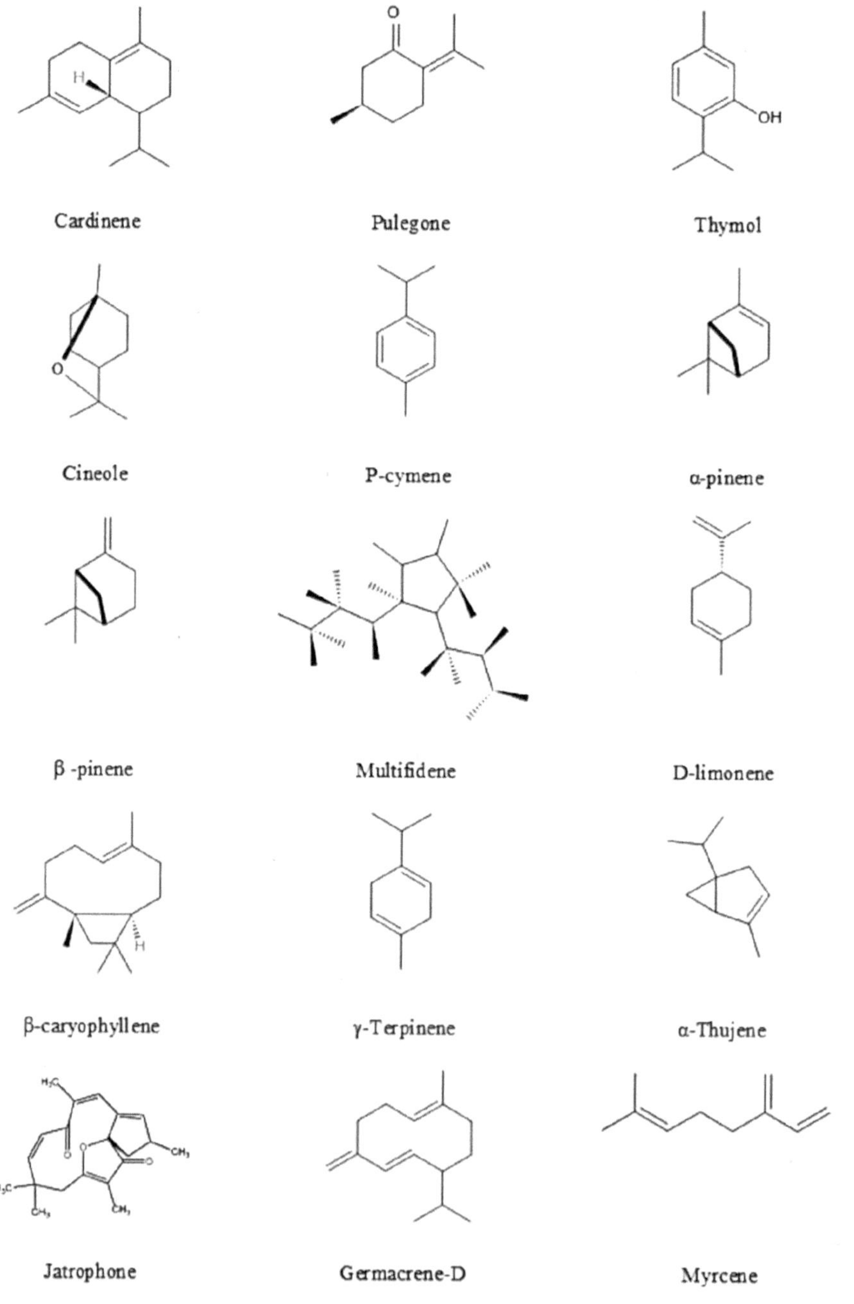

FIGURE 8.1 Chemical structures with antimicrobial activity present in species of the Euphorbiaceae family Wu et al. (2019).

from *J. curcas* and *J. dioica*. Biodiesel is obtained by transesterification of the oil contained in the plant (Pérez-Pérez et al., 2020).

J. curcas to obtain biodiesel using lipase enzymes from *Candida rugosa*. Liquid recombinant Candida rugosa lipase (CRL) isoenzymes have been investigated to convert various non-edible oils to biodiesel. These enzymes could be excellent biocatalysts to produce biodiesel from Jatropha crude oil at a low cost (Kuo et al., 2015). Therefore, the economic value of *J. curcas* largely depends on the lipid content in the seeds that are stored in the oil bodies (Liu et al., 2015). Additionally, some studies have focused on establishing in vitro cultures of *J. curcas* and *J. gossypifolia* to obtain novel compounds for pharmaceutical or industrial applications (Solís-Ramos et al., 2013). For this reason, cell cultures of *J. curcas* have been established to increase lipid production in vitro. Because to know industrial applications requires an understanding of growth conditions affecting lipid metabolism and yield (Solís-Ramos et al., 2020).

On the contrary, phorbol esters with antifungal properties have been extracted from *J. curcas* seed cake as a by-product of the biodiesel fuel industry. These esters affect *Fusarium oxysporum*, *Pythium aphanidermatum*, *Lasiodiplodia theobromae*, *Curvularia lunata*, *Fusarium semitectum*, *Colletotrichum capsici*, and *Colletotrichum gloeosporiodes*. Therefore, they could be used for agricultural applications (Saetae and and Suntornsuk, 2012). Castor oil is extracted from the seed of *Ricinus communis* L. It has a high commercial value for the manufacture of soaps, lubricants, and coatings. It is rich in hydroxy fatty acids and widely used in the cosmetic, pharmaceutical, oleochemical, and agricultural industries. The oil is a bactericidal, anti-inflammatory, and antiherpetic agent due to ricinoleic acid (Nitbani et al., 2022). The Jatropha genus has a wide range of bioactivities and it is expected that new avenues will be established to exploit its chemicals by the pharmaceutical industry to develop chemotherapeutic agents (Devappa et al., 2011).

Euphorbia cotinifolia contains enzymes that are easy to purify. Consequently, it could have applications in biophysical studies, biotechnological and industrial applications (Kumar et al., 2011). *E. tirucalli* has many bioactive properties and contains alkaloids, flavonoids, tannins, and terpenoids. In addition, the stem contains polyphenols with antioxidant activity. Scopoletin, gallic acid, and piperic acid isolated from the ethyl acetate fraction have an effect against citrus bacterial canker (Nguyen et al., 2021). On the contrary, the peroxidase *E. tirucalli* immobilized on chitosan beads with magnetic properties has been recovered to provide an industrial application. Because peroxidases can be used to treat wastewater containing phenolic compounds. In addition, they are used to the synthesis of aromatic chemicals and peroxide removal from food products, industrial waste, biosensors, and in the development of diagnostic kits (Shukla et al., 2017).

On the contrary, Euphorbiaceae species are known to synthesize antimicrobial compounds in response to invasion by phytopathogenic fungi. *H. brasiliensis* is an example that produces bioactive proteins due to Phytophthora infestation in tropical regions. These proteins protect the rubber tree from any fungal invaders; this allows us to understand some defense mechanisms and the improvement of plantations (Havanapan et al., 2016). However, the plant also contains antioxidant flavonoids and phenols that can be used in the food industry to inhibit the formation of free radicals to preserve the existing properties of food matrices. Therefore, extraction methods such as soxhlet have been implemented to recover bioactive extracts with environmentally friendly solvents (Oleinik et al., 2022).

Manihot esculenta is known as cassava and is a source of food for much of the world's population, particularly in countries in South America, Africa, and Asia (Okezie et al., 1983). It is used in cooking and is applied in industrial products such as adhesive for laundry, manufacture of paper, alcohol, butanol, dextrin, adhesive tape, textile finishing, and glue (Okezie et al., 1983). Cassava-based industries generate large amounts of waste rich in bioactive compounds, which offers the potential for obtaining value-added products (Li et al., 2017). An application of cassava starch is the formulation of biodegradable foams incorporated with grape stems because they have properties to protect foods stored with low moisture content (Engel et al., 2019). Additionally, some polymers of cassava starch, have been used to design biocomposite materials for food packaging. This material has been added with antimicrobial compounds such as cinnamaldehyde (De Souza et al., 2014). On the contrary, there are films formulated based on *M. esculenta* and carrageenan added with essential oils. These films have been used to improve the shelf life of chicken nuggets during refrigerated storage. In addition, they have advantages over conventional ones formulated based on synthetic plastics since the *M. esculenta* films are edible (Bharti et al., 2020).

8.6 CONCLUSION

The deterioration of food causes economic losses for the food industry and causes the loss of its nutritional and healthy properties. Consequently, for years, they have used methods for their conservation. However, some conventional preservation methods change the nutritional composition, while others have detrimental side effects on health, causing adverse effects such as cancer, multiple sclerosis, and heart disease. Therefore, natural preservatives of a terpenic or polyphenolic nature could reduce health risks. The plants of Euphorbiaceae are rich in this compound, and some produce polymers with properties to develop materials for food preservation since the terpenoid and polyphenolic compounds of this plant have antimicrobial and antioxidant properties. On the contrary, some Euphorbiace generate agro-industrial residues that could be used for the production of new food preservatives. Consequently, they represent a source to discover new food preservatives that replace conventional additives with higher activities in the short term. However, the formulation of new preservatives requires trials to recover bioactive compounds such as optimization of extraction methods from Euphorbiaceae. In addition, it is necessary to evaluate its effectiveness and safety.

KEYWORDS

- **euphorbiaceae**
- **antimicrobial effect**
- **terpenoids**
- **polyphenolic compounds**
- **preservative**
- **foods**
- **phytochemicals**
- **natural sources**

REFERENCES

Abban, G.; Ahiadu, B. K.; Barku, V. Y. A. Phytochemical Studies and Antioxidant Properties of Methanolic and Aqueous Extracts of the Leaves of *Mallotus oppositifolius*. *J. Basic Appl. Sci.* **2013,** 20–31.

Abdullah, A.; Asghar, M. S.; Shahid, M.; Huang, Q. Evaluating the Antimicrobial Potential of Green Cardamom Essential Oil Focusing on Quorum Sensing Inhibition of *Chromobacterium violaceum*. *J. Food Sci. Technol.* **2017,** *54*, 2306–2315.

Agyare, C.; Amuah, E.; Adarkwa-Yiadom, M.; Osei-Asante, S.; Ossei, P. P. S. Medicinal Plants Used for Treatment of wounds and Skin Infections: Assessment of Wound Healing and Antimicrobial Properties of *Mallotus oppositifolius* and *Momordica charantia*. *Int. J. Phytomed.* **2014,** *6*, 50.

Ali, A.; Yasir, M.; Jilani, M. I.; Shoaib, S.; Latif, M.; Ahmad, A.; Nazir, A. Chemical Composition an In Vitro Evaluation of Cytotoxicity, Antioxidant and Antimicrobial Activities of Essential Oil Extracted from Myristica Fragrans Houtt. *Polish J. Environ. Stud.* **2021,** *30*, 1585–1590.

Anwar, R.; Mattoo, A. K.; Handa, A. K. *Postharvest Biology and Nanotechnology. Ripening and Senescence of Fleshy Fruits*. John Wiley & Sons, Canada, Springer, **2019;** pp 15–51.

Ashraf, M. A.; Iqbal, M.; Rasheed, R.; Hussain, I.; Riaz, M.; Arif, M. S. *Environmental Stress and Secondary Metabolites in Plants: An Overview*; Elsevier Inc., 2018; pp 153–167.

Azmir, J.; Zaidul, I. S. M.; Rahman, M. M.; Sharif, K. M.; Mohamed, A.; Sahena, F.; Omar, A. K. M. Techniques for Extraction of Bioactive Compounds from Plant Materials: A Review. *J. Food Eng.* **2013,** *117*, 426–436.

Babahmad, R. A.; Aghraz, A.; Boutafda, A.; Papazoglou, E. G.; Tarantilis, P. A.; Kanakis, C.; Ouhammou, A. Chemical Composition of Essential Oil of *Jatropha curcas* L. Leaves and Its Antioxidant and Antimicrobial Activities. *Ind. Crops Prod.* **2018,** *121*, 405–410.

Bajpai, V. K.; Rahman, A.; Dung, N. T.; Huh, M. K.; Kang, S. C. In vitro inhibition of Food Spoilage and Foodborne Pathogenic Bacteria by Essential Oil and Leaf Extracts of *Magnolia liliflora* Desr. *J. Food Sci.* **2008,** *73*, 314–320.

Baran, A.; Yildirim, S.; Ghosigharehaghaji, A.; Bolat, I.; Sulukan, E.; Ceyhun, S. B. An Approach to Evaluating the Potential Teratogenic and Neurotoxic Mechanism of BHA Based on Apoptosis Induced by Oxidative Stress in Zebrafish Embryo (Danio rerio). *Hum. Exp. Toxicol.* **2021,** *40*, 425–438.

Beristain-Bauza, S. D. C.; Hernández-Carranza, P.; Cid-Pérez, T. S.; Ávila-Sosa, R.; Ruiz-López, I. I.; Ochoa-Velasco, C. E. Antimicrobial Activity of Ginger (*Zingiber officinale*) and Its Application in Food Products. *Food Rev. Int.* **2019,** *35*, 407–426.

Bharti, S. K.; Pathak, V.; Alam, T.; Arya, A.; Singh, V. K.; Verma, A. K.; Rajkumar, V. Materialization of Novel Composite Bio-Based Active Edible Film Functionalized with Essential Oils on Antimicrobial and Antioxidative Aspect of Chicken Nuggets During Extended Storage. *J. Food Sci.* **2020,** *85*, 2857–2865.

Bubonja-Šonje, M.; Knezević, S.; Abram, M. Challenges to Antimicrobial Susceptibility Testing of Plant-Derived Polyphenolic Compounds. *Arhiv za Higijenu Rada i Toksikologiju* **2020,** *71*, 300–311.

Bulama, J. S.; Dangoggo, S. M.; Mathias, S. N. Isolation and Characterization of Beta-sitosterol from Ethyl Acetate Extract of Root Bark of Terminalia Glaucescens. *Int. J. Sci. Res. Pub.* **2015,** *5*, 1–3.

Burt, S. Essential Oils: Their Antibacterial Properties and Potential Applications in Foods: A Review. *Int. J. Food Microbiol.* **2004,** *94*, 223–253.

Castro-Ríos, R.; Melchor-Martínez, E. M.; Solís-Cruz, G. Y.; Rivas-Galindo, V. M.; Silva-Mares, D. A.; Cavazos-Rocha, N. C. HPLC Method Validation for *Jatropha dioica* Extracts Analysis. *J. Chromatogr. Sci.* **2020,** *58* (5), 445–453.

Cavalcante, N. B.; da Conceição Santos, A. D.; da Silva Almeida, J. R. G. The Genus *Jatropha* (*Euphorbiaceae*): A Review on Secondary Chemical Metabolites and Biological Aspects. *Chem. Biol. Interact.* **2020,** *318*, 1–18.

Chaleshtori, F. S.; Arian, A.; Chaleshtori, R. S. Assessment of Sodium Benzoate and Potassium Sorbate Preservatives in Some Products in Kashan, Iran with Estimation of Human Health Risk. *Food Chem. Toxicol.* **2018,** *120*, 634–638.

Cucho-Medrano, J. L. L.; Mendoza-Beingolea, S. W.; Fuertes-Ruitón, C. M.; Salazar-Salvatierra, M. E.; Herrera-Calderon, O. Chemical Profile of the Volatile Constituents and

Antimicrobial Activity of the Essential Oils from *Croton adipatus*, *Croton thurifer*, and *Croton collinus*. *Antibiotics* **2021,** *10*, 1387.

Cui, H.; Zhang, C.; Li, C.; Lin, L. Antimicrobial Mechanism of Clove Oil on *Listeria monocytogenes*. *Food Control* **2018,** *94*, 140–146.

Cui, H.; Zhang, C.; Li, C.; Lin, L. The Industrial Applications of Cassava: Current Status, Opportunities and Prospects. *J. Org. Behav*. **2017,** *28*, 303–325.

Dada, E. O.; Ekundayo, F. O.; Makanjuola, O. O. Antibacterial Activities of *Jatropha curcas* (LINN) on Coliforms Isolated from Surface Waters in Akure, Nigeria. *Int. J. Biomed. Sci.* **2014,** *10*, 25–30.

Das Rocha, A. R.; Sousa, H. G.; do Vale Júnior, E. P.; De Lima, F. L.; Costa, A. S.; De Araújo, A. R.; Lago, E. C. Extracts and Fractions of *Croton* L.(*Euphorbiaceae*) Species with Antimicrobial Activity and Antioxidant Potential. *LWT—Food Sci. Technol.* **2021,** *139*, 110521.

De Araújo, A. C.; Freitas, P.; Rodrigues dos Santos Barbosa, C.; Muniz, D. F.; Esmeraldo Rocha, J.; Neto, J. B. D. A.; Melo Coutinho, H. D. Essential Oil of *Croton ceanothifolius* Baill. Potentiates the Effect of Antibiotics Against Multiresistant Bacteria. *Antibiotics* **2020,** *9*, 1–9.

De Oliveira, M. M.; Lago, A.; Dal' Magro, G. P. Food Loss and Waste in the Context of the Circular Economy: A Systematic Review. *J. Clean. Prod.* **2021,** *294*, 126–284.

De Souza, A. C.; Dias, A. M.; Sousa, H. C.; Tadini, C. C. Impregnation of Cinnamaldehyde into Cassava Starch Biocomposite Films Using Supercritical Fluid Technology for the Development of Food Active Packaging. *Carbohydr. Polym*. **2014,** *102*, 830–837.

Devappa, R. K.; Makkar, H. P.; Becker, K. *Jatropha diterpenes*: A Review. *J. Am. Oil Chem Soc*. **2011,** *88*, 301–322.

El-Saadony, M. T.; Elsadek, M. F.; Mohamed, A. S.; Taha, S. A.; Ahmed, B. M.; Saad, A. M. Effects of Chemical and Natural Additives on Cucumber Juice's Quality, Shelf Life, and Safety. *Foods*. **2020,** *9*, 1–17.

Elshamy, A. I.; Abd-ElGawad, A. M.; El Gendy, A. E. N. G.; Assaeed, A. M. Chemical Characterization of *Euphorbia heterophylla* L. Essential Oils and Their Antioxidant Activity and Allelopathic Potential on *Cenchrus echinatus* L. *Chem. Biodivers.* **2019,** *16*, 1–18.

Engel, J. B.; Ambrosi, A.; Tessaro, I. C. Development of Biodegradable Starch-Based Foams Incorporated with Grape Stalks for Food Packaging. *Carbohydr. Polym*. **2019,** *225*, 115234.

Eshilokuna, A. O.; Kasalia, A. A.; Ogunwandea, I. A.; Walkerb, T. M.; Setzerb, W. N.; Chemical Composition and Antimicrobial Studies of the Essential Oils of *Jatropha integerrima* Jacq (Leaf and Seeds). *Nat. Prod. Commun.* **2007,** *2*, 853–855.

Fabri, R. L.; De Sá, D. S.; Pereira, A. P. O.; Scio, E.; Pimenta, D. S.; Chedier, L. M. Antimicrobial, Antioxidant and Cytotoxicity Potential of *Manihot multifida* (L.) Crantz (Euphorbiaceae). *An. Acad. Bras. Cienc.* **2015,** *87*, 303–311.

Faustino, M.; Veiga, M.; Sousa, P.; Costa, E. M.; Silva, S.; Pintado, M. Agro-Food Byproducts as a New Source of Natural Food Additives. *Molecules* **2019,** *24*, 1–23.

Félix-Silva, J.; Gomes, J. A. S.; Fernandes, J. M.; Moura, A. K. C.; Menezes, Y. A. S.; Santos, E. C. G.; Tambourgi, D. V.; Silva-Junior, A. A.; Zucolotto, S. M.; Fernandes-Pedrosa, M. F. Comparison of Two *Jatropha* species (*Euphorbiaceae*) used popularly to Treat Snakebites in Northeastern Brazil: Chemical Profile, Inhibitory Activity Against *Bothrops erythromelas* Venom and Antibacterial Activity. *J. Ethnopharmacol.* **2018,** *213*, 12–20.

Floros, J. D.; Newsome, R.; Fisher, W.; Barbosa-Cánovas, G. V.; Chen, H.; Dunne, C. P.; German, J. B.; Hall, R. L.; Heldman, D. R.; Karwe, M. V.; Knabel, S. J.; Labuza, T. P.;

Lund, D. B.; Newell-McGloughlin, M.; Robinson, J. L.; Sebranek, J. G.; Shewfelt, R. L.; Tracy, W. F.; Weaver, C. M.; Ziegler, G. R. Feeding the World Today and Tomorrow: The Importance of Food Science and Technology. *Compr. Rev. Food Sci. Food Saf.* **2010,** *9*, 572–599.

Gonelimali, F. D.; Lin, J.; Miao, W.; Xuan, J.; Charles, F.; Chen, M.; Hatab, S. R. Antimicrobial Properties and Mechanism of Action of Some Plant Extracts Against Food Pathogens and Spoilage Microorganisms. *Front. Microbiol.* **2018,** *9*, 1–9.

Govea-Salas, M.; Morlett-Chávez, J.; Rodriguez-Herrera, R.; Ascacio-Valdés, J. Some Mexican Plants Used in Traditional Medicine. *Aromatic Med. Plants—Back Nat.* **2017,** 192–200.

Greay, S. J.; Hammer, K. A. Recent Developments in the Bioactivity of Mono and Diterpenes: Anticancer and Antimicrobial Activity. *Phytochem. Rev.* **2015,** *14*, 1–6.

Havanapan, P. O.; Bourchookarn, A.; Ketterman, A. J.; Krittanai, C. Comparative Proteome Analysis of Rubber Latex Serum from Pathogenic Fungi Tolerant and Susceptible Rubber Tree (*Hevea brasiliensis*). *J. Proteomics* **2016,** *131*, 82–92.

Hernández-Hernández, A. B.; Alarcón-Aguilar, F. J.; Jiménez-Estrada, M.; Hernández-Portilla, L. B.; Flores-Ortiz, C. M.; Rodríguez-Monroy, M. A.; Canales-Martínez, M. Biological Properties and Chemical Composition of *Jatropha Neopauciflora* Pax. *Afr. J. Tradit. Complement. Altern. Med.* **2016,** *14*, 32–42.

Islam, M. S.; Ara, H.; Ahmad, K. I.; Uddin, M. M. A Review on Medicinal Uses of Different Plants of Euphorbiaceae Family. *Univers. J. Pharm. Res.* **2019,** *4*, 47–51.

Joardder, M. U. H.; Masud, M. H. *Food Preservation in Developing Countries: Challenges and Solutions. A Brief History of Food Preservation*; Springer International Publishing, 2019; pp 57–66.

Keneni,Y. G.; Marchetti, J. M. Oil Extraction from Plant Seeds for Biodiesel Production. *AIMS Energy*. **2017,** 5, 316–340.

Kumar, R.; Singh, K. A.; Tomar, R.; Jagannadham, M. V. Biochemical and Spectroscopic Characterization of a Novel Metalloprotease, Cotinifolin from an Antiviral Plant Shrub: *Euphorbia cotinifolia. Plant Physiol. Biochem.* **2011,** *49*, 721–728.

Kumari, P. K.; Akhila, S.; Rao, Y. S.; Devi, B. R. Alternative to Artificial Preservatives. *Syst. Rev. Pharm.* **2019,** *10*, 99–102.

Kuo, T. C.; Shaw, J. F.; Lee, G. C. Conversion of Crude *Jatropha curcas* Seed Oil into Biodiesel Using Liquid Recombinant *Candida rugosa* Lipase Isozymes. *Bioresour. Technol.* **2015,** *192*, 54–59.

Liu, H.; Wang, C.; Chen, F.; Shen, S. Proteomic Analysis of Oil Bodies in Mature *Jatropha curcas* Seeds with Different Lipid Content. *J. Proteomics* **2015,** *113*, 403–414.

Lyu, X.; Lee, J.; Chen, W. N. Potential Natural Food Preservatives and Their Sustainable Production in Yeast: Terpenoids and Polyphenols. *J. Agric. Food Chem.* **2019,** *67*, 4397–4417.

Matara, D. N.; Nguta, J. M.; Musila, F. M.; Mapenay, I. Phytochemical Analysis and Investigation of the Antimicrobial and Cytotoxic Activities of *Croton dichogamus pax* Crude Root Extracts. *Evid. Based Complement. Altern. Med.* **2021,** *2021*.

Mazdeh, F. Z.; Aftadbari, F. E.; Moradi-Khatoonabadi, Z.; Shaneshin, M.; Torabi, P.; Ardekani, M. R. S.; Hajimahmoodi, M. Sodium Benzoate and Potassium Sorbate Preservatives in Iranian Doogh. *Food Addit. Contam. Part B Surveill.* **2014,** *7*, 115–119.

Minh, T. N.; Xuan, T. D.; Tran, H. D.; Van, T. M.; Andriana, Y.; Khanh, T. D.; Ahmad, A. Isolation and Purification of Bioactive Compounds from the Stem Bark of *Jatropha podagrica. Molecules* **2019,** *24*, 889.

Moghaddam, M.; Mehdizadeh, L. Chemistry of Essential Oils and Factors Influencing Their Constituents. *Soft Chem. Food Fermentation* **2017,** 379–419.

Mwine, J. T.; van Damme, P. Why Do Euphorbiaceae Tick as Medicinal Plants? A Review of Euphorbiaceae Family and Its Medicinal Features. *J. Med. Plants Res.* **2011,** *5*, 652–662.

Negi, P. S. Plant Extracts for the Control of Bacterial Growth: Efficacy, Stability and Safety Issues for Food Application. *Int. J. Food Microbiol.* **2012,** *156*, 7–17.

Nguyen, T. M. L.; Dang, X. C.; Pham, V. T.; Truong, N. M.; Tran, D. M.; Thuc-Huy, D.; Tran, Thi, L. M.; Vo, T. T. O. Phytochemical Screening and Evaluation of Antioxidant Properties and Antimicrobial Activity Against *Xanthomonas axonopodis* of *Euphorbia tirucalli* Extracts in Binh Thuan Province, Vietnam. *Molecules* **2021,** *26*, 2–11.

Nitbani, F. O.; Tjitda, P. J. P.; Wogo, H. E.; Detha, A. I. R. Preparation of Ricinoleic Acid from Castor Oil: A Review. *J. Oleo Sci.* **2022,** *71*, 781–793.

Okeniyi, S. O.; Adedoyin, B. J.; Garba, S. Phytochemical Screening, Cytotoxicity, Antioxidant and Antimicrobial Activities of Stem and Leave Extracts of *Euphorbia heterophylla. J. Biol. Life Sci.* **2012,** *4*, 24–31.

Okezie, B. O.; Kosikowski, F. V.; Markakis, P. Cassava as a Food. *Crit. Rev. Food Sci. Nutr.* **1983,** *17*, 259–275.

Oleinik, G.; Dario, P. P.; De Morais Gasperin, K.; Benvegnú, D. M.; Lima, F. O.; Soares, L. C.; Gallina, A. L. In Vitro Antioxidant Extracts Evaluation from the Residue of the *Hevea brasiliensis* Seed. *Sci. Rep.* **2022,** *12*, 1–14.

Özen, H.; Kamber, U.; Karaman, M.; Gül, S.; Atakişi, E.; Özcan, K.; Atakişi, O. Histopathologic, Biochemical and Genotoxic Investigations on Chronic Sodium Nitrite Toxicity in Mice. *Exp. Toxicol. Pathol.* **2014,** *66*, 367–375.

Pagare, S.; Bhatia, M.; Tripathi, N.; Pagare, S.; Bansal, Y. K. Secondary Metabolites of Plants and Their Role: Overview. *Curr. Trends Biotechnol. Pharm.* **2015,** *9*, 293–304.

Pandey, A. K.; Kumar,P.; Singh, P.; Tripathi, N. N.; Bajpai, V. K. Essential Oils: Sources of Antimicrobials and Food Preservatives. *Front. Microbiol.* **2017,** *7*, 1–14.

Pérez-Pérez, J. U.; Guerra-Ramírez, D.; Reyes-Trejo, B.; Cuevas-Sánchez, J. A.; Guerra-Ramírez, P. Actividad Antimicrobiana In Vitro de extractos de *Jatropha dioica seseé* contra bacterias fitopatógenas de tomate. *Polibotánica* **2020,** 125–133.

Pisoschi, A. M.; Pop, A.; Georgescu, C.; Turcuş, V.; Olah, N. K.; Mathe, E. An Overview of Natural Antimicrobials Role in Food. *Eur. J. Med. Chem.* **2019,** *143*, 922–935.

Rahman, M.; Sarker, S. D. Antimicrobial Natural Products. *Annu. Rep. Med. Chem.* **2020,** *55*, 77–113.

Sabandar, C. W.; Ahmat, N.; Jaafar, F. M.; Sahidin, I. Medicinal Property, Phytochemistry and Pharmacology of Several *Jatropha* Species (*Euphorbiaceae*): A Review. *Phytochemistry* **2013,** *85*, 7–29.

Saetae, D.; Suntornsuk, W. Antifungal Activities of Ethanolic Extract from *Jatropha curcas* Seed Cake. *J. Microbiol. Biotechnol.* **2012,** *20*, 319–324.

Senou, M.; Lokonon, J. E.; Abissi, E. O. T. R.; Agbogba, F.; Dehou, R. J.; Medoatinsa, E.; Baba-Moussa, L. Antibacterial Activity and Toxicity of the Sap and Aqueous Extract of the Leaves of *Jatropha multifida* Linn. *J. Biosci. Med.* **2022,** *10*, 171–182.

Sharafati Chaleshtori, F.; Taghizadeh, M.; Rafieian-kopaei, M.; Sharafati-chaleshtori, R. Effect of Chitosan Incorporated with Cumin and Eucalyptus Essential Oils As Antimicrobial Agents on Fresh Chicken Meat. *J. Food Process. Preserv.* **2016,** *40*, 396–404.

Sharif, Z. I. M.; Mustapha, F. A.; Jai, J.; Yusof, N. M.; Zaki, N. A. M. Review on Methods for Preservation and Natural Preservatives for Extending the Food Longevity. *Ing. Química* **2017,** *19*, 145–153.

Shukla, A.; Gundampati, R. K.; Jagannadham, M. V. Immobilization of *Euphorbia tirucalli* Peroxidase onto Chitosan-Cobalt Oxide Magnetic Nanoparticles and Optimization Using Response Surface Methodology. *Int. J. Biol. Macromol.* **2017,** *102*, 384–395.

Silva-Belmares, Y.; Rivas-Morales, C.; Viveros-Valdez, E.; de la Cruz-Galicia, M. G.; Carranza-Rosales, P. Antimicrobial and Cytotoxic Activities from *Jatropha dioica* Roots. *Pak. J. Biol. Sci.* **2014,** *17*, 748–750.

Solís-Ramos, L. Y.; Carballo, L. M.; Valdez-Melara, M. "Establishment of Cell Suspension Cultures of Two Costa Rican *Jatropha* Species (*Euphorbiaceae*). *Revista de Biologia Tropical* **2013,** *67*, 1095–1107.

Solís-Ramos, L. Y.; Carballo, L. M.; Valdez-Melara, M. Model-Assisted Identification of Metabolic Engineering Strategies for *Jatropha curcas* Lipid Pathways. *Plant J.* **2020,** *104*, 76–95.

Sreepian, A.; Sreepian, P. M.; Chanthong, C.; Mingkhwancheep, T.; Prathit, P. Antibacterial Activity of Essential Oil Extracted from Citrus Hystrix (Kaffir lime) Peels: An In Vitro Study. *Trop. Biomed.* **2019,** *36*, 531–541.

Sultana,A.; Hossain, M. J.; Kuddus, M. R.; Rashid, M. A.; Zahan, M. S.; Mitra, S.; Roy, A.; Alam, S.; Sarker, M. M. R.; Mohamed, I. N. Ethnobotanical Uses, Phytochemistry, Toxicology, and Pharmacological Properties of *Euphorbia neriifolia* Linn. Against Infectious Diseases: A Comprehensive Review. *Molecules* **2022,** *27*, 2–31.

Sutiany, A.; Dwiyanti, R. D.; Oktiyani, N. Inhibition Strength of Rosella (*Hibiscus sabdariffa* L.) Boiled Water on *Salmonella typhi* In Vitro. *Trop. Health Med. Res.* **2019,** *1*, 62–67.

Tatsimo, J. S. N.; Tamokou, J. D. D.; Havyarimana, L.; Kengne, I. C.; Ekom, S. E.; Lannang, A. M.; Sewald, N. Antimicrobial and Antioxidant Activities of Isolated Compounds from the Root Bark of *Croton pseudopulchellus* (Euphorbiaceae). *Investig. Med. Chem. Pharmacol.* **2020,** *13*, 1–6.

Tchangoue, Y. A. N.; Tchamgoue, J.; Lunga, P. K.; Knepper, J.; Paltinean, R.; Ibrom, K.; Schulz, S. Antibacterial Phloroglucinols Derivatives from the Leaves of *Mallotus oppositifolius* (Geisler) Müll. Arg.(*Euphorbiaceae*). *Fitoterapia* **2020,** *142*, 104527.

Thyberg, K. L.; Tonjes, D. J. Drivers of Food Waste and Their Implications for Sustainable Policy Development. *Resour. Conserv. Recycl.* **2008,** *106*, 110–123.

Ting, Y.; Jun, H.; Yan,Y.; Wen-Wen, L.; Cong-Yuan, X.; Jie-Kun, X.; Wei-Ku, Z. *Euphorbia ebracteolata* Hayata (Euphorbiaceae): A Systematic Review of Its Traditional Uses, Botany, Phytochemistry, Pharmacology, Toxicology, and Quality Control. *Phytochemistry* **2020,** *186*, 1–25.

Trasande, L.; Shaffer, R. M.; Sathyanarayana, S. Food Additives and Child Health. *Pediatrics* **2018,** *142*, 1–8.

Ughachukwu, P. O.; Ezenyeaku, C. C. T.; Ochiogu, B. C.; Ezeagwuna, D. A.; Anahalu, C. Evaluation of Antibacterial Activities of *Euphorbia heterophylla*. *IOSR J. Dent. Med. Sci.* **2014,** *13*, 69–75.

Wijesundara, N. M.; Rupasinghe, H. P. V. Essential Oils from *Origanum vulgare* and Salvia Officinalis Exhibit Antibacterial and Anti-Biofilm Activities Against *Streptococcus pyogenes*. *Microb. Pathogen.* **2018,** *117*, 118–127.

Wright, G. D. Unlocking the Potential of Natural Products in Drug Discovery. *Microb. Biotechnol.* **2019,** *12*, 55–57.

Wu, Q.; Patocka, J.; Nepovimova, E.; Kuca, K. *Jatropha gossypiifolia* L. and Its Biologically Active Metabolites: A Mini Review. *J. Ethnopharmacol.* **2019,** *234*, 197–203.

Zahan, K. A.; Kano, M. Biodiesel Production from Palm Oil, Its By-Products, and Mill Effluent: A Review. *Energies* **2018,** *11*, 2132.

CHAPTER 9

Biotechnology and Food Safety: Zoonotic Perspectives

A. ARUN PRINCE MILTON, G. BHUVANA PRIYA, SAMIR DAS, SANDEEP GHATAK, and K. SRINIVAS

ICAR Research Complex for NEH Region, Umiam, Meghalaya, India

ABSTRACT

Pork (meat of pig) is the most commonly consumed animal meat worldwide and there is increasing global demand for pork according to future projections. The pork production system has witnessed a drastic change from pasture-based rearing to the industrialized intensive system of rearing. Though pork is less attributed to foodborne diseases in comparison to other meat (poultry, beef), it is imperative to achieve pork safety owing to its large consumption. Pathogens enter the pork value chain through infected pigs, unhygienic slaughter, postharvest handling, poor processing methods, etc. Consumers acquire the infection due to the consumption of raw or improperly cooked pork and other kitchen cross-contaminations. The important challenges in achieving microbial safety in pork include cross-contamination of other foodstuffs, contaminated water, traceability, rapid on-site pathogen testing techniques, foodborne disease surveillance, implementation of hazard analysis and critical control points (HACCP), awareness of pork handlers, and consumers, etc. For effective pork safety interventions, it is decisive to understand the zoonotic pathogens transmitted via pork including their nature, control options, and prevalence in different stages of production and processing. In this chapter, pork-associated zoonotic agents discoursed are *Salmonella*, *Yersinia enterocolitica*, *Campylobacter jejuni/Campylobacter coli*, *Listeria monocytogenes*, *Clostridium perfringens*, *Staphylococcus*

Engineering Principles for Food Processing Technology and Product Realization. Leonardo Sepúlveda Torre, Porteen Kannan, A. K. Haghi, PhD (Eds.)

aureus, shigatoxigenic *Escherichia coli, Streptococcus suis, Taenia solium, Toxoplasma gondii, Trichinella spiralis*, hepatitis E virus, rotavirus, and norovirus. Attention has been given to the nature of zoonotic agents, prevalence, diseases caused in human beings, and various intervention strategies used against the pathogens in on-farm, slaughter, and postprocessing stages.

9.1 INTRODUCTION

Food safety has become an increasingly significant public health subject that has led all governments throughout the world to allocate millions of currencies to strengthen their efforts to ensure and improve food safety (Kaferstein et al., 1997). The efforts toward improving food safety can be perceived as a response to a growing number of food safety issues and rising concerns of the consumer (WHO, 2009). The Centers for Disease Control and Prevention (CDC) estimates that foodborne pathogens cause 47.8 million illness events and 3000 deaths every year (CDC, 2011). Most importantly, the foodborne disease outbreaks reported in the last decades underscore the significance of the formulation and application of preventive measures and programs relating to food safety. Particularly, a farm-to-fork tactic has been followed in all food production sectors to reduce probable biological hazards and improve hygiene (Schirone et al., 2019). Many factors associated with the increase in foodborne diseases have been described, such as changes in consumer preferences and eating habits, changes in food processing methods, inadequate sanitation and public health services including vector control and awareness building, increased global travel, the adaptation of pathogen to newer environments including the acquisition of antimicrobial resistance and virulence factors and lastly advancement in pathogen detection tools (Smith and Fratamico, 2018).

Globally, over the past decade, the growth rate of pork consumption has enhanced by 10.71% (Wu et al., 2015). World meat production is forecasted to increase by 2.2% (pork: 4.2%) to 346 million tonnes (pork: 114.4 million tonnes) (FAO, 2021). Pork (meat of pig) is the most commonly consumed animal meat worldwide and there is increasing global demand for pork according to future projections (Davies, 2010; Delgado et al., 2001). The pork production system has witnessed a drastic change from pasture-based rearing to the industrialized intensive system of rearing (Enemark et al., 2015). Although pork is less attributed to foodborne diseases in comparison to other meat (poultry, beef), it is imperative to achieve pork safety owing to its large consumption (Baer et al., 2013). As pork and pork products are extensively consumed throughout the world, it is crucial for producers (farmers), processors and other stakeholders in the value chain to be conscious of the risk factors for contamination of pork with foodborne pathogens (Baer et al., 2013). Diverse farm and husbandry systems (large/small/backyard) are associated with specific microbiological hazards, which are spread to the next stage in the pork value chain, and the subsequent stage in the value chain could also be the point of contamination (Buncic et al., 2019; Karabasil et al., 2021). The meat sector is a sector with inherent risks, so meat safety must be risk-oriented to reduce or eliminate food safety risks, with integrated and longitudinal approaches along the value chain, and incessant improvement (Karabasil et al., 2021).

9.2 ZOONOTIC PATHOGENS OF PIG AND PORK

Pigs act as a reservoir for several zoonotic pathogens of viral, bacterial, and parasitic origin (Fig. 9.1). Further, the presence of other reservoirs and vectors in the environment of pigs can increase the prevalence of zoonotic diseases and the risk for transmission to human beings (Salajpal et al., 2013; Delsart et al., 2020). Pathogens enter the pork value chain through infected pigs, unhygienic slaughter, postharvest handling, poor processing methods, etc (Fig. 9.1). The majority of pork products are prepared by smallholders and traded in classical wet markets. Nevertheless, fresh pork is exceedingly susceptible to contamination, posing health and financial burdens (Lam et al., 2020). Consumers acquire the infection due to the consumption of raw or improperly cooked pork and other kitchen cross-contamination. For effective pork safety interventions, it is decisive to understand the zoonotic pathogens transmitted via pork including their nature, control options, and prevalence in different stages of production and processing. Recent changes and intensification in the pig production system have nearly eliminated the risk of tissue-dwelling foodborne parasites such as *Trichinella spiralis*, *Taenia solium*, and *Toxoplasma gondii.* This represents a significant public health accomplishment that has gone largely underappreciated (Davies, 2011). This could have become possible as these parasites involve carnivorous transmission and effective prevention can be achieved by some preharvest interventions (Davies, 2011). However, important bacterial pathogens of concern to pork safety such as *Salmonella, Listeria, Y. enterocolitica, Campylobacter*, etc. have their basic ecological niche in the gastrointestinal tracts of healthy animals as commensals, and thus the point of contamination can be at preharvest, during harvest, postharvest, and any stages till reaching the consumers' plate. Consequently, such complex epidemiology makes them less amenable to controlling such bacterial pathogens in the pork production chain (Davies et al., 2004; Alban and Stark, 2005; Fosse et al., 2009; Oliver et al., 2009; Davies, 2011).

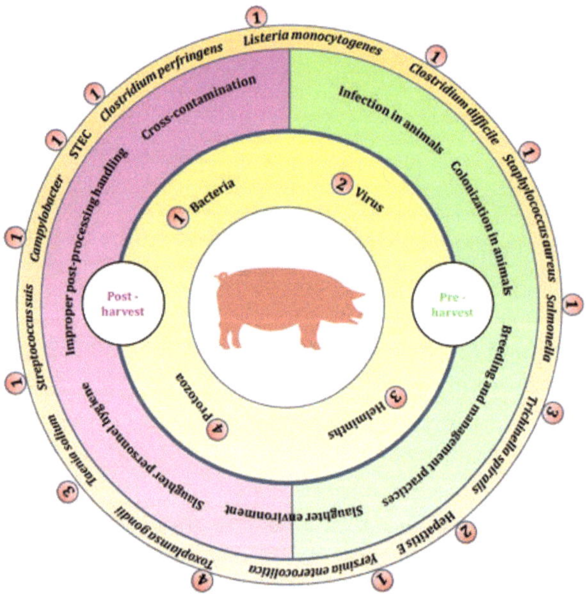

FIGURE 9.1 Zoonotic pathogens in pig value chain.

9.2.1 *SALMONELLA*

Salmonella is a Gram-negative bacteria under the family *Enterobacteriaceae,* that can survive in temperature and pH ranges of 5–45°C and 4–9, respectively (Doyle and Cliver, 1990). Some strains of *Salmonella* can also survive with a 2% NaCl concentration and low-moisture foods like dried pork products (Montville and Matthews, 2008; Baer et al., 2013). Presently, there are only 2 known species of *Salmonella* (*S. enterica* and *S. bongori*),

6 subspecies of *S. enterica*, and more than 2600 serovars of *S. enterica* subsp. *enterica* including notable serovars such as Salmonella serovars Typhimurium and *Salmonella* Enteriditis causing nontyphoidal salmonellosis (Guibourdenche et al., 2010). *Salmonella* infections in human beings are the foremost cause of hospitalization and deaths due to foodborne illnesses. It causes illnesses ranging from mild gastroenteritis and diarrhea to serious systemic infections (Scallan et al., 2011; Gupta et al., 2019). Adult pigs are frequently asymptomatic carriers of wide-ranging *Salmonella* serovars and can represent a reservoir for human infections (Gupta et al., 2019). Several studies have reported that as the growth of pigs' advances from weaning to growing and finishing stages, the prevalence of *Salmonella* is enhanced (Korsak et al., 2003; Dorr et al., 2009; Wilkins et al., 2010). Nevertheless, the prevalence of *Salmonella* was also reported to decrease as pigs advance through the production chain (Kranker et al., 2003; Rowe et al., 2003). Notwithstanding, the prevalence of *Salmonella* in pigs is not unusual in any of the stages of production and processing, and the prevalence rates vary from 0.7 to 77.8 (Baer et al., 2013). In a study conducted by USDA in two commercial slaughter establishments, it was found that 91% of prescald carcasses were carrying *Salmonella* (Schmidt et al., 2012). In another study, it was reported that 37% of carcasses harboring *Salmonella* contamination contained *Salmonella* counts between 1 and 3.9 log CFU per 100 cm^2. Hence, there is a chance of cross-contamination with other carcasses and processing equipment (Dickson et al., 2002). Several studies have indicated the prevalence of *Salmonella* in retail fresh pork, pork products, and ready-to-eat pork products and the reasons noticed were contaminated slaughter environment, complacence in hygiene, under processing, contamination from food handlers, raw materials like spices, and pests inside the facility such as rodents, insects, and birds (Baer et al., 2013).

9.2.2 *YERSINIA ENTEROCOLITICA*

Y. enterocolitica is a Gram-negative psychrotrophic salt-tolerant bacteria that is extensively distributed in nature, in animal and aquatic reservoirs, with pigs representing the main known reservoir for zoonotic strains of *Y. enterocolitica* (Bottone, 1997). The pathogenicity of *Y. enterocolitica* is usually linked with its bio-serotype. The biotypes, 1B, 2, 3, 4, and 5 and serogroups, O:3, O:5,27, 42 O:8, and O:9, are pathogenic to human beings. The bio-serotype 4/O:3 is predominant in human and pork samples (Martins et al., 2020). The usual symptoms of *Y. enterocolitica* infection in human beings are fever, diarrhea, and abdominal pain. However, reactive arthritis and nodular erythema are some uncommon symptoms (Wang et al., 2021). Yersiniosis in humans is the fourth and fifth most common foodborne zoonosis generally transmitted through pork and water in the European Union and the United States, respectively (EFSA, 2019; Tack et al., 2019). In the European Union, from 2005 to 2018, pork and pork products were among the most commonly described food types causing yersiniosis of foodborne origin (EFSA and ECDC, 2019). Fosse et al. (2008) assessed that 77% of human yersiniosis cases are linked to the consumption of pork and cold storage might guard *Y. enterocolitica* as it is psychrotrophic in nature (Martins et al., 2020). Pork chitterlings were also linked to a yersiniosis outbreak (Anonymous, 1990). It is assessed that 35–70% of pig herds and 45–100% of individual pigs carry Y. enterocolitica (Jakobsen et al., 2019). A higher prevalence of *Y. enterocolitica* was reported in conventional intensive pig farms (Laukkanen et al., 2009; Fondrevez et al., 2014; Zdolec and Kiš, 2021). In Norway, a considerable (25%) reduction in the incidence of *Y. enterocolitica* prevalence after the sealing-off of the rectum with a plastic bag indicates that the fecal material of pigs acts as an important source of contamination for pork and pork products (Nesbakken and Skjerve, 1996). However, the tonsils of pigs are considered as an important source of *Y. enterocolitica* than fecal and intestinal contents (Fondrevez et al., 2014). The reasons for the contamination of pig carcasses and products are unhygienic slaughtering and processing, and poor handling of possibly contaminated sites including tonsils and mesenteric lymph nodes (Martins et al., 2020).

9.2.3 CAMPYLOBACTER

Campylobacter is a Gram-negative microaerophilic bacteria that can survive in temperature ranges and pH of 30–47°C and 5.8, respectively (Doyle, 1990). The *C. jejuni/C. coli* group of the pathogen has come out from obscurity to rank among the foremost causes of diarrhea of bacterial origin in humans (Butzler and Skirrow, 1979). *C. jejuni* is predominantly recovered from gastroenteritis cases and is widely distributed in the animal kingdom with the majority isolations from poultry, wild birds, exotic zoo animals, and domestic animals like pigs, dogs, cats, cattle, sheep, etc (Milton et al., 2020a). *C. coli* is the second most frequent and neglected cause of human campylobacteriosis (Kelley et al., 2020). *C. coli* cause an illness in humans that is identical to the one caused by *C. jejuni* (Karmali et al., 1983). Despite pig being its major reservoir, *C. coli* is also found in other animals, but with an occurrence considerably less than that of *C. jejuni*. Retail raw pork and products are one of the important food vehicles (Noormohamed and Fakhr, 2013; Quintana-Hayashi and Thakur 2012; Beier et al. 2018). The widely held *Campylobacter* infections in humans are self-limiting, which set hurdles for the determination of its true incidence (Hanninen et al., 2000). Human campylobacteriosis caused by *C. jejuni/C. coli* is typically manifested by diarrhea (often bloody), vomiting, abdominal cramps, and fever and after acute illness, it has been associated with long-term complications such as Guillain–Barre syndrome, irritable bowel syndrome, and reactive arthritis (Milton et al., 2020a). Swine get colonized with *Campylobacter* immediately (<1 week) after birth and as they advance through various production phases, the prevalence of *Campylobacter* usually increases (Schuppers et al., 2005). A study reported a 15% increase in the prevalence of *Campylobacter* from the growing to finishing stages of pigs (Farzan et al., 2010). In the finishing stage, most of the pigs tested positive for *Campylobacter*, and across all the stages of production, *C. coli* is found to be more prevalent than *C. jejuni* (Schuppers et al., 2005). The occurrence of *Campylobacter* at the production farm translates to the presence in the slaughter process and thereby in pork. However, it can be reduced by following certain hygienic slaughter procedures such asscalding, singeing, dehairing, polishing, etc. followed by chilling as campylobacters are sensitive to low temperature and drying (Pearce et al., 2003).

9.2.4 *LISTERIA MONOCYTOGENES*

Listeria is a Gram-positive, psychrotrophic, ubiquitously present bacteria with common inhabitation in soil, water, and intestines of animals (Bahk and Marth, 1990). Under the genus *Listeria, L. monocytogenes* is the only zoonotic human pathogen representing the highest reasons for the hospitalizations and death rates among infections caused by foodborne pathogens (Scallan et al., 2011). The common clinical symptoms of *L. monocytogenes* in humans are fever, chills, and back pain and mortality are usually due to encephalitis, meningitis, and sepsis in 20–30% of infected cases, mostly immunocompromised ones (Schlech, 2000). Listeriosis causes abortion and fetal infection in pregnant women with around 1500 cases occurring in the US alone and 99% are of foodborne origin (Scallan et al., 2011). Although

L. monocytogenes covers 13 serotypes, the serotypes 4b (28%), 1/2a (24%), and 1/2c (22%) are commonly reported in slaughtered retail pork (Farber and Peterkin, 1991; Kanuganti et al., 2002). Sows carry more *L. monocytogenes* in feces than piglets (Fenlon et al., 1996). Apart from wet feed (or silage), bedding or litter, and boots are all potential sources of *Listeria* contamination to live pigs on the farm (Baer et al., 2013). As the environment acts as an important source of *L. monocytogenes* to pigs, improper hygienic management and biosecurity measures can lead to its enhanced prevalence on farms (Hellstrom et al., 2010). Through serotype tracing, it was found that infected live animals and cross-contamination of the carcass are the two important routes of contamination to pork by *L. monocytogenes* (Hellstrom et al., 2010). Higher levels of *L. monocytogenes* prevalence in tonsils and isolations even after chilling of carcasses are the two noteworthy observations made earlier (FSIS, 1996; Autio et al., 2000; Hellstrom et al., 2010). Ready-to-eat pork products such as deli meats, frankfurters, etc., and home-stored meats are known to harbor *L. monocytogenes*, and are associated with numerous foodborne outbreaks (Schlech, 2000; de Valk et al., 2001; Baer et al., 2013). These are also the indicators of cold insensitivity of *L. monocytognes* and environmental contamination (Baer et al., 2013).

9.2.5 *CLOSTRIDIUM PERFRINGENS* AND *CLOSTRIDIUM DIFFICILE*

C. perfringens and *C. difficile* are Gram-positive anaerobic spore-forming bacteria associated with economic and public health concerns in swine production (Baker et al., 2010). These pathogens are normal colonizers of the intestinal tract of various food animals mainly pigs (Songer and Uzal, 2005). Clostridial infections generally occur in piglets immediately (within the first week) after birth and are associated with under-developed normal microbiota, scouring, and antibiotic management (Yaeger et al., 2002; Baker et al., 2010). Clinically, *C. perfringens* (type A and C) causes enteric and systemic diseases in human beings including food poisoning, gas gangrene, and enterocolitis (pigbel) (Milton et al., 2020b). *C. difficile* infections in human beings can cause antibiotic-associated diarrhea to a life-threatening infection such as severe pseudomembranous colitis (Milton et al., 2020b). Both the pathogenic species of *Clostridium* have been reported from pork and the environment of pig slaughterhouses (Rho et al., 2001; Metcalf et al., 2010; Chan et al., 2012; Curry et al., 2012; Rodriguez et al., 2013; Mooyottu et al., 2015; Wu et al., 2017). Nevertheless, although *C. perfringens* grades among the most significant pathogens of food-borne diseases, the sorting of *C. difficile* as a foodborne pathogen is still contentious (Milton et al., 2020b). However, spotting identical strains of *C. difficile* in food animals, meat, and humans put forward the possible transmission of *C. difficile* through foods as well as pork products (Milton et al., 2020b).

9.2.6 *STAPHYLOCOCCUS AUREUS*

S. aureus is a Gram-positive, facultative anaerobic bacteria whose common habitat is the upper respiratory tract and skin of humans and animals (Bergdoll, 1990). As the primary reservoir of *S. aureus* seems to be skin, the pork usually sold along with the skin makes them an important source if inadequately scalded and singed. The pathogen is zoonotic and the foodborne illness caused by them involves clinical symptoms of vomiting, nausea, diarrhea, and abdominal cramps after 1–6 h of ingestion (Bergdoll, 1990). The pathogen releases enterotoxins that are the triggering factors of illness and are accountable for the 14% hospitalization rate (Noskin et al., 2007). However, for adequate enterotoxin levels to trigger illness, at least 10^5 *S. aureus* cells/g must be present in the food (Bahtia and Zahoor, 2007). As *S. aureus* is not a sturdy competitor compared to other bacterial flora present in food, the contamination is typically linked to the postprocessing handling of meat products when the competition has been removed (Argudin et al., 2010). Despite the relatively low-risk attribute of *S. aureus*, the worldwide emergence of methicillin-resistant *S. aureus* (MRSA) has become an important public health concern (Baer et al., 2013). Varied prevalence of *S. aureus* ranging from 42.2 to 76.7% has been reported in pork and products (Atanassova et al., 2001; van Loo et al., 2007; O'Brien et al., 2012; Savariraj et al., 2018). Around 3.1–10.7% of fresh grounded and chopped pork were described to be contaminated with MRSA (van Loo et al., 2007; de Boer et al., 2009). Poor personal hygiene of slaughter and processing personnel, unhygienic and unscientific abattoir, and the processing practices contaminate the pork with a high count of *S. aureus* and consequent release of enterotoxins (Savariraj et al., 2018).

9.2.7 SHIGATOXIGENIC *ESCHERICHIA COLI* (STEC)

E. coli is a Gram-negative, facultative anaerobic bacteria under the family *Enterobacteriaceae* that is normally found in the intestines of warm-blooded animals. *E. coli* is classified by its links with human disease, namely, enterohemorrhagic or shiga toxigenic (EHEC/STEC), enterotoxigenic (ETEC), enteropathogenic (EPEC), enteroinvasive (EIEC), enteroaggregative (EAEC), diffuse adhering (DAEC), and extraintestinal pathogenic *E. coli* (ExPEC) (Logue et al., 2017). STEC possesses more zoonotic concerns among all the pathogenic *E. coli* and pigs are important reservoirs of STEC (Colello et al., 2016). STEC can cause severe foodborne diseases leading to life-threatening complications like hemolytic uremic syndrome (HUS) and bloody diarrhea due to colitis (Paton and Paton, 1998). The important cause of acute kidney injury and acquired chronic kidney disease in children are due to the damages caused by HUS through the cytotoxic actions of Stx1 and Stx2. Normally, STEC does not cause disease in animals except edema disease in pigs caused due to Stx2e subtype of STEC (Niewerth et al., 2001). Pork and pork products have been incriminated in human STEC-associated diarrhoeic and HUS infections (Sonntag et al., 2005; Kaufmann et al., 2006; Trotz-Williams et al., 2012). A family outbreak of STEC (O157) infection was reported to be due to the consumption of dry fermented pork salami (Conedera et al., 2007). A varied prevalence of STEC in pigs ranging from 2 to 31% was documented (Parma et al., 2000; Kaufmann et al., 2006; Meng et al., 2014). A prevalence range of 0.2–26% of STEC was reported earlier in pig carcasses during slaughtering (Leung et al., 2001; Bouvet et al., 2002; Bohaychuk et al., 2011; Koláčková et al., 2014). In pig slaughtering establishments, diverse STEC serogroups (O91, O121, and O157) that are associated with human illnesses were detected earlier (Colello et al., 2016). Slaughterhouse operations such as skinning, handling, and evisceration are found to be the more likely reason for the contamination of pork (Koohmaraie et al., 2005; Etcheverría et al., 2010).

9.2.8 *STREPTOCOCCUS SUIS*

S. suis is a Gram-positive facultative anaerobic bacteria whose common habitat is the nasal cavities, tonsils, and alimentary and genital tracts of pigs thus pig becomes the natural reservoir of the pathogen (Lun et al., 2007). *S. suis* serotype 2 is a zoonotic pathogen largely of occupational origin and most frequently associated with diseases in human beings and pigs (Lun et al., 2007; Dutkiewicz et al., 2017). Human *S. suis* infections are clinically manifested as pus-filled meningitis, the other described syndromes comprise septicemia, streptococcal toxic shock-like syndrome with multiple organ failure, cellulitis, endocarditis, rhabdomyolysis, arthritis, peritonitis, spondylodiscitis, pneumonia, uveitis and endophthalmitis (Dutkiewicz et al., 2017). The carriage rates of *S. suis* may vary between swine herds and can range from 0% to up to 80–100% (Dutkiewicz et al., 2017). In the majority of Western countries, the transmission of *S. suis* is considered occupational affecting persons working in the pig or pork industries. But, in Southeast Asian countries, the main risk is due to close contact with pigs or ingestion of uncooked or partially cooked pork products (Wertheim et al., 2009; Segura, 2020). The percutaneous route of infection associated with handling pigs or pork has also been reported multiple times (Chattopadhyay, 1979; Watkins et al., 2001; Vilaichone et al., 2002; Wangkaew et al., 2006).

9.2.9 *TAENIA SOLIUM*

T. solium is a tapeworm (cestode) that dwells only in the intestines of human beings (definitive host). Pigs acquire the infection by ingesting the eggs/proglottids shed in human feces and play the role of intermediate hosts and the parasite resides in the muscles and other tissues of pigs in the cyst form. The human population acquires intestinal infections (taeniasis) of tapeworm by consuming undercooked cyst-infested pork, that is, measly pork (Phiri et al., 2003). In addition, human beings can also get infected with cystic form infection (cysticercosis) if encounter infested human feces either directly or indirectly. The cysts are formed throughout the body, but predominantly in the eyes and brain leading to neurocysticercosis condition (Davies, 2011). Neurocysticercosis accounts for 50% of epileptic cases in developing countries. Apart from seizures, it causes raised intracranial pressure, headaches, focal neurologic deficits, and psychiatric and ocular symptoms (Rajshekhar et al., 2006). It was estimated by the Food and Agriculture Organization (FAO) that worldwide 50 million people harbor adult *T. solium* worm and up to 50,000 mortalities annually are attributable to neurocysticercosis (Eddi et al., 2003; Davies, 2011). *T. solium* prevails where there is poor sanitation and traditional free-ranging or scavenging type of pig rearing is practiced (Phiri et al., 2003; Davies, 2011). The elimination of *T. solium* from North America and Europe was basically due to economic development and adaptation of the modern intensive system of pig production (Gonzalez et al., 2003). The United States's pig industry has acknowledged that open defecation in swine farms by the farmworkers is the sole potential concern and education efforts can eliminate the risk (Davies, 2011). Presently, the *T. solium* infection is endemic in South and Central America, Asia, sub-Saharan Africa, and some parts of the Pacific regions. The prevalence of cysticercosis infection in pigs in Africa, South/Central America, and Asia are 64%, 65.4%, and 26%, respectively (Djurković-Djaković et al., 2013). The prevalence of *T. solium* infection in human beings in Africa, South/Central America, and Asia are 10%, 22%, and 1%, respectively (Djurković-Djaković et al., 2013). Nevertheless, increasing economic immigration from endemic countries to developed countries and increased tourism in developing nations indicate *T. solium* as a pathogen with global health implications and the ability to spread to nonendemic zones (Pawlowski et al., 2005).

9.2.10 *TOXOPLASMA GONDII*

T. gondii is a coccidian parasite (protozoan) of felids which is the definitive host. The life cycle of *T. gondii* comprises three infective stages, namely, sporozoites, tachyzoites, and bradyzoites. Cats shed sporozoites (reproducing stage) as oocysts in their feces and it remains in the environment and contaminates soil and water, and infects other intermediate hosts such as food animals (such as pigs), human beings, and wildlife. After ingestion of oocysts by intermediate vertebrate hosts, the parasite assumes tachyzoites and bradyzoites (nonreproducing) stages and develops cysts in the tissues such as muscle, heart, and brain (Cliver 1990; Baer et al., 2013). Human beings acquire infection mainly through the consumption of raw or improperly cooked infested meat and through oocyst-contaminated water (Foroutan-Rad et al., 2016; Sher et al., 2016). Severe clinical manifestations are usually noticed in immunosuppressed individuals such as AIDS patients, and pregnant ladies. In pregnant women, it causes stillbirths and abortion and if the fetus survives, *T. gondii* is passed to the neonate and results in a congenital form of the disease with major abnormalities (Cliver, 1990; Carvalheiro et al., 2005). However, throughout the world, a large number of people have antibodies to the parasite, indicating *T. gondii* is an important public health concern. In the United States, it was reported that out of 225000 cases, 50% were of foodborne origin (Mead et al., 1999). *T. gondii* was described to be responsible for 8% of hospitalizations and 24% of deceases from foodborne diseases (Scallan et al., 2011). The prevalence of *T. gondii* in pigs in different countries has been well recognized (Baer et al., 2013). The important risk factors for *T. gondii* infection in pigs were described as free-range production system, increased contact with wild animals or cats, improper discarding of pig carcasses, and contaminated water or feed (Dubey et al., 2002; Gebreyes et al., 2008; Hill et al., 2010; Baer et al., 2013). As *T. gondii* can endure infectivity in the tissues of pigs for not less than 171 days, its presence in pork is a zoonotic concern (Dubey et al., 1984). In a US-based study, it was found that pork was the only retail meat reported to be positive for the parasite (Dubey et al., 2005). The same study has estimated that people in the northeastern US have a 78% chance of procuring pork contaminated with *T. gondii* (Dubey et al., 2005). As *T. gondii* can be destructed by several preservation and processing methods such as heating (min. 67°C), freezing (min. –12.4°C), etc., improper processing or cooking remains a major risk factor (Dubey et al., 1990; Lunden and Uggla, 1992).

9.2.11 *TRICHINELLA SPIRALIS*

T. spiralis is a nematode (roundworm) that causes trichinellosis in human beings due to the consumption of undercooked meat containing encysted larvae. In addition to the predominant and classical agent *T. spiralis*, a few rare species of *Trichinella* implicated in human disease are *Trichinella pseudospiralis*, *Trichinella britovi*, *Trichinella murelli*, *Trichinella nativa*, *Trichinella nelson,* and *Trichinella papuae (CDC, 2019)*. Human beings who consume infested meat contract the disease with clinical manifestations such as abdominal pain, fever, and periorbital edema. The larvae are released from the cysts and invade the mucosa of the small intestine and develop into adult worms and the adult female further releases larvae that migrate to striated muscles and encyst. This leads to myalgia, joint pain, and muscle contraction inefficiency (Cliver, 1990). It can also cause serious disease in elderly individuals where serious complications such as encephalitis or myocarditis can lead to death (Djurković-Djaković et al., 2013). Roughly, 150 trichinellosis cases are reported annually in the US and all of those cases are of foodborne origin (Kennedy et al., 2009; Scallan et al., 2011). A survey conducted in 198 countries revealed that 21.9% of the countries had *Trichinella* spp. in pig herds (Pozio, 2006). A CDC report on trichinellosis revealed that during 2002–2007, 19% of trichinellosis in the US was caused by the consumption of pork, whereas 50% were owing to the consumption of wild game meat (Kennedy et al., 2009). The important risk factors for *T. spiralis* infection in pigs were described as a free-range production system, feeding of garbage and wildlife carcasses, and improper discarding of pig carcasses

(Cliver, 1990; Murrell and Pozio, 2000; van der Giessen, et al., 2007). A study conducted between 1999 and 2004 in China revealed a prevalence of 1.6–3.7% of *T. spiralis* in retail meat (Cui et al., 2006). As thorough cooking remains the best option to inactivate *Trichinella,* improper cooking becomes the major risk factor in the transmission of *T. spiralis* to human beings via the consumption of pork (Baer et al., 2013).

9.3 HEPATITIS E VIRUS

Hepatitis E virus (HEV) under the genus *Hepevirus* is an RNA virus recognized as the leading cause of acute viral hepatitis. Among the diverse HEV genotypes (GT) that affect human beings, GT3 and GT4 are zoonotic, have been reported in various animal species (Zhou et al., 2019; Li et al., 2020). Pigs serve as the main reservoir for the zoonotic HEV strains. Antibodies against HEV have been extensively detected in both domestic and wild pigs (Baechlein et al., 2010; Geng et al., 2010; Fredriksson-Ahomaa et al., 2020). Though HEV infection is generally asymptomatic or self-limiting in healthy people, GT3 and GT4 HEV infections in organ transplant recipients are more likely to develop chronic hepatitis (Kamar et al., 2008). However, asymptomatic individuals in endemic areas have been recognized to be potential reservoirs shedding the virus, permitting its ecological persistence (Christou and Kosmidou, 2013). Worldwide, it is estimated that around 939 million equivalent to 1 in 8 persons have ever encountered HEV infection (Li et al., 2020). The route of transmission of HEV from pigs to humans is frequently through direct contact with pigs and indirectly through the consumption of contaminated pork products and environmental exposure. Nevertheless, the extensive consumption of pork products is certainly posing a great risk of HEV transmission through food. HEV infections associated with the consumption of undercooked pork and wild boar meat have been repeatedly reported (Matsuda et al., 2003; Yazaki et al., 2003). It was estimated that globally, close to 60% of domestic pigs and 27% of wild boars have ever been HEV infected. Approximately 13% of domestic and 9.5% of wild pigs are undergoing an active infection. In addition, globally, 10% of commercial pork products are positive for HEV RNA (Li et al., 2022).

9.4 OTHER PATHOGENS IN THE CONTEXT OF PORK SAFETY

Shigella, *Bacillus cereus*, rotavirus, norovirus, and *Sarcocystis*, are other pathogens associated with pork-associated foodborne diseases. *Shigella* is a Gram-negative bacterial contaminant of food that is estimated to cause 164.7 million foodborne illnesses (dysentery) and 1.1 million deaths worldwide (Thiem et al., 2004). Foodborne outbreaks of shigella are widely reported and 7.5% of *Shigella* outbreaks were attributed to meat and meat products including pork (Pires et al., 2012). *B. cereus* is a Gram-positive bacterial contaminant of food that is estimated to cause 1.4–12% of foodborne outbreaks worldwide and 8.7% of *B. cereus* outbreaks were attributed to pork products (Pires et al., 2012; Grutsch et al., 2018). Rotavirus and norovirus are RNA viruses that are predominantly responsible for diarrheal illnesses worldwide (Hakim et al., 2018). Globally rotavirus accounts for 6% of diarrheic episodes and is responsible for 35,2000–592,000 deaths in children below 5 years of age (Parashar et al., 2003). Rotavirus has been previously detected in retail pork and products (Soares et al., 2022). Norovirus is estimated to be associated with 18% of diarrheal diseases and responsible for 212,000 deaths annually worldwide (Pires et al., 2015). Between 1988 and 2007, 2% of norovirus outbreaks were attributed to pork products (Greig and Ravel, 2009). *Sarcocystis* is a coccidian parasite and three of its species, namely, *Sarcocystis suihominis*, *Sarcocystis miescheriana*, and *Sarcocystis porcifelis* have been recognized in pigs. However, only *S. suihominis* can cause human infection upon consumption of raw pork (Djurković-Djaković et al., 2013). A foodborne outbreak of *S. suihominis* in children who consumed undercooked pork was reported in India (Banerjee et al., 1994).

9.5 CONCLUSION

This chapter has thoroughly outlined all the zoonotic pathogens and their significance in the pork industry. As the urge for intensive and alternative (outdoor) pig production is equally increasing across the world, comprehensive tailor-made preharvest plans specific to various situations or farms have to be delineated and strictly followed to ensure the safety of pork. Further hygienic slaughter, HACCP implementation, minimum handling, complete processing, proper packaging, appropriate personal hygiene of handlers, pest-proof food processing area, etc. are some of the postharvest strategies that ensure safe pork on the consumer's plate.

KEYWORDS

- **pork**
- **safety**
- **Zoonoses**
- **pathogen**
- **foodborne diseases**

REFERENCES

Alban, L.; Stark, K. D. Where Should the Effort be Put to Reduce the Salmonella Prevalence in the Slaughtered Swine Carcass Effectively? *Prev. Vet. Med.* **2005,** *68,* 63–79.

Anonymous. Topics in Minority Health *Yersinia enterocolitica* Infections during the Holidays in Black Families—Georgia. Morbid. *Mortal. Weekly Rep.* **1990,** *39,* 819–820.

Argudın, M. A.; Mendoza, M. C.; Rodicio, M. R. Food Poisoning and *Staphylococcus aureus* Enterotoxins. *Toxins.* **2010,** *2,* 1751–1773.

Atanassova, V.; Meindl, A.; Ring, C. Prevalence of *Staphylococcus aureus* and Staphylococcal Enterotoxins in Raw Pork and Uncooked Smoked Ham. *Int. J. Food Microbiol.* **2001,** *68,* 105–113.

Autio, T.; Sateri, T.; Fredriksson-Ahomaa, M.; Rahkio, M.; Lunden, J.; Korkeala, H. *Listeria monocytogenes* Contamination Pattern in Pig Slaughterhouses. *J Food Prot.* **2000,** *63,* 1438–1442.

Baechlein, C.; Schielke, A.; Johne, R.; Ulrich, R. G.; Baumgaertner, W.; Grummer, B. Prevalence of Hepatitis E Virus-Specific Antibodies in Sera of German Domestic Pigs Estimated by Using Different Assays. *Vet Microbiol.* **2010,** *144,* 187–191.

Baer, A. A.; Miller, M. J.; Dilger, A. C. Pathogens of Interest to the Pork Industry: A Review of Research on Interventions to Assure Food Safety. *Compr. Rev. Food Sci. Food Saf.* **2013,** *12,* 183–217.

Bahk, J.; Marth, E. H. *Listeriosis and Listeria monocytogenes. Foodborne Diseases*; Academic Press Inc.: San Diego, California, 1990; pp 247–257.

Bahtia, A.; Zahoor, S. *Staphylococcus aureus* Enterotoxins: A Review. *J. Clin. Diagn. Res.* **2007,** *3,* 188–197.

Baker, A. A.; Davis, E.; Rehberger, T.; Rosener, D. Prevalence and Diversity of Toxigenic *Clostridium perfringens* and Clostridium Difficile among Swine Herds in the Midwest. *Appl. Environ. Microbiol.* **2010,** *76,* 2961–2967.

Banerjee, P. S.; Bhatia, B. B.; Pandit, B. A. *Sarcocystis suihominis* Infection in Human Beings in India. *Vet. Parasitol.* **1994,** *8,* 57–58.

Beier, R. C.; Harvey, R. B.; Hernandez, C. A.; Hume, M. E.; Andrews, K.; Droleskey, R. E. Interactions of Organic Acids with *Campylobacter coli* from Swine. *PLoS One* **2018,** *13,* 1–17.

Bergdoll, M. S. *Staphylococcal Food Poisoning. Foodborne Diseases*; Academic Press, 1990; pp 85–106.

Bohaychuk, V. M.; Gensler, G. E.; Barrios, P. R. Microbiological Baseline Study of Beef and Pork Carcasses from Provincially Inspected Abattoirs in Alberta. Canada. *Can. Vet. J.* **2011,** *52,* 1095–1100.

Bottone, E. J. *Yersinia enterocolitica*: The Charisma Continues. *Clin. Microbiol. Rev.* **1997,** *10,* 257–276.

Bouvet, J.; Montet, M.; Rossel, R.; Le Roux, A.; Bavai, C.; Ray-Gueniot, S.; et al. Prevalence of Verotoxin-Producing *Escherichia coli* (VTEC) and *E. coli* O157:H7 in French Pork. *J. Appl. Microbiol.* **2002,** *93,* 7–14.

Buncic, S.; Alban, L; Blagojević, B. From Traditional Meat Inspection to Development of Meat Safety Assurance Programs in Pig Abattoirs—the European Situation. *Food Control.* **2019,** *106,* 1–9.

Butzler, J. P.; Skirrow, M. B. *Campylobacter* enteritis. *Clin. Gastroenterol.* **1979,** *8,* 737–765.

Carvalheiro, C. G.; Mussi-Pinhata, M. M.; Yamamoto, A. Y.; Souza, C. B. S. D.; Maciel, L. M. Z. Incidence of Congenital Toxoplasmosis Estimated by Neonatal Screening: Relevance of Diagnostic Confirmation in Asymptomatic Newborn Infants. *Epidemiol. Infect.* **2005,** *133,* 485–491.

CDC, CDC Estimates of Foodborne Illness in the United States. Center for Disease Control [Online] 2011. http://www.cdc.gov/foodborneburden/2011-foodborne-estimates.html

Centers for Disease Prevention and Control. Parasites—Trichinellosis (also known as trichinosis) [Online], 2019. https://www.cdc.gov/parasites/trichinellosis/biology.html (accessed May 17, 2022).

Chan, G.; Farzan, A.; Soltes, G.; Nicholson, V. M.; Pei, Y.; Friendship, R.; et al. The Epidemiology of *Clostridium perfringens* Type A on Ontario Swine Farms, with Special Reference to cpb2-Positive Isolates. *BMC Vet Res*. **2012,** *8,* 1–8.

Chattopadhyay, B. Group R Streptococcal Infection amongst Pig Meat Handlers. A Review. *Public Health.* **1979,** *93,* 140–142.

Christou, L.; Kosmidou, M. Hepatitis E Virus in the Western World—A Pork-Related Zoonosis. *Clin. Microbiol. Infect.* **2013,** *19,* 600–604.

Cliver, D. O. *Parasites. Foodborne Diseases*; Academic Press Inc.: San Diego, Calif., 1990; PP 296–297.

Colello, R.; Cáceres, M. E.; Ruiz, M. J.; Sanz, M.; Etcheverría, A. I.; Padola, N. L. From Farm to Table: Follow-up of Shiga Toxin-Producing *Escherichia coli* Throughout the Pork Production Chain in Argentina. *Front. Microbiol.* **2016,** *7,* 1–7.

Conedera, G.; Mattiazzi, E.; Russo, F.; Chiesa, E.; Scorzato, I.; Grandesso, S.; et al. A Family Outbreak of *Escherichia coli* O157 Haemorrhagic Colitis Caused by Pork Meat Salami. *Epidemiol Infect.* **2007,** *135,* 311–314. doi: 10.1017/s0950268806006807. PMID: 17291366; PMCID: PMC2870571.

Cui, J.; Wang, Q.; Hu, D. S. The Epidemiology of Swine Trichinellosis in China during 1999–2004. *Helminthologia.* **2006,** *43,* 21–26.

Curry, S. R.; Marsh, J. W.; Schlackman, J. L.; Harrison, L. H. Prevalence of Clostridium Difficile in Uncooked Ground Meat Products from Pittsburgh, Pennsylvania. *Appl. Environ. Microbiol.* **2012,** *78,* 4183–4186.

Davies P. R. Pork Safety: Achievements and Challenges. *Zoonoses Public Health* **2010,** *57,* 1–5.

Davies, P. R. Intensive Swine Production and Pork Safety. *Foodborne Pathog Dis.* 2011 *8,*189–201. Doi: 10.1089/Fpd.2010.0717.

Davies, P. R.; Hurd, H. S.; Funk, J. A.; Fedorka-Cray, P. J.; Jones, F. T. The Role of Contaminated Feed in the Epidemiology and Control of *Salmonella enterica* in Pork Production. *Foodborne Pathog. Dis.* **2004,** *1,* 202–215.

De Boer, E.; Zwartkruis-Nahuis, J. T.; Wit, B.; Huijsdens, X. W.; De Neeling, A. J.; Bosch, T.; et al. Prevalence of Methicillin-Resistant *Staphylococcus aureus* in Meat. *Int. J. Food Microb.* **2009,** *134,* 52–56.

de Valk, H.; Vaillant, V.; Jacquet, C.; Le Querrec, F.; Stainer, F.; Quelquejeu, N.; Pierre, O.; Pierre, V.; Desenclos, J. C.; Goulet, V. Two Consecutive Nationwide Outbreaks of Listeriosis in France, October 1999–February 2000. *Am. J. Epidemiol.* **2001,** *154,* 944–995.

Delgado, C. L.; Rosegrant, M. W.; Meijer, S. *Livestock to 2020: The Revolution Continues*; Intl. Trade Research Consortium (IATRC); Auckland, New Zealand, 2001.

Delsart, M.; Pol, F.; Dufour, B.; Rose, N.; Fablet, C. Pig Farming in Alternative Systems: Strengths and Challenges in Terms of Animal Welfare, Biosecurity, Animal Health and Pork Safety. *Agriculture* **2020,** *10,* 1–34. doi: 10.3390/agriculture10070261

Dickson, J.; Hurd, H. S.; Rostagno, M. H. Salmonella in the Pork Production Chain. Natl. Pork Board/Pork Information Gateway Factsheet [Online], 2002. http://www.pork.org/filelibrary/factsheets/porksafety/pork%20production%20chain.pdf

Djurković-Djaković, O.; Bobić, B.; Nikolić, A.; Klun, I, Dupouy-Camet J. Pork as a Source of Human Parasitic Infection. *Clin. Microbiol. Infect.* **2013,** *19,* 586–594.

Dorr, P.; Tadesse, D. A.; Zewde, B.; Fry, P.; Thakur, S.; Gebreyes, W. A. Longitudinal Study of *Salmonella* Dispersion and the Role of Environmental Contamination in Commercial Swine Production Systems. *Appl. Environ. Microbiol.* **2009,** *75,* 1478–1486.

Doyle, M. P. *Campylobacter jejuni. Foodborne Diseases*; Academic Press Inc.; San Diego, California, 1990; pp 217–228.

Doyle, M. P.; Cliver, D. O. *Salmonella. Foodborne Diseases*; Academic Press Inc.: San Diego, California, 1990; pp 185–204.

Dubey, J. P.; Gamble, H. R.; Hill, D.; Sreekumar, C.; Romand, S.; Thulliez, P. High Prevalence of Viable *Toxoplasma gondii* Infection in Market Weight Pigs from a Farm in Massachusetts. *J. Parasitol.* **2002,** *88,* 1234–1238.

Dubey, J. P.; Hill, D. E.; Jones, J. L.; Hightower, A. W.; Kirkland, E.; Roberts, J. M.; et al. Prevalence of Viable *Toxoplasma gondii in* Beef, Chicken and Pork from Retail Meat Stores in the United States: Risk Assessment to Consumers. *J. Parsitol.* **2005,** *91,* 1082–1093.

Dubey, J. P.; Kotula, A. W.; Sharar, A.; Andrews, C. D.; Lindsay, D. S. Effect of High Temperature on Infectivity of *Toxoplasma gondii* Tissue Cysts in Pork. *J. Parsitol.* **1990,** *76,* 201–204.

Dubey, J. P.; Murrell, K. D.; Fayer, R. Persistence of encysted *Toxoplasma gondii in* tissues of pigs fed oocysts. *Am. J. Vet. Res.* **1984,** *45,* 1941–1943.

Dutkiewicz, J.; Sroka, J.; Zając, V.; Wasiński, B.; Cisak, E.; Sawczyn, A.; Kloc, A.; Wójcik-Fatla, A. *Streptococcus suis*: A Re-emerging Pathogen Associated with Occupational Exposure to Pigs or Pork Products. Part I—Epidemiology. *Ann. Agric. Environ. Med.* **2017,** *24,* 683–695.

Eddi, C.; Nari, A.; Amanfu, W. *Taenia solium* Cysticercosis/Taeniosis: Potential Linkage With FAO Activities; FAO Support Possibilities. *Acta Trop.* **2003,** *87,* 145–148.

EFSA. The European Union One Health 2018 Zoonoses Report. *EFSA J.* **2019,** *17,* 1–275

Enemark, H. L.; Johansen, M. V.; Alban, L. Transmission Dynamics of Foodborne Parasites in Pork (Pig and Wild Boar). In *Parasite Transmission Dynamics in Food Sources*; Gajadhar, A., Ed.; Woodhead Publishing Ltd, 2015; pp 259–292.

Etcheverría, A.; Padola, N.; Sanz, M.; Polifroni, R.; Krüger, A.; Passucci, J.; et al. Occurrence of Shiga Toxin-Producing *E. coli* (STEC) on Carcasses and Retail Beef Cuts in the Marketing Chain of Beef in Argentina. *Meat Sci.* **2010,** *86,* 418–421.

European Food Safety Authority and European Centre for Disease Prevention and Control (EFSA and ECDC). The European Union One Health 2018 Zoonoses Report. *EFSA J.* **2019,** *17,* E05926.

FAO. Food Outlook—Biannual Report on Global Food Markets (Rome: FAO) [Online] 2021. http://www.fao.org/3/cb4479en/cb4479en.pdf

Farber, J. M.; Peterkin, P. I. *Listeria monocytogenes*, a Food-Borne Pathogen. *Microbiol. Rev.* **1991,** *55,* 476–511.

Fenlon, D. R.; Wilson, J.; Donachie, W. The Incidence and Level of *Listeria monocytogenes* Contamination of Food Sources at Primary Production and Initial Processing. *J. Appl. Bacteriol.* **1996,** *81,* 641–50.

Fondrevez, M.; Minvielle, B.; Labbé, A.; Houdayer, C.; Rose, N.; Esnault, E.; Denis, M. Prevalence of Pathogenic *Yersinia enterocolitica* in Slaughter-Aged Pigs During a One-Year Survey 2010–2011 France. *Int. J. Food Microbiol.* **2014,** *174,* 56–62.

Foroutan-Rad, M.; Majidiani, H.; Dalvand, S.; Daryani, A.; Kooti, W.; Saki, J.; et al. Toxoplasmosis in Blood Donors: A Systematic Review and Meta-Analysis. *Transfus. Med. Rev.* *30,* 116–122.

Fosse, J.; Seegers, H.; Magras, C. Foodborne Zoonoses Due to Meat: A Quantitative Approach for A Comparative Risk Assessment Applied to Pig Slaughtering in Europe. *Vet. Res.* **2008,** *39,* 1–16.

Fosse, J.; Seegers, H.; Magras, C. Prevalence and Risk Factors for Bacterial Food-Borne Zoonotic Hazards in Slaughter Pigs: A Review. *Zoonoses Public Health* **2009,** *56,* 429–454.

Fredriksson-Ahomaa, M.; London, L.; Skrzypczak, T.; Kantala, T.; Laamanen, I.; et al. Foodborne Zoonoses Common in Hunted Wild Boars. *Ecohealth* **2020,** *17,* 512–522.

FSIS. Nationwide Pork Microbiological Baseline Data Collection Program: Market Hogs (April 1995-March 1996). Food Safety and Inspection Service [Online] 1996. http://www.fsis.usda.gov/science/baseline_data_market_hogs/index.asp

Gebreyes, W. A.; Bahnson, P. B.; Funk, J. A.; Mckean, J. D.; Patchanee, P. Seroprevalence of *Trichinella*, *Toxoplasma*, and *Salmonella in* Antimicrobial-Free and Conventional Swine Production Systems. *Foodborne Pathog. Dis.* **2008,** *5,* 199–203.

Geng, Y.; Wang, C.; Zhao, C.; Yu, X.; Harrison, T. J.; Tian, K.; et al. Serological Prevalence of Hepatitis E Virus in Domestic Animals and Diversity of Genotype 4 Hepatitis E Virus in China. *Vector Borne Zoonotic Dis.* **2010,** *10,* 765–770.

Gonzalez, A. E.; García, H. H.; Gilman, R. H.; Tsang, V. C. Control of *Taenia solium*. *Acta Trop.* **2003,** *87,* 103–109.

Greig, J. D.; Ravel, A. Analysis of Foodborne Outbreak Data Reported Internationally for Source Attribution. *Int. J. Food Microbiol.* **2009,** *130,* 77–87.

Grutsch, A. A.; Nimmer, P. S.; Pittsley, R. H.; Kornilow, K. G.; Mckillip, J. L. Molecular Pathogenesis of Bacillus spp., with Emphasis on the Dairy Industry. *Fine Focus*. **2018,** *4,* 203–222.

Guibourdenche, M.; Roggentin, P.; Mikoleit, M.; Fields, P. I.; Bockemuhl, J.; Grimont, P. A.; et al. Supplement 2003–2007 (No. 47) to the White-Kauffmann-Le Minor Scheme. *Res. Microbiol.* **2010,** 26–29. PMID:19840847.

Gupta, S. K.; Sharma, P.; Mcmillan, E. A.; Jackson, C. R.; Hiott, L. M.; Woodley, T.; et al. Genomic Comparison of Diverse Salmonella Serovars Isolated From Swine. *PLoS One* **2019,** *14,* 1–24.

Hakim, M. S.; Nirwati, H.; Aman, A. T.; Soenarto, Y.; Pan, Q. Significance of Continuous Rotavirus and Norovirus Surveillance in Indonesia. *World J. Pediatr.* **2018,** *14,* 4–12.

Hanninen, M. L.; Perko-Makela, P.; Pitkala, A; Rautelin, H. A Three Year Study of *Campylobacter jejuni* Genotypes in Humans with Domestically Acquired Infections and in Chicken Samples from the Helsinki Area. *J. Clin. Microbiol.* **2000,** *38,* 1998–2000.

Hellstrom, S.; Laukkanen, R.; Siekkinen, K. M.; Ranta, J.; Maijala, R.; Korkeala, H. *Listeria monocytogenes* can Originate from Farms. *J. Food Prot.* **2010,** *73,* 641–648.

Hill, D. E.; Haley, C.; Wagner, B.; Gamble, H. R.; Dubey, J. P. Seroprevalence of and Risk Factors for *Toxoplasma gondii* in the US Swine Herd Using Sera Collected During the National Animal Health Monitoring Survey (Swine 2006). *Zoonoses Public Health.* **2010,** *57,* 53–59.

Jakobsen, A. M.; Bahl, M. I.; Buschhardt, T.; Hansen, T. B.; Al-Soud, W. A.; Brejnrod, A. D.; Sørensen, S. J.; Nesbakken, T.; Aabo, S. Bacterial Community Analysis for Investigating Bacterial Transfer from Tonsils to the Pig Carcass. *Int. J. Food Microbiol.* **2019,** *295,* 8–18.

Kaferstein, F. K.; Motarjemi, Y.; Bettcher, D.W. Foodborne Disease Control: A Transnational Challenge. *Emerg. Infect. Dis.* **1997,** *3,* 503–510.

Kamar, N.; Selves, J.; Mansuy, J. M.; Quezzani, L.; Peron, J. M.; Guitard, J.; et al. Hepatitis E Virus and Chronic Hepatitis in Organ-transplant Recipients. *N. Engl. J. Med.* **2008,** *358,* 811–817.

Kanuganti, S. R.; Wesley, I. V.; Reddy, P. G.; Mckean, J.; Hurd, H. S. Detection of *Listeria monocytogenes* in Pigs and Pork. *J. Food Prot.* **2002,** *65,* 1470–1474.

Karabasil, N.; Boskovic, T.; Vasilev, D.; Betic, N.; Dimitrijevic, M. Pork Safety—Challenges and Opportunities. *IOP Conf. Ser.: Earth Environ. Sci.* **2021,** *854,* 1–5.

Karmali, M. A.; Penner, J. L.; Fleming, P. C.; Williams, A.; Hennessy, J. N. The Serotype and Biotype Distribution of Clinical Isolates of *Campylobacter jejuni* and *Campylobacter coli* Over A Three-Year Period. *J. Infect. Dis.* **1983,** *147,* 243–246.

Kaufmann, M.; Zweifel, C.; Blanco, M.; Blanco, J.; Blanco, J.; Beutin, L.; et al. *Escherichia coli* O157 and non-O157 Shiga Toxin–Producing *Escherichia coli* in Fecal Samples of Finished Pigs at Slaughter in Switzerland. *J. Food Prot.* **2006,** *69,* 260–266. doi: 10.4315/0362–028x.jfp-10-579.

Kelley, B. R.; Ellis, J. C.; Large, A.; Schneider, L. G.; Jacobson, D.; Johnson, J. G. Whole-Genome Sequencing and Bioinformatic Analysis of Environmental, Agricultural, and Human *Campylobacter jejuni* Isolates From East Tennessee. *Front. Microbiol.* **2020,** *11,* 1–17.

Kennedy, E. D.; Hall, R. L.; Montgomery, S. P.; Pyburn, D. G.; Jones, J. L. Trichinellosis Surveillance: United States, 2002–2007. *MMWR Surveill Summ.* **2009,** *58,* 1–7.

Kolackova, I.; Hazova, K.; Skockova, A.; Karpiskova, R. Occurrence of Shiga Toxigenic *Escherichia coli* Strains in Pigs and Cattle at Slaughterhouses in the Czech Republic in 2013. *Klin. Mikrobiol. Infekc. Lek.* **2014,** *20,* 36–39.

Koohmaraie, M.; Arthur, T.; Bosilevac, J.; Guerini, M.; Shackelford, S.; Wheeler, T. Post-harvest Interventions to Reduce/Eliminate Pathogens in Beef. *Meat Sci.* **2005,** *71,* 79–91.

Korsak, N.; Jacob, B.; Groven, B.; Etienne, G.; China, B.; Ghafir, Y.; Daube, G. Salmonella Contamination of Pigs and Pork in an Integrated Pig Production System. *J. Food Prot.* **2003,** *66,* 1126–1133.

Kranker, S.; Alban, L.; Boes, J.; Dahl, J. Longitudinal Study of *Salmonella enterica* Serotype Typhimurium Infection in Three Danish Farrow-to-Finish Swine Herds. *J. Clin. Microbiol.* **2003,** *41,* 2282–2288.

Lam, S.; Nguyen, H. T. T.; Nguyen-Viet, H. Unger, F. Mapping Pathways toward Safer Pork in Vietnam. *ILRI Policy Brief* [Online] 2020, *95.* https://cgspace.cgiar.org/bitstream/handle/10568/108768/research_brief_95.pdf?sequence=5&isallowed=y (accessed May 13, 2022).

Laukkanen, R.; Martínez, P. O.; Siekkinen, K. M.; Ranta, J.; Maijala, R.; Korkeala, H. Contamination of Carcasses with Human Pathogenic *Yersinia enterocolitica* 4/O:3 Originates from Pigs Infected on Farms. *Foodborne Pathog. Dis.* **2009,** *6,* 681–688.

Leung, P.; Yam, W.; Ng, W.; Peiris, J. The Prevalence and Characterization of Verotoxin-Producing *Escherichia coli* Isolated from Cattle and Pigs in an Abattoir in Hong Kong. *Epidemiol. Infect.* **2001,** *126,* 173–179.

Li, P.; Ji, Y.; Li, Y.; Ma, Z.; Pan, Q. Estimating the Global Prevalence of Hepatitis E Virus in Swine and Pork Products. *One Health* 2022; *14,* 100362.

Li, P.; Liu, J.; Li, Y.; Su, J.; Ma, Z.; Bramer, W. M.; et al. The Global Epidemiology of Hepatitis E Virus Infection: A Systematic Review and Meta-analysis, *Liver Int.* **2020,** *40,* 1516–1528.

Logue, C. M.; Barbieri, N. L.; Nielsen, D. W. Pathogens of Food Animals: Sources, Characteristics, Human Risk, and Methods of Detection. *Adv. Food Nutr. Res.* **2017,** *82,* 277–365. doi: 10.1016/bs.afnr.2016.12.009.

Lun, Z. R.; Wang, Q. P.; Chen, X. G.; Li, A. X.; Zhu, X. Q. *Streptococcus suis*: An Emerging Zoonotic Pathogen. *Lancet. Infect. Dis.* **2007,** *7,* 201–209.

Lunden, A.; Uggla, A. Infectivity of *Toxoplasma gondii in* Mutton Following Curing, Smoking, Freezing or Microwave Cooking. *Int. J. Food. Microbiol.* **1992,** *15,* 357–363.

Martins, B. T. F.; De Azevedo, E. C.; Yamatogi, R. S.; Call, D. R.; Nero, L. A. Persistence of *Yersinia enterocolitica* Bio-Serotype 4/O:3 in A Pork Production Chain in Minas Gerais, Brazil. *Food Microbiol.* **2020,** *94,* 103660.

Matsuda, H.; Okada, K.; Takahashi, K.; Mishiro, S. Severe Hepatitis E Virus Infection after Ingestion of Uncooked Liver from a Wild Boar. *J. Infect. Dis.* **2003,** *188,* 944.

Mead, P. S.; Slutsker, L.; Dietz, V.; Mccaig, L. F.; Bresee, J. S.; Shapiro, C.; et al. Food-Related Illness and Death in the United States. *Emerg. Infect. Dis.* **1999,** *5,* 607–625.

Meng, Q.; Bai, X.; Zhao, A.; Lan, R.; Du, H.; Wang, T.; et al. Characterization of Shiga Toxin-Producing *Escherichia coli* Isolated from Healthy Pigs in China. *BMC Microbiol.* **2014,** *14,* 1–14.

Metcalf, D.; Reid-Smith, R. J.; Avery, B. P.; Weese, J. S. Prevalence of *Clostridium difficile* in Retail Pork. *Can. Vet. J.* **2010,** *51,* 873–876.

Milton, A. A. P.; Priya, G. B.; Angappan, M.; Momin, K. M.; Ghatak, S.; Kannan, P. Campylobacteriosis: Emerging Foodborne Zoonosis. In *Food Product Optimization For Quality and Safety Control: Process, Monitoring, and Standards*, 1st ed.; **Contreras-Esquivel,** J. C., **Badwaik, L. S.,** Kannan, **P.,** Hagi, A. K., Eds.; CRC Press: United States, 2020a; pp165–187.

Milton, A. A. P.; Priya, G. B.; Momin, K. M.; Angappan, M.; Ghatak, S.; Kannan, P. Foodborne Pathogenic Anaerobes. In *Food Product Optimization For Quality and Safety Control:*

Process, Monitoring, and Standards, 1st ed.; **Contreras-Esquivel,** J. C., **Badwaik, L. S.,** Kannan, **P.,** Hagi, A. K., Eds.; CRC Press: United States, 2020b; pp 31–52.

Montville, T. J.; Matthews, K. R. *Food Microbiology: An Introduction*, 2nd ed.; ASM Press: Washington, D.C.; 2008.

Mooyottu, S.; Flock, G.; Kollanoor-Johny, A.; Upadhyaya, I.; Jayarao, B.; Venkitanarayanan, K. Characterization of a Multidrug Resistant *C. difficile* Meat Isolate. *Int. J. Food Microbiol.* **2015,** *192,* 111–116.

Murrell, K. D.; Pozio, E. Trichinellosis: the Zoonosis that Won't Go Quietly. *Int. J. Parasitol.* **2000,** *30,* 1339–1349.

Nesbakken, T.; Skjerve, E. Interruption of Microbial Cycles in Farm Animals from Farm to Table. *Meat Sci.* **1996,** *43,* S47–S57.

Niewerth, U.; Frey, A.; Voss, T.; Le Bouguénec, C.; Baljer, G.; Franke, S.; et al. The AIDA Autotransporter System is Associated With F18 and stx2e in Escherichia Coli Isolates from Pigs Diagnosed with Edema Disease and Postweaning Diarrhea. *Clin. Diagn. Lab. Immunol.* **2001,** *8,* 143–149.

Noormohamed, A.; Fakhr, M. K. A Higher Prevalence Rate of Campylobacter in Retail Beef Livers Compared to Other Beef and Pork Meat Cuts. *Int. J. Environ. Res. Public Health.* **2013,** *10,* 2058–2068. doi: 10.3390/ijerph10052058

Noskin, G. A.; Rubin, R. J.; Schentag, J. J.; Kluytmans, J.; Hedblom, E. C.; Jacobson, C.; Smulders, M.; Gemmen, E.; Bharmal, M. National Trends in *Staphylococcus aureus* Infection Rates: Impact on Economic Burden and Mortality Over a 6-Year Period (1998–2003). *Clin. Infect. Dis.* **2007,** *45,* 1132–1140.

O'Brien, A. M.; Hanson, B. M.; Farina, S. A.; Wu, J. Y.; Simmering, J. E., Wardyn, S. E.; et al. MRSA in Conventional and Alternative Retail Pork Products. *PLoS One*. **2012,** *7,* 1–6.

Oliver, S. P.; Patel, D. A.; Callaway, T. R.; Torrence, M. E. Developments and Future Outlook for Preharvest Food Safety. *J. Anim. Sci.* **2009,** *87,* 419–437.

Parashar, U. D.; Hummelman, E. G.; Bresee, J. S.; Miller, M. A.; Glass, R. I. Global Illness and Deaths Caused by Rotavirus Disease in Children. *Emerg. Infect. Dis.* **2003,** *9,* 565–572.

Parma, A.; Sanz, M.; Blanco, J.; Blanco, J.; Viñas, M.; Blanco, M.; et al. Virulence Genotypes and Serotypes of Verotoxigenic *Escherichia coli* Isolated from Cattle and Foods in Argentina. *Eur. J. Epidemiol.* **2000,** *16,* 757–762.

Paton, J. C.; Paton, A. W. Pathogenesis and Diagnosis of Shiga Toxin-Producing *Escherichia coli* Infections. *Clin. Microbiol. Rev.* **1998,** *11,* 450–479.

Pawlowski, Z.; Allan, J.; Sarti, E. Control of *Taenia solium* Taeniasis/Cysticercosis: from Research Towards Implementation. *Int. J. Parasitol.* **2005,** *35,* 1221–1232.

Pearce, R. A.; Wallace, F. M.; Call, J. E.; Dudley, R. L, Oser, A.; Yoder, L.; Sheridan, J. J.; Luchansky, J. B. Prevalence of *Campylobacter* Within a Swine Slaughter and Processing Facility. *J. Food Prot.* **2003,** *66,* 1550–1556.

Phiri, I. K.; Sikasunge, C. S.; Siziya, S.; Gabriel, S.; Dorny, P.; Willingham, A. L. 3rd. *The Prevalence and Risk Factors of Porcine Cysticercosis in Zambia, Proceedings of the 5th International Symposium On the Epidemiology and Control of Foodborne Pathogens in Pork*, Heraklion, Greece, 2003, pp 68–69.

Pires, S. M.; Fischer-Walker, C. L.; Lanata, C. F.; Devleesschauwer, B.; Hall, A. J.; Kirk, M. D.; et al. Aetiology-Specific Estimates of the Global and Regional Incidence and Mortality of Diarrhoeal Diseases Commonly Transmitted Through Food. PLoS One. **2015,** *10,* 1–17.

Pires, S. M.; Vieira, A. R.; Perez, E.; Lo Fo Wong, D.; Hald, T. Attributing Human Foodborne Illness to Food Sources and Water in Latin America and the Caribbean Using Data from Outbreak Investigations. *Int. J. Food Microbiol.* **2012,** *152,* 129–138.

Pozio, E.; Kapel, C. M.; Gajadhar, A. A.; Boireau, P.; Dupouy-Camet, J.; Gamble, H. R. Trichinella in Pork: Current Knowledge on the Suitability of Freezing as a Public Health Measure. *Euro. Surveill.* **2006,** *11,* 3079.

Quintana-Hayashi, M. P.; Thakur, S. Longitudinal Study of the Persistence of Antimicrobial-Resistant *Campylobacter* Strains in Distinct Swine Production Systems on Farms, at Slaughter, and in the Environment. *Appl. Environ. Microbiol.* **2012,** *78,* 2698–2705.

Rajshekhar, V.; Raghava, M. V.; Prabhakaran, V.; Oommen, A.; Muliyil, J. Active Epilepsy as an Index of Burden of Neurocysticercosis in Vellore District, India. *Neurology.* **2006,** *67,* 2135–2139.

Rho, M. J.; Chung, M. S.; Lee, J. H.; Park, J. Monitoring of Microbial Hazards at Farms, Slaughterhouses, and Processing Lines of Swine in Korea. *J. Food Prot.* **2001,** *64,* 1388–1391.

Rodriguez, C.; Avesani, V.; Van Broeck, J.; Taminiau, B.; Delmée, M.; Daube, G. Presence of *Clostridium difficile* in Pigs and Cattle Intestinal Contents and Carcass Contamination at the Slaughterhouse in Belgium. *Int. J. Food Microbiol.* **2013,** *166,* 256–262.

Rowe, T.; Leonard, F. C.; Kelly, G.; Lynch, P.; Egan, J.; Quirke, A.; Quinn P. Salmonella Serotypes Present on a Sample of Irish Pig Farms. *Vet. Rec.* **2003,** *153,* 453–456.

Salajpal, K.; Karolyi, D.; Luković, Z. Sanitary Aspects of Outdoor Farming Systems. In *Proceedings of the Acta Argiculturae Slovenica*; Slovenia: Ljubljana, 109–117.

Savariraj, W. R.; Ravindran, N. B.; Kannan, P.; Paramasivam, R.; Senthilkumar, T. M. A.; Kumarasamy, P.; et al. Prevalence, Antimicrobial Susceptibility and Virulence Genes of *Staphylococcus aureus* Isolated from Pork Meat in Retail Outlets in India. *J. Food Saf.* **2018,** *39,* 1–8.

Scallan, E.; Hoekstra, R. M.; Angulo, F. J.; Tauxe, R. V.; Widdowson, M. A.; Roy, S. L.; et al. Foodborne Illness Acquired in the United States—Major Pathogens. *Emerg. Infect. Dis.* **2011,** *17,* 7–15.

Schirone, M.; Visciano, P.; Tofalo, R.; Suzzi, G. Editorial: Foodborne Pathogens: Hygiene and Safety. *Front. Microbiol.* **2019,** *10,* 1974.

Schlech, W. F. Foodborne Listeriosis. *Clin Infect Dis,* **2000,** *31,* 770–775.

Schmidt, J.; Brichta-Harhay, D.; Kalchayanand, N.; Bosilevac, J.; Shackelford, S.; Wheeler, T.; Koohmaraie, M. Prevalence, Enumeration, Serotypes, and Antimicrobial Resistance Phenotypes of *Salmonella enterica* Isolates from Carcasses at Two Large United States Pork Processing Plants. *Appl. Environ. Microbiol.* **2012,** *78,* 2716–2726.

Schuppers, M. E.; Stephan, R.; Ledergerber, U.; Danuser, J.; Bissig-Choisat, B.; Stark, K. D. C.; Regula, G.. Clinical Herd Health, Farm Management and Antimicrobial Resistance in *Campylobacter coli* on Finishing Pig Farms in Switzerland. *Prev. Vet. Med.* **2005,** *69,*189–202.

Segura, M. *Streptococcus suis* Research: Progress and Challenges. *Pathogens.* **2020,** *9,* 1–8.

Sher, A.; Tosh, K.; Jankovic, D. Innate Recognition of *Toxoplasma gondii* in Humans Involves A Mechanism Distinct from that Utilized by Rodents. *Cell. Mol. Immunol.* **2016,** *14,* 36–42.

Smith, J. L.; Fratamico, P. M. Emerging and Re-emerging Foodborne Pathogens. *Foodborne Pathog. Dis.* **2018,** *15,* 737–757.

Soares, V. M.; Dos Santos, E. A. R.; Tadielo, L. E.; Cerqueira-Cézar, C. K.; Da Cruz Encide Sampaio, A. N.; Eisen, A. K. A.; et al. Detection of Adenovirus, Rotavirus, and Hepatitis E

Virus in Meat Cuts Marketed in Uruguaiana, Rio Grande Do Sul, Brazil. *One Health.* **2022,** *14,* 100377.

Songer, J. G.; Uzal, F. A. Clostridial Enteric Infections in Pigs. *J. Vet. Diagn. Invest.* **2005,** *17,* 528–536.

Sonntag, A.-K.; Bielaszewska, M.; Mellmann, A.; Dierksen, N.; Schierack, P.; Wieler, L. H.; et al. Shiga Toxin 2e-Producing *Escherichia coli* Isolates From Humans and Pigs Differ in Their Virulence Profiles and Interactions With Intestinal Epithelial Cells. *Appl. Environ. Microbiol.* **2005,** *71,* 8855–8863.

Tack, D. M.; Marder, E. P.; Griffin, P. M.; Cieslak, P. R.; Dunn, J., Hurd, S.; et al. Preliminary Incidence and Trends of Infections with Pathogens Transmitted Commonly Through Food—Foodborne Diseases Active Surveillance Network, 10 U.S. Sites, 2015–2018. *Mortal. Week. Rep.* **2019,** *68,* 369–373.

Thiem, V. D.; Sethabutr, O.; Von, S. L.; Van, T. B.; Chien, T.; Lee, H.; et al. Detection of Shigella by A PCR Assay Targeting the IpaH Gene Suggests Increased Prevalence of Shigellosis in Nha Trang, Vietnam. *J. Clin. Microbiol.* **2004,** *42,* 2031–2035.

Trotz-Williams, L. A.; Mercer, N. J.; Walters, J. M.; Maki, A. M.; Johnson, R. P. Pork Implicated in A Shiga Toxin-Producing *Escherichia coli* O157: H7 Outbreak in Ontario, Canada. *Can. J. Public Health.* **2012,** *103,* E322–E326.

Van Der Giessen, J.; Fonville, M.; Bouwknegt, M.; Langelaar, M.; Vollema, A. Seroprevalence of *Trichinella spiralis and Toxoplasma gondii in* Pigs From Different Housing Systems in the Netherlands. *Vet. Parasitol.* **2007,** *148,* 371–374.

Van Loo, I. H. M.; Diederen, B. M. W.; Savelkoul, P. H. M.; Woudenberg, J. H. C.; Roosendaal, R.; Van Belkum, A. Methicillin-Resistant *Staphylococcus aureus* in Meat Products, the Netherlands. *Emerg. Infect. Dis.* **2007,** *13,* 1753–1755.

Vilaichone, R. K.; Vilaichone, W.; Nunthapisud, P.; Wilde, H. *Streptococcus suis* Infection in Thailand. *J. Med. Assoc. Thai.* **2002,** *85,* S109–S117.

Wang, J.; Liu, M.; Wang, H.; Wu, Q.; Xu, T.; Ma, G.; et al. Occurrence, Molecular Characterization, and Antimicrobial Susceptibility of *Yersinia enterocolitica* Isolated from Retail Food Samples in China. *LWT* **2021,** *150,* 111876.

Wangkaew, S.; Chaiwarith, R.; Tharavichitkul, P.; Supparatpinyo, K. *Streptococcus suis* Infection: A Series of 41 Cases from Chiang Mai University Hospital. *J. Infect.* **2006,** *52,* 455–460.

Watkins, E. J.; Brooksby, P.; Schwelger, M. S.; Enright, S. M. Septicaemia in A Pig-Farm Worker. *Lancet,* **2001,** *357,* 38.

Wertheim, H. F.; Nghia, H. D.; Taylor, W.; Schultsz, C. *Streptococcus suis*: An Emerging Human Pathogen. *Clin. Infect. Dis.* **2009,** *48,* 617–625.

WHO. *Food Safety and Foodborne Illnesses. Fact Sheet 237*; WHO's Health & Safety Information Services: Geneva, Revised March 2007, 2009.

Wilkins, W.; Rajic, A.; Waldner, C.; Mcfall, M.; Chow, E.; Muckle, A.; Rosengren L. Distribution of *Salmonella* Serovars in Breeding, Nursery, and Grow-to-Finish Pig, and Risk Factors for Shedding in Ten Farrow-to-Finish Swine Farms in Alberta and Saskatchewan. *Can. J. Vet. Res.* **2010,** *74,* 81–90.

Wu, L. H.; Wang, S. X.; Zhu, D.; Hu, W. Y.; Wang, H. S. Chinese Consumers' Preferences and Willingness to Pay for Traceable Food Quality and Safety Attributes: the Case of Pork. *China Econ. Rev.* **2015,** *35,* 121–136.

Wu, Y. C.; Chen, C. M.; Kuo, C. J.; Lee, J. J.; Chen, P. C.; Chang, Y. C.; Chen, T. H. Prevalence and Molecular Characterization of *Clostridium difficile* Isolates from a Pig Slaughterhouse, Pork, and Humans in Taiwan. *Int. J. Food Microbiol.* **2017,** *242,* 37–44.

Yaeger, M.; Funk, N.; Hoffmann, L. A Survey of Agents Associated with Neonatal Diarrhea in Iowa Swine Including *Clostridium difficile* and Porcine Reproductive and Respiratory Syndrome Virus. *J. Vet. Diagn. Invest.* **2002,** *14,* 281–287.

Yazaki, Y.; Mizuo, H.; Takahashi, M.; Nishizawa, T.; Sasaki, N.; Gotanda, Y.; et al. Sporadic Acute or Fulminant Hepatitis E in Hokkaido, Japan, may be Food-Borne, as Suggested by the Presence of Hepatitis E Virus in Pig Liver as Food. *J. Gen. Virol.* **2003,** *84,* 2351–2357.

Zdolec, N.; Kiš, M. Meat Safety from Farm to Slaughter—Risk-Based Control of *Yersinia enterocolitica and Toxoplasma gondii. Processes* **2021,** *9,* 1–13.

Zhou, J. H.; Li, X. R.; Lan, X.; Han, S. Y.; Wang, Y. N.; Hu, Y.; et al. The Genetic Divergences of Codon Usage Shed New Lights on Transmission of Hepatitis E Virus from Swine to Human. *Infect. Genet. Evol.* **2019,** *68,* 23–29.

SECTION 3
CASE STUDIES

CHAPTER 10

Functional Properties of Agroindustrial Wastes from Pomegranate Peel (*Punica granatum*)

ANA L. IZÁBAL-CARVAJAL[1], CRISTIAN TORRES-LEÓN[2], LEONARDO SEPÚLVEDA[1], MÓNICA L. CHÁVEZ GONZÁLEZ[1], CRISTÓBAL N. AGUILAR[1], and J. A. ASCACIO-VALDÉS[1]

[1]*Bioprocesses and Bioproducts Group, Food Research Department, School of Chemistry, Universidad Autónoma de Coahuila, Saltillo, Coahuila, México*

[2]*Research Center and Ethnobiological Garden, Universidad Autónoma de Coahuila, Unidad Torreón, Viesca, Coahuila, México*

ABSTRACT

Fruits are rich sources of bioactive components, which is mainly attributed to various medicinal properties and beneficial effects on health (Kalaycıoğlu and Erim, 2017). Pomegranate (*Punica granatum* L.) is being investigated because it is a source of various beneficial phytochemicals for health since it contains anthocyanins, ellagitannins, gallotannins, flavonoids, among other polyphenolic compounds (Bonzanini et al., 2009; García-Villalba et al., 2015; N. Seeram et al., 2005; Verotta et al., 2018).

The increase in food processing has greatly contributed to the increase in agro-industrial waste, which is produced in all phases of the agricultural production cycle, such as industrial manufacturing, processing, and distribution (Kumar et al., 2017). In recent years, the use of agroindustrial wastes has become a global concern due to the scarcity of natural resources and environmental protection; following this line of research,

Engineering Principles for Food Processing Technology and Product Realization. Leonardo Sepúlveda Torre, Porteen Kannan, A. K. Haghi, PhD (Eds.)

some groups have carried out studies to value the residues of plant origin (Devesa-Rey et al., 2011). The valorization of these wastes can be accomplished by extracting fibers, flavoring compounds, and phytochemicals (Baiano, 2014).

For this reason, the use of pomegranate peel has been studied, due to its high content of bioactive compounds with biological properties of great relevance (antioxidant, antimicrobial, antiproliferative, etc.).

10.1 INTRODUCTION

According to the botanist N. I. Vavilov, pomegranate (*Punica granatum* L.) is a very ancient plant and fruit first cultivated in Egypt, Italy, and Greece, and then propagated to many regions as Asia, North Africa, and especially grown in Mediterranean countries of Europe (Evreinoff, 1957). Pomegranate (*Punica granatum* L.) is an important fruit crop adaptable to a wide range of agro-climatic conditions. It is a fruit-bearing deciduous shrub or a small tree that belongs to Punicaceae family (Singh et al., 2018). The term *Punica* is the feminized Roman name for Carthage, an ancient city in Tunisia where the best fruits came to Italy. The term *Punica granatum* is derived from Latin word *pōnum*, which means apple, and the word *grānātus* which means granulated. At the United States, the common name means "seedy apple" (Stover and Mercure, 2007). This fruit can be grown in arid areas due to its tolerance to drought conditions. However, the main producers of pomegranate are India, Iran, Turkey China, United States of America, Argentina, Brazil, Chile, Peru, Australia, and Italy (Kahramanoglu, 2019a). Pomegranate has been considered for cultivating in arid zones due to the high tolerance to drought (Kandylis and Kokkinomagoulos, 2020).

Currently, the main producers and exporters of pomegranate in the world are India, Iran, China, Turkey, the United States, South Africa, Spain, Peru, Chile, and Argentina. The world production of this fruit is approximately 3.8 million tons (Kahramanoglu, 2019b). The pomegranate fruit is called by various names depending on the region where it's cultivated, because the color of the rind and the aril can vary considerably depending on the growing region. These differences mainly affect fruit size and peel color (red, yellow, purple, and pink) (Pareek et al., 2015). Figure 10.1 shows the fruit's different parts, which can be divided into some anatomical origins like peel, seeds, and arils (Viuda-Martos et al., 2010).

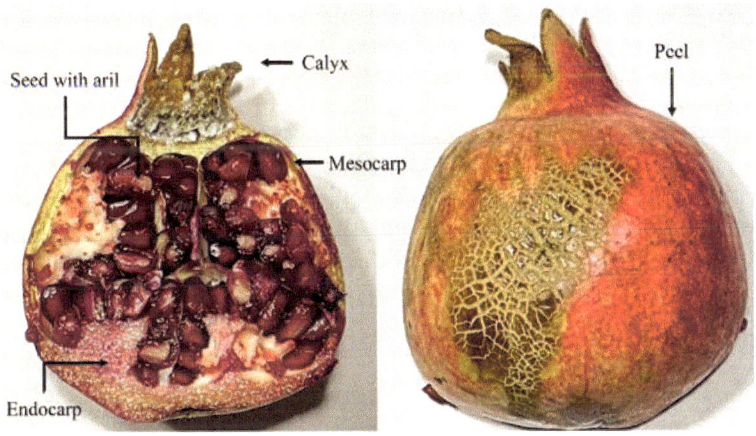

FIGURE 10.1 Pomegranate (*Punica granatum*).

Pomegranate (*Punica granatum* L.) has been investigated for being a dietary source with a high content of various phytochemicals beneficial to health because it contains anthocyanins, ellagitannins, gallotanins, flavonoids, and some other polyphenolic compounds (Bonzanini et al., 2009; García-Villalba et al., 2015; N. Seeram et al., 2005).

Ripe pomegranates are traditionally consumed fresh, because it's difficult to extract the arils from the fruit, for this reason processed fruits such as jellies, syrups, juices, among others, are preferred (Ferrara et al., 2011). Food processing has increased the production of agro-industrial residues or by-products, which are produced throughout the food supply chain such as agricultural production, industrial manufacturing, processing, and distribution (Kumar et al., 2017). In recent years, the use of agro-industrial wastes has become a global concern due to environmental protection; following this, some research groups have carried out studies with the aim to investigate the value of the residues of plant origin (Devesa-Rey et al., 2011), which can be achieved by extracting high-value compounds which can be reused as functional and nutraceutical ingredients, for example, proteins, polysaccharides, fibers, flavoring compounds, and phytochemicals (Baiano, 2014).

Bioactive compounds found in food waste can be extracted and used to develop nutraceuticals and functional foods that (Baiano, 2014) can have beneficial effects on health, such as the prevention of chronic diseases (Kumar et al., 2017).

The aim of this chapter is to study the chemical composition of the pomegranate and the use of agro-industrial wastes as a source of purification of ellagitannins, biological activities, and bioactive compounds as nutraceutical and functional foods.

10.2 CHEMICAL COMPOSITION OF POMEGRANATE

The interior of the pomegranate is composed of 40% arils and 10% seeds. Aryls contain 85% water, 10% total sugars such as fructose and glucose, 1.5% pectin, phenolic compounds, and flavonoids, mainly anthocyanins; organic acids like ascorbic acid, citric acid, and malic acid (Viuda-Martos et al., 2010) (Fig. 10.2). Also, several studies have reported that the main phenolic compounds in pomegranate juice are anthocyanins, while the main phenolic compounds found in the mesocarp and pericarp of the fruit are hydrolyzable tannins (Bonesi et al., 2019), which are classified as ellagitannins and gallatonins, since their hydrolysis produces ellagic acid and gallic acid (Arapitsas, 2012).

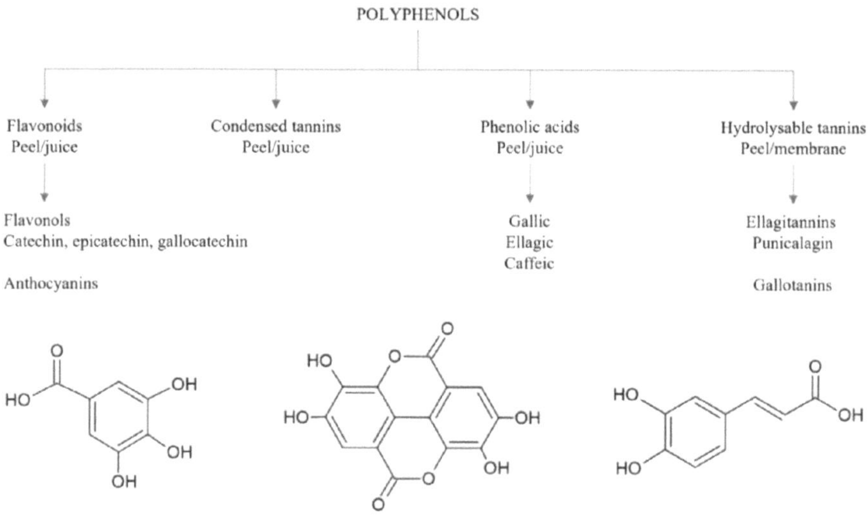

1 2 3

FIGURE 10.2 Major polyphenols found in pomegranate (*Punica granatum*). [Gallic acid (1), ellagic acid (2), caffeic acid (3).]

10.3 PHENOLIC COMPOUNDS IN POMEGRANATE PEEL (PP)

The pomegranate peel (PP) represents approximately 50% of the total weight of the fruit. It is a source of bioactive compounds such as phenols, flavonoids, ellagitannins, proanthocyanidin compounds; minerals such as potassium, nitrogen, calcium, phosphorus, magnesium, and sodium; and complex polysaccharides (Sreekumar et al., 2014). The entire fruit is rich in large polyphenolic compounds like punicalagin isomers, ellagic acid derivatives, and anthocyanins (delphinidin, cyanidin, and pelargonidin 3-gluco-sides and 3,5-diglucosides); however, the peel contains a pool of phenolics (hydrolyzable tannins) compared their concentration in another part of the fruit.

PP is rich in hydrolyzable tannins, mainly punicalin, pedunculagin, and punicalagin (N. Seeram et al., 2005). PP has a high antioxidant value. Several studies have demonstrated the high concentration of phenolic compounds of the peel (Akhtar et al., 2015).

The ellagitannins are one of the main groups of hydrolyzable tannins; they are esters of hexahydroxydiphenic acid (HHDP) bound to a polyol, usually glucose and quinic acid. When exposed to strong acids or bases, the ester bonds hydrolyze, and HHDP spontaneously reorganizes into water-insoluble dilactone, obtaining ellagic acid (2,3,7,8-tetrahydroxchromene [5,4,3-cde] chromene-5,10-dione, according to IUPAC). Ellagic acid is a thermodynamically stable molecule; it has four rings that perform the lipophilic domain, four phenolic groups (hydroxyl), and two lactone groups that represent the hydrophilic part, where they can proceed as donors or acceptors of hydrogen bonds (Aguilera-Carbo et al., 2008a; Bala et al., 2006).

The ellagitannins are found mostly in the pomegranate's pericarp, seeds, and flowers, while the gallotanins are found at the leaves and seldom in other parts of the pomegranate. The hydrolysis of ellagitannins is used to recognize and quantify ellagitannins in food (Brighenti et al., 2017). The punicalagin is the most abundant of the ellagitannins in the pomegranate (Figure 10.3), the total antioxidant activity of the juice is attributed to this compound present in the fruit peels (Tzulker et al., 2007).

10.4 AGRO-INDUSTRIAL WASTES AS A SOURCE OF BIOACTIVE COMPOUNDS

After the extraction of sugar, oil, juice, and starch, there remains some residues such as peelings, stems, trimmings, seeds, bran, and shells. The recovered biocompounds and outgrowth can be applied to produce basic foods (Baiano, 2014).

According to Kumar et al. (2017), the bioactive compounds present in agro-industrial waste can be recovered using various techniques. According

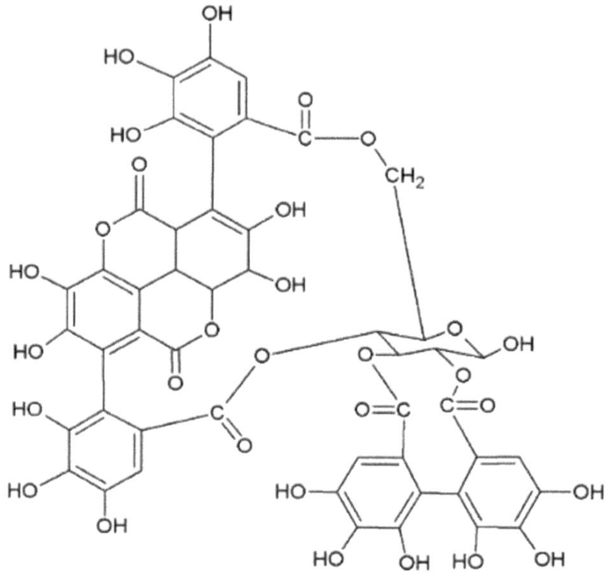

FIGURE 10.3 Punicalagin chemical structure.

to the literature, the extraction methods for bioactive compounds are in basic based on:

- Subcritical water extraction (SWC),
- Supercritical fluid extraction (SFE),
- Solvent extraction (SE),
- Microwaves,
- Ultrasounds, and
- Use pf anzymes

These technique's availability provides an opportunity for the recovery of specific compounds.

The extraction of bioactive compounds, mainly polyphenols, is an alternative for recovering agro-industrial waste. However, the extraction methods currently used have some disadvantages; for example, the use of toxic solvents generates damage to the environment, and green technologies are expensive, complex, and difficult to implement. Also, some traditional techniques use high temperatures that can generate negative effects on the biological activity of the extracts. Moreover, the techniques currently used produce an incomplete release of phenolic compounds, due to the impossibility of recovering the bound polyphenols (Bhanja Dey et al., 2016) through the strong interactions of these compounds with the plant matrix (Torres-León et al., 2019).

Recently emerging techniques for the extraction of ellagitannin have been made since it seeks to reduce the use of polluting solvents and obtain better extraction yields. Emerging technologies include pressurized liquid extraction, ultrasound-assisted extraction, pressurized hot water extraction, simultaneous distillation, microwave-assisted dual extraction, and supercritical fluid extraction (Domínguez-Rodríguez et al., 2017).

Furthermore, solid state fermentation has great potential to obtain bioactive compounds such as phenolic compounds (Martínez-Ávila et al., 2012) being that microorganisms like fungi naturally produce enzymes that degrade the cell wall, generating hydrolysis reactions (Jamal et al., 2011) and the mobilization of compounds unto the extraction solvent (Torres-León et al., 2019). Additionally, this bioprocess is economical and simple to perform; it requires small equipment and a reduction in operating costs because it doesn't need sterilization, aeration, or agitation processes (Soccol et al., 2017).

During these bioprocesses, it must be considered the choice of microorganisms and the substrates, since filamentous fungi are suitable solid state fermentation as the technique simulates their natural habitat: in this condition, they can synthesize considerable amounts of enzymes and some other metabolites (Farinas, 2015). These extracellular enzymes are produced and simultaneously utilized for the extraction/release of phenolics from the matrix of substrates in a single process, along with the production of new phenolic compounds by the secondary metabolism of microorganisms (Bhanja Dey et al., 2016).

This bioprocess allows the use of solid agro-industrial waste as a substrate that is considered a pollutant and produce less wastewater production. The use of agro-industrial waste reduces production costs, and the extracts high biological activities guarantee the bioprocess's economic profitability (Acosta-Estrada et al., 2014).

Ascacio-Valdés et al. (2013) reported that ellagitanase is directly related to the degradation and accumulation of ellagitannins, since the ester bonds between the HHDP group and the glycosides are degraded by this enzyme, which has high specificity.

Vattem and Shetty (2003) mentioned that the β-glucosidase enzyme is associated with the degradation of ellagitannins present in blueberry residues through solid state fermentation with the fungus Lentinus edodes to produce ellagic acid. Huang et al. (2008) co-cultivated *Aspergillus oryzae* and *Trichoderma reesei* using acorn crown extract with a content higher than 62% ellagitannins as a substrate to produce ellagic acid with high levels of enzymes such as hydrolase, cellulase, and xylanase. The results indicated that the mixed culture of T. reesei and A. oryzae is an efficient method to produce an enzymatic system for the degradation of ellagitannins.

Aguilera-Carbo et al. (2008) reported on the production of ellagic acid by cultivating fungi in the solid state using polyurethane foam (PUF) as a support and an aqueous extract obtained from the peel of pomegranate (*Punica granatum*) as a source of carbon and energy. In this study, *Aspergillus niger* GH1 consumed ellagitannins during the first 36 h of culture with a maximum concentration of ellagic acid reached at 48 h. The authors attributed the biodegradation of ellagitannins to a new tannase, which is probably different from acyl hydrolase.

10.5 PURIFICATION OF ELLAGITANNINS

To purify ellagitannins, several steps must be followed—first, a column chromatography must be performed using chromatographic fillers such as Amberlite XAD-16, Sephadex LH-20 and macroporous resin H-103 (Aguilar-Zárate et al., 2017; Seeram et al., 2005). The most reported techniques are high-performance liquid chromatography (HPLC), centrifugal partition chromatography (CPC), high-speed countercurrent chromatography (HSCC), and preparative HPLC (Aguilar-Zárate et al., 2017).

It is recommended to perform a HPLC analysis since they are commonly used to detect and quantify compounds. Some methods have been reported based on reverse phase C18 column technology, with solvents such as methanol, ethanol, and acetonitrile, mixed with acidified water (acetic or phosphoric acid) (Aguilera-Carbo et al., 2008b).

10.6 BIOLOGICAL ACTIVITIES

Fruits play a main part in the maintenance of a well-balanced diet. They contain macro- and micronutrients, and also bioactive compounds that promote health. Plenty of studies indicate the importance of fruit consumption for the preventing of health-associated risks and diseases (Kandylis, 2020).

Pomegranate is a rich in polyphenols and have shown many health-related properties, so it has become in the object of study in medicine. Various studies have established that pomegranate has antibacterial, anti-inflammatory, antiviral, and anticancer properties (Kalaycıoğlu and Erim, 2017). These investigations have concluded that the regular consumption of pomegranate fruit, juice, or even its added compounds in other food products, acts in a beneficial way for health and can even protect or improve health against various diseases such as obesity, diabetes, cardiovascular diseases, and even some types of cancer (Kandylis et al., 2020).

The fruit's health-promoting properties are mainly due to punicalagin's presence and, slightly, to other metabolites like anthocyanins and lavonols (Syed et al., 2007).

The punicalagin, HHDP-hexosides, and ellagic acid are derived from the partial or complete hydrolysis of punicalagin and have generated interest because of the human health effects (Aguilar-Zárate et al., 2017). Punicalagin is found in high amounts in PP. The powerful bioactivity of this compound has been demonstrated in several studies.

10.6.1 ANTIOXIDANT ACTIVITY

Phenolic compounds have an aromatic ring bearing on one or more hydroxyl groups and their structures may vary from a simple phenolic molecule to a complex high-molecular-weight polymer. Flavonoids, have C6–C3–C6 structure, and account for more than half of the over 8000 different phenolic compounds (Balasundram et al., 2006). Pomegranate's phenolic content is higher than those reported for cranberry, strawberry, red wine, orange juice, and green tea (Gil et al., 2000).

The use of synthetic antioxidants has been increasingly restricted due to the secondary effects they may produce; the determination of the antioxidant capacity of pomegranate compounds and derivates it's been studied for use as natural additives to replace synthetic antioxidants (Viuda-Martos et al., 2010).

Wang et al. (2013) studied a large-scale separation method of punicalin from PP using medium pressure liquid chromatography (MLPC) and they obtained large amounts of punicalin with high purity of 95.9%. Also, they evaluated the antioxidant and protein precipitating capacities of the purified punicalin; the results showed that punicalin has strong antioxidant potential according to the DPPH and hydroxyl radical scavenging assay and exhibited a low affinity for protein, suggested that would not be masked by tannin protein precipitation.

10.6.2 ANTIMICROBIAL ACTIVITY/PREBIOTIC EFFECT

One of the oldest techniques to control the growth of microorganisms is using chemical or synthetic agents as inhibitors, growth reducers, or even inactivators. Several studies have demonstrated the antibacterial activity of pomegranate against various pathogens and organisms resistant to antibiotics (Viuda-Martos et al., 2010).

An important function of pomegranate and its derivatives is it's effect exerts on the intestinal microbiota and its antimicrobial activity (Acosta-Estrada et al., 2014). Ellagitannins are hydrolyzed into ellagic acid in the intestine before being metabolized by the colon microbiota to form urolithin A and B (Finegold et al., 2014), which has been associated with the prebiotic potential of pomegranate and its products (Kandylis, 2020).

Betanzos-Cabrera et al. (2015) evaluated the antimicrobial activity of fresh pomegranate juice in clinical isolates of multi-resistant *Staphylococcus* in epidermal strains. They observed inhibition in the growth kinetics of the strain at different concentrations of the juice, which can be attributed to the high content of polyphenols and the antioxidant capacity of the pomegranate. Šavikin et al. (2018) studied PP as an antimicrobial agent and a reinforcing agent was utilized in developing starch-based films like hydroxypropyl high-amylose starch plasticized by water and glycerol. The results demonstrated that PP inhibited gram-positive (*S. aureus*) and gram-negative (*Salmonella*) bacteria growth. Meanwhile, PP also increased Young's modulus, tensile strength, and stiffness of the starch-based films.

Prebiotics can improve selected or randomized growth of colon microbiota. Pomegranate extracts inhibit pathogens and initiate the growth of beneficial microbiota in human guts. Probiotic lactobacilli were relatively unaffected by pomegranate chemical constituents: ellagic acid slightly reduced the growth of lactobacilli (*Lactobacillus pentosus, Lactobacillus rhamnosus, and Lactobacillus acidophilus*) to approximately 10–20% of controls (Bialonska et al., 2009).

10.6.3 ANTI-INFLAMMATORY ACTIVITY

Various investigations have been reported in the literature on the anti-inflammatory activity of pomegranate extracts through in vitro and in vivo studies. Macrophages function as immune system control switches maintaining the balance between pro and anti-inflammatory activities; they can become different subsets with positive inflammatory phenotypes, classical polarization (M1), and activation of macrophages in powerful pro-inflammatory cells characterized by the generation of species reactive oxygen and nitrogen, and the alternative activation profile (M2) characterized by the secretion of anti-inflammatory cytokines (Bouhlel et al., 2007).

Hollebeeck et al. (2012) evaluated the extract's anti-inflammatory properties of the PP, rich in Punicalagin, using an in vitro model of the epithelium of the human intestine. They concluded that the extracts showed anti-inflammatory properties in intestinal cells, acting on the transcription of the pro-inflammatory gene and protein levels, indicating that the latter phenomenon is possibly due to a direct molecular capture of the ellagitannins from the shell, helping to prevent inflammation in chronic intestinal.

Ellagic acid has been shown to be used to treat pain and inflammation through central and peripheral sites. Mansouri et al. (2015) examined the interaction between ellagic acid and venlafaxine (VLF) in intraperitoneal acetic acid-induced pain in mice using an isobolographic analysis, which indicated a synergistic antinociceptive and anti-inflammatory interaction when these drugs were combined.

10.7 BIOACTIVE COMPOUNDS AS NUTRACEUTICALS AND FUNCTIONAL FOODS

Bioactive compounds are known to lower the risk of developing diseases like cancer, Alzheimer's, cataracts, and Parkinson.

The idea of basic foods is complex and can refer to various possible aspects, including foods obtained through any process, whose main characteristic is that one or more of its components, even if that component is a nutrient, affect the function of the organism of in a particular way, by helping the physiological or psychological effect far away the nutritional. The pomegranate fruit conforms to the definition of functional food, although establishing any function implies identifying the bioactive components to specify their possible beneficial effects on health (Viuda-Martos et al., 2010).

ACKNOWLEDGMENTS

Autor Izábal-Carvajal A.L. thanks the Mexican Council for Science and Technology (CONACYT) for the financial support for the development of the project (PhD) in Food Science and Technology offered by the Autonomous University of Coahuila.

KEYWORDS

- **agroindustrial waste**
- **pomegranate**
- **bioactive compound**
- **environmental protection**
- **valorization**
- **phytochemicals**

REFERENCES

Acosta-Estrada, B. A.; Gutiérrez-Uribe, J. A.; Serna-Saldívar, S. O. Bound Phenolics In Foods, A Review. *Food Chem.* **2014,** *152,* 46–55. https://doi.org/10.1016/j.foodchem.2013.11.093

Aguilar-Zárate, P.; Wong-Paz, J. E.; Michel, M.; Buenrostro-Figueroa, J.; Díaz, H. R.; Ascacio, J. A.; Contreras-Esquivel, J. C.; Gutiérrez-Sánchez, G.; Aguilar, C. N. Characterisation of Pomegranate-Husk Polyphenols and Semi-Preparative Fractionation of Punicalagin. *Phytochem. Anal.* **2017,** *28* (5), 433–438. https://doi.org/10.1002/pca.2691

Aguilera-Carbo, A.; Augur, C.; Prado-Barragan, L. A.; Favela-Torres, E.; Aguilar, C. N. Microbial Production of Ellagic Acid and Biodegradation of Ellagitannins. *Appl. Microbiol. Biotechnol.* **2008**a, *78* (2), 189–199. https://doi.org/10.1007/s00253-007-1276-2

Aguilera-Carbo, A. F.; Augur, C.; Prado-Barragan, L. A.; Aguilar, C. N.; Favela-Torres, E. Extraction and Analysis of Ellagic Acid from Novel Complex Sources. *Chem. Papers* **2008b,** *62* (4), 440–444. https://doi.org/10.2478/s11696-008-0042-y

Akhtar, S.; Ismail, T.; Fraternale, D.; Sestili, P. Pomegranate Peel and Peel Extracts: Chemistry and Food Features. *Food Chem.* **2005,** *174,* 417–425. https://doi.org/10.1016/j.foodchem.2014.11.035

Arapitsas, P. Hydrolyzable Tannin Analysis in Food. *Food Chem.* **2012,** *135* (3), 1708–1717. https://doi.org/10.1016/j.foodchem.2012.05.096

Ascacio-Valdés, J.; Burboa, E.; Aguilera-Carbo, A. F.; Aparicio, M.; Pérez-Schmidt, R.; Rodríguez, R.; Aguilar, C. N. Antifungal Ellagitannin Isolated from Euphorbia Antisyphilitica Zucc. *Asian Pac. J. Trop. Biomed.* **2013,** *3* (1), 41–46. https://doi.org/10.1016/S2221-1691(13)60021-0

Baiano, A. Recovery of Biomolecules from Food Wastes—A Review. *Molecules* **2014,** *19* (9), 14821–14842. https://doi.org/10.3390/molecules190914821

Bala, I.; Bhardwaj, V.; Hariharan, S.; Kumar, M. N. V. R. Analytical Methods for Assay of Ellagic Acid and Its Solubility Studies. *Jo. Pharm. Biomed. Anal.* **2006,** *40* (1), 206–210. https://doi.org/10.1016/j.jpba.2005.07.006

Balasundram, N.; Sundram, K.; Samman, S. Phenolic Compounds in Plants and Agri-Industrial By-Products: Antioxidant Activity, Occurrence, and Potential Uses. *Food Chem.* **2006,** *99* (1), 191–203. https://doi.org/10.1016/j.foodchem.2005.07.042

Betanzos-Cabrera, G.; Montes-Rubio, P. Y.; Fabela-Illescas, H. E.; Belefant-Miller, H.; Cancino-Diaz, J. C. Antibacterial Activity of Fresh Pomegranate Juice Against Clinical Strains of *Staphylococcus epidermidis*. *Food Nut. Res.* **2015,** *59*. https://doi.org/10.3402/fnr.v59.27620

Bhanja Dey, T.; Chakraborty, S.; Jain, K. K.; Sharma, A.; Kuhad, R. C. Antioxidant Phenolics and Their Microbial Production by Submerged and Solid-state Fermentation Process: A Review. *Trends Food Sci. Technol.* **2016,** *53,* 60–74. https://doi.org/10.1016/j.tifs.2016.04.007

Bialonska, D.; Kasimsetty, S. G.; Schrader, K. K.; Ferreira, D. The Effect of Pomegranate (Punica Granatum l.) Byproducts and Ellagitannins on the Growth of Human Gut Bacteria. *J. Agric. Food Chem.* **2009,** *57* (18), 8344–8349. https://doi.org/10.1021/jf901931b

Bonesi, M.; Tundis, R.; Sicari, V.; Loizzo, M. R. The Juice of Pomegranate (*Punica granatum L.*): Recent Studies on Its Bioactivities. In *Quality Control in the Beverage Industry: Volume 17: The Science of Beverages*; Elsevier, 2019; pp 459–489. . https://doi.org/10.1016/B978-0-12-816681-9.00013-8

Bonzanini, F.; Bruni, R.; Palla, G.; Serlataite, N.; Caligiani, A. Identification and Distribution of Lignans in *Punica granatum L.* Fruit Endocarp, Pulp, Seeds, Wood Knots and Commercial Juices by GC-MS. *Food Chem.* **2009,** *117* (4), 745–749. https://doi.org/10.1016/j.foodchem.2009.04.057

Bouhlel, M. A.; Derudas, B.; Rigamonti, E.; Dièvart, R.; Brozek, J.; Haulon, S.; Zawadzki, C.; Jude, B.; Torpier, G.; Marx, N.; Staels, B.;Chinetti-Gbaguidi, G. PPARγ Activation Primes Human Monocytes into Alternative M2 Macrophages with Anti-inflammatory Properties. *Cell Metabolism.* **2007,** *6* (2), 137–143. https://doi.org/10.1016/j.cmet.2007.06.010

Brighenti, V.; Groothuis, S. F.; Prencipe, F. P.; Amir, R.; Benvenuti, S.; Pellati, F. Metabolite Fingerprinting of *Punica granatum L.* (Pomegranate) Polyphenols by Means of High-performance Liquid Chromatography with Diode Array and Electrospray Ionization-mass Spectrometry Detection. *J. Chromatogr. A* **2017,** *1480*, 20–31. https://doi.org/10.1016/j.chroma.2016.12.017

Devesa-Rey, R.; Vecino, X.; Varela-Alende, J. L.; Barral, M. T.; Cruz, J. M.; Moldes, A. B. Valorization of Winery Waste vs. the Costs of Not Recycling. *Waste Manag.* **2011,** *31* (11), 2327–2335. https://doi.org/10.1016/j.wasman.2011.06.001

Domínguez-Rodríguez, G.; Marina, M. L.; Plaza, M. Strategies for the Extraction and Analysis of Non-extractable Polyphenols from Plants. *J. Chromatograp. A* **2017,** *1514,* 1–15. https://doi.org/10.1016/j.chroma.2017.07.066

Evreinoff, V. A. Contribution à l'étude du Grenadier. *J. Agric. Trop. Bot. Appl.* **1957,** *4* (3), 124–138. https://doi.org/10.3406/jatba.1957.2380

Farinas, C. S. Developments in Solid-state Fermentation for the Production of Biomass-degrading Enzymes for the Bioenergy Sector. *Renew. Sustain. Ener. Rev.* **2015,** *52,* 179–188. https://doi.org/10.1016/j.rser.2015.07.092

Ferrara, G.; Cavoski, I.; Pacifico, A.; Tedone, L.; Mondelli, D. Morpho-pomological and Chemical Characterization of Pomegranate (*Punica granatum L.*) Genotypes in Apulia Region, Southeastern Italy. *Sci. Hortic.* **2011,** *130* (3), 599–606. https://doi.org/10.1016/j.scienta.2011.08.016

Finegold, S. M.; Summanen, P. H.; Corbett, K.; Downes, J.; Henning, S. M.; Li, Z. Pomegranate Extract Exhibits Invitro Activity Against Clostridium Difficile. *Nutrition* **2014,** *30* (10), 1210–1212. https://doi.org/10.1016/j.nut.2014.02.029

García-Villalba, R.; Espín, J. C.; Aaby, K.; Alasalvar, C.; Heinonen, M.; Jacobs, G.; Voorspoels, S.; Koivumäki, T.; Kroon, P. A.; Pelvan, E.; Saha, S.; Tomás-Barberán, F. A. Validated Method for the Characterization and Quantification of Extractable and Nonextractable Ellagitannins after Acid Hydrolysis in Pomegranate Fruits, Juices, and Extracts. *J. Agric. Food Chemi.* **2015,** *63* (29), 6555–6566. https://doi.org/10.1021/acs.jafc.5b02062

Gil, M. I.; Tomas-Barberan, F. A.; Hess-Pierce, B.; Holcroft, D. M.; Kader, A. A. Antioxidant Activity of Pomegranate Juice and Its Relationship with Phenolic Composition and Processing. *J. Agric. Food Chemi.* **2000,** *48* (10), 4581–4589. https://doi.org/10.1021/jf000404a

Hollebeeck, S.; Winand, J.; Hérent, M. F.; During, A.; Leclercq, J.; Larondelle, Y.; Schneider, Y. J. Anti-inflammatory Effects of Pomegranate (*Punica granatum L.*) Husk Ellagitannins in Caco-2 Cells, an In Vitro Model of Human Intestine. *Food Func.* **2012,** *3* (8), 875–885. https://doi.org/10.1039/c2fo10258g

Huang, W.; Niu, H.; Li, Z.; He, Y.; Gong, W.; Gong, G. Optimization of Ellagic Acid Production from Ellagitannins by Co-culture and Correlation Between Its Yield and activities of Relevant Enzymes. *Biores. Technol.* **2008,** *99* (4), 769–775. https://doi.org/10.1016/j.biortech.2007.01.032

Jamal, P.; Idris, Z. M.; Alam, M. Z. Effects of Physicochemical Parameters on the Production of Phenolic Acids from Palm Oil Mill Effluent Under Liquid-state Fermentation by Aspergillus Niger IBS-103ZA. *Food Chemi.* **2011,** *124* (4), 1595–1602. https://doi.org/10.1016/j.foodchem.2010.08.022

Kahramanoglu, I. *International Journal of Agriculture, Forestry and Life Sciences Trends in Pomegranate Sector: Production, Postharvest Handling and Marketing*; 2019a.

Kahramanoglu, I. *Trends in Pomegranate Sector: Production, Postharvest Handling and Marketing*; 2019b.

Kalaycıoğlu, Z.; Erim, F. B. Total Phenolic Contents, Antioxidant Activities, and Bioactive Ingredients of Juices from Pomegranate Cultivars Worldwide. *Food Chem.* **2017,** *221,* 496–507). https://doi.org/10.1016/j.foodchem.2016.10.084

Kandylis, P.; Kokkinomagoulos, E. Food Applications and Potential Health Benefits of Pomegranate and Its Derivatives. *Foods* **2020,** *9* (2), 122. https://doi.org/10.3390/foods9020122

Kumar, K.; Yadav, A. N.; Kumar, V.; Vyas, P.; Dhaliwal, H. S. Food Waste: A Potential Bioresource for Extraction of Nutraceuticals and Bioactive Compounds. *Bioresour. Bioprocess.* **2017,** *4* (1). https://doi.org/10.1186/s40643-017-0148-6

Mansouri, M. T.; Naghizadeh, B.; Ghorbanzadeh, B. Ellagic Acid Enhances the Antinociceptive Action of Venlafaxine in Mouse Acetic Acid-induced Pain: An Isobolographic Analysis. *Pharmacol. Rep.* **2015,** *67* (3), 473–477. https://doi.org/10.1016/j.pharep.2014.11.004

Martínez-Ávila, G. C.; Aguilera-Carbó, A. F.; Rodríguez-Herrera, R.; Aguilar, C. N. Fungal Enhancement of the Antioxidant Properties of Grape Waste. *Ann. Microbiol.* **2012,** *62* (3), 923–930. https://doi.org/10.1007/s13213-011-0329-z

Pareek, S.; Valero, D.; Serrano, M. Postharvest Biology and Technology of Pomegranate. *J. Sci. Food Agric.* **2015,** *95* (12), 2360–2379. https://doi.org/10.1002/jsfa.7069

Šavikin, K.; Živković, J.; Alimpić, A.; Zdunić, G.; Janković, T.; Duletić-Lašević, S.; Menković, N. Activity Guided Fractionation of Pomegranate Extract and Its Antioxidant,

Antidiabetic and Antineurodegenerative Properties. *Ind. Crops Prod.* **2018,** *113*, 142–149. https://doi.org/10.1016/j.indcrop.2018.01.031

Seeram, N.; Lee, R.; Hardy, M.; Heber, D. Rapid Large-scale Purification of Ellagitannins from Pomegranate Husk, A By-product of the Commercial Juice Industry. *Sep. Purif. Technol.* **2005,** *41* (1), 49–55. https://doi.org/10.1016/j.seppur.2004.04.003

Seeram, N. P.; Adams, L. S.; Henning, S. M.; Niu, Y.; Zhang, Y.; Nair, M. G.; Heber, D. In Vitro Antiproliferative, Apoptotic and Antioxidant Activities of Punicalagin, Ellagic Acid and a Total Pomegranate Tannin Extract are Enhanced in Combination with Other Polyphenols as Found in Pomegranate Juice. *J. Nutr. Biochem.* **2005,** *16* (6), 360–367. https://doi.org/10.1016/j.jnutbio.2005.01.006

Singh, B.; Singh, J. P.; Kaur, A.; Singh, N. Phenolic Compounds as Beneficial Phytochemicals in Pomegranate (*Punica granatum L.*) Peel: A Review. *Food Chemi.* **2018,** *261*, 75–86. https://doi.org/10.1016/j.foodchem.2018.04.039

Soccol, C. R.; da Costa, E. S. F.; Letti, L. A. J.; Karp, S. G.; Woiciechowski, A. L.; de Vandenberghe, L. P. S. Recent Developments and Innovations in Solid State Fermentation. *Biotechnol. Res. Innova.* **2017,** *1* (1), 52–71. https://doi.org/10.1016/j.biori.2017.01.002

Sreekumar, S.; Sithul, H.; Muraleedharan, P.; Azeez, J. M.; Sreeharshan, S. Pomegranate Fruit as a Rich Source of Biologically Active Compounds. *BioMed Res. Int.* **2014,** *2014*, 686921. https://doi.org/10.1155/2014/686921

Stover, E.; Mercure, E. W. *The Pomegranate: A New Look at the Fruit of Paradise*; n.d.

Syed, D. N.; Afaq, F.; Mukhtar, H. Pomegranate Derived Products for Cancer Chemoprevention. *Semin. Cancer Biol.* **2007,** *17* (5), 377–385. https://doi.org/10.1016/j.semcancer.2007.05.004

Torres-León, C.; Ramírez-Guzmán, N.; Ascacio-Valdés, J.; Serna-Cock, L.; dos Santos Correia, M. T.; Contreras-Esquivel, J. C.; Aguilar, C. N. Solid-state Fermentation with Aspergillus Niger to Enhance the Phenolic Contents and Antioxidative Activity of Mexican Mango Seed: A Promising Source of Natural Antioxidants. *LWT* **2019,** *112*. https://doi.org/10.1016/j.lwt.2019.06.003

Tzulker, R.; Glazer, I.; Bar-Ilan, I.; Holland, D.; Aviram, M.; Amir, R. Antioxidant Activity, Polyphenol Content, and Related Compounds in Different Fruit Juices and Homogenates Prepared from 29 Different Pomegranate Accessions. *J. Agric. Food Chemi.* **2007,** *55* (23), 9559–9570. https://doi.org/10.1021/jf071413n

Vattem, D. A.; Shetty, K. Ellagic Acid Production and Phenolic Antioxidant Activity in Cranberry Pomace (*Vaccinium macrocarpon*) Mediated by Lentinus Edodes Using a Solid-state System. *Proc. Biochem.* **2003,** *39* (3), 367–379. https://doi.org/10.1016/S0032-9592(03)00089-X

Verotta, L.; Panzella, L.; Antenucci, S.; Calvenzani, V.; Tomay, F.; Petroni, K.; Caneva, E.; Napolitano, A. Fermented Pomegranate Wastes as Sustainable Source of Ellagic Acid: Antioxidant Properties, Anti-inflammatory Action, and Controlled Release under Simulated Digestion Conditions. *Food Chemi.* **2018,** *246,* 129–136. https://doi.org/10.1016/j.foodchem.2017.10.131

Viuda-Martos, M.; Fernández-Lóaez, J.; Pérez-álvarez, J. A. Pomegranate and its Many Functional Components as Related to Human Health: A Review. *Compr. Rev. Food Sci. Food Saf.* **2010,** *9* (6), 635–654. https://doi.org/10.1111/j.1541-4337.2010.00131.x

Wang, Y.; Zhang, H.; Liang, H.; Yuan, Q. Purification, Antioxidant Activity and Protein-precipitating Capacity of Punicalin from Pomegranate Husk. *Food Chemi.* **2013,** *138* (1), 437–443. https://doi.org/10.1016/j.foodchem.2012.10.092

CHAPTER 11

Foodborne Zoonoses: Current Status and Control Strategies

M. SUMAN KUMAR[1], HIMANI DHANZE[1], S. S. REDDY VAKAMALLA[1], A. R. CHANDNI[1], and S. WILFED RUBAN[1,2]

[1]*Division of Veterinary Public Health, ICAR-IVRI, Izatnagar, Uttar Pradesh, India*

[2]*Veterinary College, KVA&FSU, Hebbal, Bengaluru, Karnataka, India*

ABSTRACT

The aim of this chapter is to review and report the latest control strategies that is available on food borne diseases to public health considering the socioeconomic development.

11.1 INTRODUCTION

The World Food Summit held at Rome in 1996 reaffirmed that gaining secure and healthy food is the fundamental right of every person, in accordance with the integrity to sufficient food and fair to be free from starvation. Foodborne zoonotic diseases pose a continuous hazard to public well-being with implications on socioeconomic development and international trade worldwide. Zoonoses refer to those illnesses and epidemic that is normally passed on among vertebrate animals and human. More than 200 familiar illnesses are transferred from food by pathogenic microorganisms such as bacteria and their toxins, viruses, and parasites. The World Health Organization's (WHO) action to measure the world hardship of foodborne illness considered illness caused by 31 foodborne hazards and reported over 600 million illnesses and

Engineering Principles for Food Processing Technology and Product Realization. Leonardo Sepúlveda Torre, Porteen Kannan, A. K. Haghi, PhD (Eds.)

4,20,000 deaths in the year 2010 alone (WHO, 2015). The disease burden was highest in Africa (43%), followed by South-East Asia (24%) and Eastern Mediterranean (16%) sub regions. Around 40% of the foodborne diseases (FBD) were reported from children less than 5 years. The highest rates of under-5 mortality due to diarrhea were in sub-Saharan Africa and South Asia with India considered for many loss of life in 2015 (Troeger et al., 2017). Based on the CDC, 48 million people are affected by food-borne illness each year, 128,000 are hospitalized, and around 3000 die as a result (CDC, 2021). Norovirus is the most common cause of FBD, followed by Campylobacter, Salmonella, and Listeria monocytogenes. The observed trend was mostly comparable to that of the EU, the United States of America (USA), and Australia. However, pathogenic E. coli caused the most cases in Korea, and L. monocytogenes was not as lethal. The number of Campylobacter, norovirus, and *C. perfringens*-related infections in Japan was quite high (Lee and Yoon, 2021). FBD cost India about $28 billion (Rs. 1,78,100 crore) or around 0.5% of the country's gross domestic product (GDP) every year (Kristkova et al., 2017). It is estimated that the current FBD burden in India represents about 100 million cases per year. This corresponds to one in 12 people falling ill, which might be well underestimated as not all cases are reported or recorded.

The risk of foodborne illness has been rising during the last two decades due to several factors including pathogen behavior, rapid growth of population, and an increased open global market trade in foods and farm animals from countries where appropriate microbiological security strategy are not followed or implemented. The improved transport logistics and conditions facilitates some microorganisms to survive on food products for longer periods and reach the consumer in a viable form. The change in eating habits of consumer, such as the consumption of lightly cooked food and a shift from low-to-high protein-based diets (e.g., meat and fish products) has also increased the probability of occurrence of foodborne illnesses. The increased movement of human population also facilitates easy spreading of pathogens worldwide. These factors have impacted the severity of FBD and their impacts on health of communities. Therefore, preventing illness and deaths due to FBD remains a major public health challenge across the globe. Lack of accurate epidemiological data on incidence and cost of FBD, particularly in the developing countries, has delayed the setting up of suitable policies and allocation of resources to mitigate the ills of FBD. Further, alarming rise in antimicrobial resistance among bacterial pathogens particularly in the food chain has also impacted the public health to a greater extent. Effective control involves a thorough understanding of the essential aspects of epidemic dynamics, such as the spatiotemporal distribution and common origins, as well as strong identification of outbreaks from baseline sporadic individuals (Deng et al., 2016; Ronholm et al., 2016). Studies that gather information regarding FBDs are vitally necessary to protect health. Greater understanding of biological agents, foods, and the sources of contamination that are frequently implicated in FBD outbreaks allows agencies to design effective preventative measures, thereby minimizing the incidence of outbreaks.

11.2 MAJOR BACTERIAL FOODBORNE PATHOGENS

Human diarrheal disease are mostly caused by bacterial foodborne infections (64%) followed by viral (24%) and protozoan (12%) infections. During the last two decades, *Salmonella* spp., *Campylobacter* spp., and *E. coli* were reported to be the major foodborne bacterial agents. Recent WHO estimates revealed that *Campylobacter* spp. (27%), enterotoxigenic *E. coli* (25%), non-typhoidal *Salmonella enterica* (23%), *Shigella* spp. (15%), and enteropathogenic *E. coli* (7%) are major diarrheal disease pathogens. *Campylobacter* spp. has been estimated to cause 96 million cases of illness in year 2010 alone. Globally, about 4.6 billion cases of infectious diarrhea and 1.6 million deaths are reported annually due to contaminated food and water (Pires et al., 2015).

11.2.1 CAMPYLOBACTERIOSES

Campylobacter is the leading cause of bacterial gastroenteritis worldwide—one of four key global causes of FBD with an annual incidence of 4.4–9.3/1000 cases leading to 30,931 deaths (WHO, 2012). A report of Food Borne Epidemiology Reference Group of World Health Organization (WHO, 2015) estimates global median of 95,613,970 cases of campylobacteriosis. In Southeast Asian Region D (SEAR-D) (in which India is included), the median 16,379,442 cases have been estimated. Global median deaths estimated due to campylobacteriosis are 21,374 and in SEAR-D deaths are 5410. *Campylobacter* foodborne illness is a concern and a monetary burden for the human population, accounting for 8.4% of all diarrhea cases worldwide (Connerton and Connerton, 2017). Campylobacteriosis is increasing at a rate more than twice that of salmonellosis (Mace et al., 2015). Most campylobacteriosis outbreaks have been endemic, with asymptomatic symptoms, and *C. jejuni,* preceded by *C. coli*, has been found to be responsible for nearly 80% of outbreaks in most EU countries and the USA (Kaakoush et al., 2015). The public health impact of *Campylobacter* and its prevalence in India is firmly established. However, the figures for actual prevalence and economic losses incurred in India are sketchy. Sporadic studies conducted in different parts of the country have revealed the presence of *Campylobacter* organisms in foods of plant origin (Kumar et al., 2001), various species of animals, birds, and foods of animal origin including milk and from cases of human diarrhea (Kumar et al., 2015). Campylobacteriosis occurs in people of all ages (HSE, 2017), with the highest incidence of notification in children aged 0–5, and the largest percentage of patients hospitalized and dying in adults aged 65 and older (Facciola et al., 2017). Troeger et al. (2017) estimated nearly 10,211 mean number of deaths in children younger than 5 years due to campylobacterioses in India alone. The complications of infection include meningitis, urinary tract infections, reactive arthritis, and Gullian Barré syndrome, a nerve disorder that causes muscle weakness and paralysis of the limbs. Major sources of infection include undercooked poultry meat, pets (especially young puppies), water, and raw milk. Several studies in India have estimated the occurrence of *C. jejuni* and *C. coli* in the range of 4.42–28% in chicken meat samples, 1.3–4.4% in cattle milk, 3.6–25% in meat samples, 13.52–22.64% in water samples, 2.35–25% in human diarrheal cases, and 0.56–8% in healthy human fecal samples (OPZD Annual Reports 2014–2019). Infections due to chicken eating account for 50–70% of all Campylobacter infections worldwide (Seliwiorsto et al., 2015). Macrolides and fluoroquinolones are the drug of choice for treatment of human diarrheal cases. However, increasing instances of antimicrobial resistance have been reported of late against multiple drugs. Different studies in India have shown an increasing trend of multidrug resistance of 2.2% (Prasad et al., 1994), 30.6% (Jain et al., 2005), and 51.33% (Kumar, 2016) in *Campylobacter* spp. isolates from diverse origin. Campylobacter spp. also has the ability to form biofilms and survives in the environment. A nation-wide surveillance of campylobacteriosis along with typing of strains isolated from both human and animal cases is required. Preventive measures such as rearing of campylobacter-free poultry flocks and the good hygienic practices throughout the entire production chain can limit the transmission of the pathogen. Withholding feed from chickens for around 12 h before slaughter is another important preventive action that will help decrease the amount of these bacteria (Hansson et al., 2018). Campylobacter infections can be prevented in humans in a variety of ways, including improved sewage sanitation, the availability of clean drinking water, public awareness of the importance of pasteurization of milk, proper cooking of animal-sourced foods, vaccination, and the use of therapeutics in cases of infection (Hansson et al., 2018).

11.2.2 SALMONELLOSIS

Salmonella is the second leading causative agent of foodborne illness in most developed countries and most common reason for foodborne zoonosis in other developing countries. *Salmonella enterica* subsp. *enterica* causes 99% of the human and animal infections. It is estimated that approximately 78.7 million cases of foodborne illness cases occur due to non-typhoidal *Salmonella* globally, leading to 59,000 deaths (WHO, 2015). *Salmonella* is one of the most frequent foodborne infections in the world, responsible for approximately 93.8 million foodborne illnesses and 155,000 fatalities per year (Eng et al., 2015). In 2018, the European Union (EU) member states reported 5146 outbreaks of FBD, giving rise in the infections of 48,365 persons. Salmonella was solely responsible for 33% of these outbreaks. *Salmonella* spp. has wide range of livestock species (poultry, cattle, and pigs) as host and the infected animals act as a reservoir of zoonotic salmonellae. Salmonellosis is mostly a foodborne illness; however, reptiles are responsible for 7% salmonellosis occurrences in humans (CFSPH, 2013). Contaminated eggs, poultry meat, pork, and other meats and products are common vehicles of salmonellosis to humans. The Salmonella epidemics that resulted in 1581 cases in Poland, Slovakia, and Spain were directly connected to infected eggs (Munck et al., 2020). Historically, plants have not been regarded as hosts for human infections for *Salmonella*; nevertheless, in the last several decades, the habitats for these species have shifted (Lenzi et al., 2021). Devleesschauwer et al. (2017) observed that despite the fact that salmonellosis outbreaks caused by fruits and vegetables have been widely documented, their frequency remains intermittent. In India, foods of animal origin are considered as the main source with contributing factors such as inadequate cooking, lack of refrigeration, slow cooling of food, and inadequate reheating before serving. Various studies in India have documented prevalence of *Salmonella* spp. in foods of animal origin including meat (4–12.4%), carabeef (1.25%), pork (10%), water (3.4–14.89%), chicken (14.89%), and in poultry litter (1.73–28.33%) (OPZD Annual Reports 2014–2019). Serotypes Typhi and Paratyphi A of typhoidal Salmonella are more frequent in Southeast Asia, while non-typhoidal Salmonella (NTS) is more prominent in Africa (WHO, 2016). The non-typhoidal *Salmonella* serovars, especially Typhimurium, Enteritidis, Heidelberg, and Newport have been reported in many outbreaks of human salmonellosis around the globe which have been linked with consumption of Salmonella-contaminated foods of animal origins. Furthermore, the emergence of several clones of Salmonella resistant to multiple antimicrobials worldwide underscores a significant food safety hazard (Jajere, 2019). A detailed surveillance followed by risk assessment studies is imperative to control salmonellosis.

11.2.3 E. COLI INFECTION

E. coli has wide range of host and so far at least seven different groups of *E. coli* strains have been isolated from gastrointestinal diseases, out of which foodborne infections are mostly associated with enterotoxigenic *E. coli* (ETEC) and enteropathogenic *E. coli* (EPEC) strains although Shiga-toxigenic *E. coli* (STEC) causes more severe disease leading to mortality. Among the STEC strains, *E. coli* O157:H7 is widely recognized as an important cause of foodborne illness. The clinical outcomes of STEC infections would be mild-to-severe diarrhea and hemolytic uremic syndrome. Cattle are the principal reservoirs of STEC (Bibbal et al., 2015) and beef has been implicated in several global outbreaks. ETEC strains have also been isolated from carabeef, dung, feces of diarrhea patients, and river water in India. The presence of antimicrobial-resistant *E. coli* other than STECs in foods of animal origin is also an emerging public health concern. Of particular, public health concern among antimicrobial-resistant *E. coli* are strains that produce extended-spectrum β-lactamases (ESBLs). ESBL-producing *E. coli* have been documented in feces of healthy poultry and retail meat, from several parts of the world including India (Kar et al., 2015). A pooled prevalence of 15% was estimated for *E.coli* in the food products of livestock origin in Ethiopia. The high prevalence may be due to unhygienic slaughtering of food animals and consumption of raw meat in the area concerned (Assefa and Bihon, 2018). Recently, Odo et al. (2021) reviewed different prevalence studies of *E. coli* O157:H7 in Nigeria. Chicken and beef samples from slaughterhouses of southwest Nigeria when tested showed an overall prevalence of 17.1% for multidrug resistant *E coli* O157:H7. This bacterial strain could also be detected in fermented milk products from Nasarawa state of Nigeria. Spatial epidemiology study of *E. coli* O157:H7 in Netherlands revealed the increased incidence being associated with exposure to small ruminants rather than cattle (Mulder et al., 2020). The total number of confirmed cases of foodborne ETEC infection in the EU were above 6000 in 2016 and 2017 and increased to 8167 in 2018 while the number reduced to below 5000 during 2020 (EFSA-ECDC, 2020). The statistical analysis to identify the major sources contributing to different foodborne pathogens showed that STEC was largely transmitted through direct contact with infected animals. Nevertheless, among the foodborne infection pathways, beef/lamb accounted for 65% of the infections (Mughini-Gras et al., 2022). In India, the occurrence of *E. coli* have been documented in cattle milk (1–11%), water (2.6–33.3%), meat and meat products (24.10%), poultry eggs (10%), carabeef (9%), and chicken meat and giblets (1.33%). In a study conducted on street-vended foods, a higher presence of *E. coli* was observed in raw food samples (60.62%), followed by environmental samples (46.73%) and ready-to-eat foods (RTE) (42.99%), of which 31 (26.49%), 3 (6.97%), and 7 (15.21%) samples, respectively, harbored ESBL producers. ESBL *E. coli* were detected in finished food products like salad, chutney, and paneer (Sivakumar, 2016).

Escherichia albertii, a close relative of *E. coli*, is considered as an emerging zoonotic pathogen which is often misunderstood as other members of *Enterobacteriaceae*. The organism has been isolated from meat and meat products of chicken, pig, and duck and from the environmental sources. Diarrhea outbreaks due to infection with *E. albertii* have been reported in different areas across the globe including countries like China, Brazil, Germany, Bangladesh, Alaska, USA, Poland, Japan, Iran, Mexico, and Nigeria (Muchaamba et al., 2022). The presence of the organism could be detected in a recent study in retail oysters with a prevalence of 1.9% which may be accounted to the contamination of the water bodies with the bacteria (Arai et al., 2022).

11.2.4 LISTERIOSIS

Listeriosis is a serious, but preventable and treatable zoonotic disease. *Listeria monocytogenes* is one of the most virulent foodborne pathogens, with 20–30% case fatality. According to WHO (2015) estimates, annually 14,169 cases of foodborne illness and 3175 deaths occur globally due to listeriosis. *L. monocytogenes* has been discovered to be a common pathogen in several nations, including the United Kingdom, Mexico, Canada, Australia, and USA

(Yang et al., 2020). Although it is considered uncommon in comparison to other foodborne infections, listeriosis has a fatality rate of 20–30% in its most severe form (Li et al., 2020). The survival capability of *L. monocytogenes* in wide range of pH, salt concentration, and other processing and refrigeration conditions poses potential threat to public health. Processed, refrigerated, and RTE of animal origin are the major sources often implicated in Listeria outbreaks. Cooking or other antimicrobial treatments are not performed between food production and consumer consumption in these circumstances, which increases their risk (Jordan and McAuliffe, 2018; Gray et al., 2021). Listeriosis is directly responsible for the deaths of 260 individuals per year in the USA because of consuming contaminated food (Centers for Disease Control and Prevention, 2015). A widespread outbreak of listeriosis resulted to 937 cases and 216 fatalities in South Africa in 2018 due to the eating of tainted RTE-processed beef (Thomas et al., 2020). Pregnant, elderly, or immuno-compromised persons are at a greater risk of infection. *L. monocytogenes* has also been reported from cases of spontaneous abortion in humans (Kaur et al., 2007). Occurrence rates of 3.5–15.7% in water and 5.21–20.5% in meat have been recorded for *L. monocytogenes* in different studies (OPZD Annual Reports 2014–2019). Many studies have reported occurrence of *L. monocytogenes* in animal origin foods such as milk and its products (1–23%), chevon (0–17%) poultry meat (8–15%), and pork (1.3–2.8%) (Barbuddhe et al. 2012; Shakuntala et al. 2019). Existence of multidrug-resistance and biofilm-forming capacity of *L. monocytogenes* isolates is also a cause for public health concern. The clinical outcomes of listeriosis include perinatal and non-perinatal-related symptoms of septicaemia, meningitis, neurological consequences, stillborn, and death. *Listeria* spp. isolates from food have been found to be resistant to antibiotics currently used in the treatment of human listeriosis, such as penicillin, ampicillin, tetracycline, and gentamicin (Olaimat et al., 2018). Further research is needed on alternative therapies to treat listeriosis and better understanding about bacterial persistence and survival over a wide range of temperature outside mammalian hosts.

11.3 OTHER IMPORTANT BACTERIAL PATHOGENS CAUSING FBDS

11.3.1 VIBRIO SPP.

Among more than 100 species of *Vibrio*, *Vibrio cholera* (*V. cholera*), *Vibrio parahemolyticus* (*VP*), and *Vibrio Vulnificus* (*V. Vulnificus*) are involved in foodborne infection. *VP* is the most common kind of bacterial gastroenteritis that is linked to seafood consumption all over the world (Baker-Austin, 2018). The reservoirs include normal flora of brackish water, aquatic flora (macrophytes, microphytes), and fauna (zooplanktons, crustaceans), humans, and animals. Raw oysters, crustaceans, crabs, and oysters or undercooked mussels are common sources of the disease in man. *V. cholera* is found in oysters for weeks under refrigeration and for more than 2 weeks in dairy products, including soft deserts and cakes. Complications of infection include severe dehydration, cardiac complications, and circulatory failure, which occur due to loss of potassium, septicemia, wound infection, ear infection, cellulitis, peritonitis, necrotizing fascitis, cholecystitis, endophthalmitis, and meningitis. Various studies have documented occurrence of *V. cholera* from 3.45% to 57.38% and that of *V. vulnificus* as 2.3–38.5% in marine fish, clams, shell fish, water, brackish water fish, and aquatic environment. In a joint report published by WHO and FAO, it was documented that there were between 500 and 800 outbreaks of *VP* gastroenteritis recorded in Japan each year, affecting more than 10,000 persons. *VP* is one of the most common causes of foodborne gastroenteritis in Japan (WHO and FAO, 2011). It is believed that this bacterium is responsible for 50% of the infections that are caused by consuming contaminated food in Asian nations (Chen et al., 2018). Occurrence of *VP* was found to be between 6.9% and 56.09% in crab, shrimps, fish, and shell fish in coastal regions of India. Although the data regarding burden of cholera in India is well documented (Ali et al., 2017), the foodborne burden due to other *Vibrio* spp. is sketchy and focused studies are required to estimate the true burden.

11.3.2 AEROMONAS

Aeromonas species, which are sometimes referred to as "jack of all trades," have drawn growing interest as human pathogens in recent years. *Aeromonas* spp. are ubiquitous aquatic bacteria, recognized as increasingly important human pathogen due to the frequent prevalence in all types of foods, particularly in seafood. The reservoirs include wide range of fish, environment, and water. This organism has been isolated from meat, raw milk, poultry, fish, shellfish, and vegetables. Additionally, salt or sodium chloride (NaCl) is a usual preservative for meat products and raw fish; nevertheless, *Aeromonas* may thrive in NaCl concentration as high as 4%. pH is another factor that regulates bacterial growth in food. Aeromonas has been shown to be able to live at pH-5 (Figueras et al., 2014; John et al., 2019). In humans, the symptoms range from acute gastroenteritis to life-threatening complications such as septicemia and, in extremely rare instances, meningitis (Dallagassa et al., 2018). Seafood products are more commonly contaminated because of the widespread occurrence of these bacteria in the marine environment. *Aeromonas* spp. can enter the food chain through cross-contamination during processing from biofilms on food-contacted surfaces. Batra et al. (2016) estimated 76 cases per million population in Asia. Occurrence rate of *Aeromonas* spp. in various samples were 22.2–50% in fresh water fish, 18.34% in prawns, 46.67% in frozen fish, 10–22.5% in chicken, and 11–16.52% in different water sources (Sharma et al., 2009; Smita and Brahmabhatt, 2011; Das et al., 2012; Didugu et al., 2016).

11.3.3 BRUCELLOSIS

Brucellosis is a re-emerging foodborne bacterial zoonosis which is considered as a public health concern and is an important economic burden with losses occurring in the human, livestock, and wildlife populations. Each year around 5 lakhs human cases of brucellosis are reported globally. Countries like Canada, Denmark, Germany, Austria, Australia, France, New Zealand, Sweden, and USA are considered to be free from bovine brucellosis. Disinfectants like 70% ethanol, formaldehyde, hypochlorite, and iodophores can easily destroy the organisms while pasteurization of milk remains the important method to kill the organisms in milk. Consumption of unpasteurized milk is the major route of acquiring the infection by humans, but direct contact with infected animals, consumption of infected raw or undercooked meat, and in rare cases blood transfusion may also result in infection. A seroprevalence of 9% was recorded in hunted wild boars in Finland in 2016 (Fredriksson-Ahomaa et al., 2020), which hints to the probability of acquiring brucellosis from wild animals. Disease outbreaks in humans have been documented by different authors in countries including Mexico, Turkey, Saudi Arabia, and South Korea of which most of them were due to consumption of raw or unpasteurized milk and milk products (Pal, 2018). Epidemiological study of human cases in Dalian and Liaoning of China between 2014 and 2019 could report 571 total cases with no fatality (Yucun et al., 2021). In India, brucellosis caused a median loss of USD 3.4 billion to the livestock sector of which 96% was to the dairy sector alone (Singh et al., 2015). In India, the annual median losses due to human brucellosis were estimated to be INR 627.5 million (Singh et al., 2018). The main source of infection in Indian population is dairy products. Cheese prepared from unpasteurized milk from endemic areas is a particular risk for tourists. The milk of infected sheep, goats, or cattle may contain large numbers of viable organisms, which become concentrated in products such as yogurt, paneer, and cheese (Ongor et al., 2006). In dairy animals, the organism localizes in the supra-mammary lymph nodes and mammary glands of 80% of infected animals, and these continue to excrete the pathogen in milk throughout their lives acting as carriers but intermittently (OIE, 2016). Various studies have reported occurrence rate of 4.35–55.6% in cattle milk samples in India. The prevalence rate was found to be higher in unorganized farms. Recent studies have indicated resistance toward commonly used drugs such as rifampicin and co-trimoxazole, which is matter of concern and calls for judicious use of antibiotics (Doimari et al., 2018). Although numerous studies have been conducted on sero-prevalence of brucellosis in various livestock, very few have focused on the occurrence of *Brucella* spp. in foods. Studies on estimating risk of brucellosis through foods need to be carried out.

11.3.4 STAPHYLOCOCCUS AUREUS

Staphylococcus aureus (*S. aureus*) is the predominant agent associated with food poisoning and is present in the soil, water, air, and on everyday objects and surfaces, in humans and animals. Human hands and nasal tracts are common areas of *S. aureus* inhabitation. *S. aureus* has also been identified from a variety of animal-based foods and was described as the third-most prevalent cause of food-borne disease worldwide (Hachemi et al., 2019). *S. aureus* food poisoning occurs due to preformed heat stable enterotoxins produced in foods. Among the 21 Staph enterotoxin (SE) or enterotoxin-like proteins (SEls) identified, SEA is the most common toxin responsible for food poisoning. Consuming food infected with toxins generated by *S. aureus* can result in staphylococcal food poisoning, which can induce severe gastroenteritis, nausea, vomiting, diarrhea, and abdominal discomfort within 1–6 h (crago et al., 2012). Foods frequently associated include meat and meat products; poultry and egg products; and salads, such as egg, tuna, chicken, bakery products, and milk and dairy products. It has been reported that the prevalence of *S. aureus* in milk and milk products produced in developing nations is higher than in industrialized nations (EFSA, 2009), for instance, in India wherein milk products such as sweets are a common source for infection. Methicillin-resistant *S. aureus* (MRSA), when associated with human or animal infections, poses severe public health concern (Silva et al., 2013). The transmission of MRSA occurred in clinical settings, but its prevalence in the population, outside of hospitals, has been observed in recent decades (Oniciuc et al., 2015). Recent research by Caggiano et al. (2016) evaluated healthy persons who worked in the food sector and found that the presence of *S. aureus* and MRSA posed a threat to public health. LA-MRSA strains were detected in pork, poultry, and raw turkey products in the USA (Smith, 2015). Caggiano et al. (2016) conclude that the spread of *S. aureus* and MRSA in non-hospital settings, such as communities and animals, require ongoing and vigilant surveillance. Costa et al. (2015) performed *S. aureus* isolation and MRSA identification in which 9.5% of the samples contained MRSA. The high frequency of MRSA in meat, particularly in food prepared for consumption, highlights the need for the best food handling techniques in hospitals as well as the best animal handling methods.

Several reports on the presence of *S. aureus* in different foods such as meat, poultry, milk, and their products with prevalence in different products ranging from 10% to 100%. The description of several IDSP report outbreaks indicates toward staph poisoning.

11.3.5 *BACILLUS CEREUS*

It is believed that *Bacillus cereus* (*B. cereus*) is responsible for between 1.4% and 12% of foodborne outbreaks (FBOs) globally (Grutsch et al., 2018). In general, the incidence of outbreaks attributed to bacterial toxins increased, and in 2018, *B. cereus* was the fifth most common cause of FBOs (EFSA, 2019). After *S. aureus* , *B. cereus* is the second-most commonly detected causal agent in FBOs in France (Glasset et al., 2016). It causes acute gastroenteritis due to *B. cereus* intoxication and is commonly transmitted through consumption of rice, spices, and dairy products. A wide variety of foods such as meats, milk, vegetables, and fish are implicated. In the Chinese market, dairy products such as baby formula have produced a significant number of *B. cereus*-induced FBOs (Liu et al., 2020). Moreover, diarrheal-type illness has been recorded more often in Northern Europe, which includes Finland and Norway, and emetic-type disease has been observed more frequently in Japan and the United Kingdom, posing a significant issue in both industrialized and emerging nations (Bennett et al., 2013; Jessberger et al., 2020). In India, prevalence in meat and poultry products varies from 10% to 70%. Sporadic reports are available from the country. Although several IDSP outbreak descriptions from rice/biryani/noodles have been documented, no further reports on aetiology are available. Even though *B. cereus* contributes just a small percentage to the total number of FBOs, it is essential to highlight that these figures are underestimations. This is because the majority of *B. cereus*-caused food poisoning outbreaks are often minor and do not require medical treatment. Consequently, the majority of instances are not reported to health authorities (Jovanovic et al., 2021)

11.3.6 *CLOSTRIDIUM PERFRINGENS*

Foodborne illnesses caused by *Clostridium perfringens* (*C. perfringens*) are always linked with meat and poultry products. During slaughtering, animal products can get infected with this disease by contact with a contaminated surface or faeces (Leung et al., 2017; EFSA, 2017). After *Salmonella*, *C. perfringens-type A* food poisoning is the second-most widespread bacterial food poisoning in the USA, with an estimated 1 million cases each year (Scallan et al., 2011). The number of *C. perfringens*-related food poisoning occurrences in EU member states was estimated to be roughly 5 million per year (Felicio et al., 2015). *C. perfringens* is responsible for 5% of FBD outbreaks, 10% of illnesses, and 4% of hospitalizations, according to CDC epidemiology monitoring statistics (Dewey-Mattia et al., 2018). Heat tolerance is connected with the production of spores that may germinate at temperatures between 15 and 55 °C. Toxins of types A and C have been proven to cause human illness. Type A causes the majority of food poisoning and diarrheal diseases linked with *C. perfringens*. Important pathogen of both humans and animals are present in soil, water, dust, and the intestinal tracts of man and animals. Among the five types, type A is associated with food poisoning and harbors enterotoxin genes. Meat and poultry products are common sources. Inadequate heating, survival of spores, and post-production temperature abuse especially in large-scale preparation of foods, leads to human outbreaks. Prevalence in meats in India has been estimated to be 40–85% in carabeef, 35–75% in chevon, 30–75% in poultry, and 20–85% in RTE meat and poultry products.

11.3.7 *CLOSTRIDIUM BOTULINUM*

Foodborne botulism is a potentially lethal condition caused by the ingestion of neurotoxins generated by *Clostridium botulinum*. Foodborne botulism is due to preformed toxin botulinum neurotoxins (BoNTs), in food which are secreted by *Clostridium* organisms known as BoNT-releasing *clostridia*. Food sources include improperly canned foods, meat products, and seafoods. Foodborne botulism usually develops 18–36 h (range 6–8 days) following exposure (Peck et al., 2017). *C. botulinum* produces neurotoxins A, B, C, D, E, F, and G among which toxin A is the most potent toxin known with a toxic dose of 0.001 µg/kg BW. Inadequately heated and packed foods are common sources. Honey is most commonly implicated in infant botulism. During 2004–2020, a total of 80 foodborne botulism outbreaks occurred in China, involving 386 illnesses and 55 deaths; most outbreaks were reported between June and August, with a sharp peak in January (Li et al., 2022). In Italy, botulism is a public health issue with 355 confirmed cases of botulism involving 515 people from 1986 to 2019 (Benevenia et al., 2022). Between 2011 and 2015, an average of 162 yearly cases of botulism were recorded in the USA. Foodborne botulism accounts for 1%–20% of all botulism cases as documented by Jeffery and Karim (2017).

Lalitha and Surendran (2002) reported occurrence in foods such as cured fishes and salt dried shrimps in retail market of Cochin with an overall prevalence of 19%.

11.3.8 *MYCOBACTERIUM TUBERCULOSIS* AND *M. BOVIS* (BOVINE TUBERCULOSIS)

According to WHO, about 30% of total global tuberculosis (TB) cases are found in India. *M. bovis* (bovine TB) plays a crucial role in India, where 70% of the human population resides in rural areas. Consumption of unpasteurized raw milk is the major risk factor for transmission. Srinivasan et al. (2018) estimated a prevalence of 7.3% indicating that there may be approx. 21.8 million infected cattle in India affected with bovine TB. In another study to estimate the occurrence of bovine TB in three high-risk populations, *M. bovis* was detected in 11.4% of samples from farmers, dairy workers, and livestock keepers; 8.9% zookeepers and animal handlers; and in 12.6% of residents of TB endemic area (Bapat et al., 2017). In all these cases, previous contact with active TB cases and raw milk consumption were found to be important determinants of bovine TB. Though zoonotic TB organisms are widely present in cattle herds of different regions across the globe, a surveillance based on one health perspective is lacking in 89.9% of the WHO signatory countries (De Macedo Couto et al., 2022). In 2020, EU recorded 88 cases of TB in humans of which 86 were caused by *M. bovis* while the remaining two cases reported in Germany was due to *M. caprae* (EFSA-ECDC, 2020). In the USA, 1.4% of the total TB cases in humans are due to *M. bovis* infection. Some studies have proven that mortality is more for *M. bovis*-infected patients than those infected with *M. tuberculosis*. Global incidence of zoonotic TB estimated by WHO in 2010 came around 1.2 lakhs and in countries like Africa, about 70,000 cases are being reported annually (Olea-Popelka et al., 2017). Many studies have documented presence of *M. tuberculosis* in milk at the occurrence rate of 4.8–52.17% depending on the endemicity of region (OPZD Annual Reports 2014–2019). There is an urgent need to study the contribution of *M. bovis* toward human TB burden in India.

11.3.9 PARATUBERCULOSIS

M. avium subspp. *Paratuberculosis* (MAP) is responsible for causing paratuberculosis and is a major pathogen of the livestock in India and globally. The animals under sub-clinical to clinical stage continue to shed MAP bacilli in their milk and feces. Among the 26,000 domestic livestock screened, the bio-load of MAP in was found in 43% cattle, 36% buffaloes, 23% goats, and 41% sheep. Further, mass screening of 28,291 human samples from 2008 to 2016 also revealed a high bio-load of 33.7% (Chaubey et al., 2017). It is believed that 90% of the herds in USA are possessing at least few animals which are infected with MAP. A study conducted in disease-suspected animals from selected abattoirs in Mexico revealed a seroprevalence of 2.08% (Osuna-Chávez et al., 2021). There are reports that MAP infection in humans can trigger the development of many chronic and autoimmune diseases including type 1 diabetes mellitus. Some studies could discuss that the children who were seropositive for MAP antibodies were more prone to develop the disease than the healthy controls who were otherwise considered to be at risk (Ozana et al., 2022). The risk of MAP transmission exists even in pasteurized milk and its products owing to the ability of the organism to forms clumps in milk, which can lead to its survival during pasteurization. When the risk assessment for acquiring infection from different sources of milk and milk products were done using statistical models, it was found that the risk was very high for the people who consume raw milk from infected farms while pasteurization can decrease the risk substantially but not completely (Knific et al., 2022). Live bacilli have also been recovered from meat products and the environment thus illustrating the potential of MAP as a pathogen of public health concern. The organism is known to survive for 3 months in soft cheese and 10 months in hard cheese. MAP causes chronic inflammatory bowel disease in humans and is similar to Jhone's disease in ruminants. Due importance must be given toward the transmission of MAP from milk and its products with research to ensure zero-survival during pasteurization.

11.3.10 YERSINIOSIS

Yersinia enterocolitica is a widespread zoonotic pathogen found in soil, water, animals, and numerous foods (Ahmed et al., 2019; Bari et al., 2013). In some countries (including New Zealand), the intestinal condition yersiniosis caused by *Yersinia enterocolitica* or *Y. pseudotuberculosis* is among the most frequently reported foodborne infections, posing a significant public health crisis (Pattis et al., 2019). Because of the effort required in pig farming and the close interaction between dogs and humans, pigs and dogs are a major source of *Y. enterocolitica* infections in humans (Duan et al., 2014). Acute gastroenteritis is the most common word for Yersiniosis, and the severity of the illness is mostly influenced by the age of the person who becomes infected. (El Qouqa et al., 2011). *Y. enterocolitica* is an increasingly diverse foodborne zoonotic pathogen that may cause significant morbidity and death, particularly in newborns and young children (Pal, 2018). It is also an important FBD, mostly spread by direct contact between humans and animals or through the environment (Bancerz-Kisiel and Szweda, 2015). Six biotypes of *Y. enterocolitica* exist: 1A, 1B, 2, 3, 4, and 5. Biotype 1B was previously considered extremely pathogenic, whereas biotype 1A was deemed to be non-pathogenic in humans; other biotypes are considered to be of low pathogenicity in humans (Valentin-Weigand et al., 2015). As recently calculated, the prevalence of Y. enterocolitica in instances of human gastroenteritis is 1.97% (95% confidence interval 1.32–2.74%), with serotype O:3 being the most prevalent (Riahi et al., 2021). Reports released by EFSA and the ECDC Zoonoses Report for 2020 show the number of human cases of this disease in Europe has increased to 5668, with very little data available on the meat-producing supply chain (EFSA, 2021). Furthermore, just 0.2% of the 2351 pigs examined in six European nations tested positive for *Y. enterocolitica. Y. enterocolitica* infection is more prevalent in nations with temperate climates than in tropical or subtropical locations. Europe had the most severe epidemic of enteric yersiniosis in 2019, and it was the third-most frequent zoonotic illness in Europe (EFSA, 2017). This foodborne bacteria is thought to be the cause of around 87,000 cases per year throughout the world (Pal et al., 2013).

11.3.11 FOODBORNE ZOONOTIC VIRUSES

Virology research is rapidly expanding leading to discovery of new viruses at a comparatively higher pace. Out of the 22 families of viruses which are found to infect humans, at least 10 are documented to cause foodborne transmission. In the last decade, we have witnessed outbreaks due to consumption of food contaminated with emerging viruses like Nipah virus, avian influenza virus, and SARS coronavirus. Though those outbreaks were mainly associated with a particular food habit, they have raised the query if consumption of food contaminated during preparation, handling, slaughtering, etc., could be one of the routes by which emerging viruses can infect the human population. Foodborne viruses are mainly transmitted via the fecal-oral route through ingestion of contaminated food and water, and/or by person-to-person contact. Though burden of foodborne viruses is largely unknown, according to WHO (2015) estimates, 15% of foodborne related deaths due to diarrheal agents were because of Norovirus and 23.6% of foodborne-related deaths due to invasive disease agents were due to Hepatitis A virus. FAO and WHO have considered the noroviruses, hepatitis A virus, and the group A rotaviruses as the priority viral foodborne hazards. The genomic studies of these viruses have shown that they are chiefly transmitted through the food handlers. Besides these, other viruses like enteroviruses, sapoviruses, rotaviruses, astroviruses, and adenoviruses have also been implicated in food- and/or water-borne illness. Several challenges including development of reliable diagnostics to detect viruses in food and generation of reliable surveillance data needs to be addressed for devising adequate preventive and control measures.

11.3.12 NOROVIRUS

Worldwide, Norovirus is the leading causative agent for acute gastroenteritis, particularly in countries where rotavirus vaccination is implemented. Norovirus infection can cause explosive vomiting, abdominal pain, and watery diarrhea. The virus belongs to the family *Caliciviridae* and is reported to cause infection mainly in humans, while its presence is detected in cattle, sheep, dog, pig, and mice, too. Incubation period for the disease is usually 12–48 h and it usually lasts for 1–3 days (Velebit et al., 2019). Severe long-drawn-out diarrhea can be observed in elderly and immuno-compromised individuals, where they shed the virus for extended time period indicating active viremia phase but slow viral clearance. The virus concentration in the feces and vomitus (as high as 1010 viruses per gram) was found to be independently associated with the protracted diarrhea (Lee et al., 2007). Norovirus causes 124.8 million cases of foodborne illness and 34,929 deaths (WHO, 2015). There are seven genogroups of norovirus, of which five are responsible for human gastroenteritis. Infective dose of this virus is as low as 10 organisms. The low infective dose, the ability to survive in harsh environment, several transmission routes, and short-lived immunity against the virus has hiked its impact in global population.

A meta-analysis report on a global scale regarding selected gastroenteritis cases showed that 16% of the total cases were due to norovirus infection. The prevalence of gastroenteritis produced by norovirus was found to be maximum in South America with 22%, followed by Europe and Asia with 16% each while Africa showed a prevalence of 15% which was the least (Liao et al., 2021). Globally, GII.4 and GII.3 were found to be the most commonly occurring genotypes with highest frequency being reported in children below 1 year of age (Farahmand et al., 2022). Asymptomatic infections with a prevalence of 7% was estimated by a meta-analysis and this should be overlooked as those individuals can obviously contribute to the disease transmission (Qi et al., 2018). In Southern India, out of 1856 diarrheal episodes, 207 (11.2%) were associated with norovirus (Menon et al., 2016). NoV positivity varied between 6.3% and 12.6% in different cities of Western India with the predominance of GII (96.6%) (Chhabra et al., 2009). First instance of inter-genogroup recombination has been reported from Kolkata, having polymerase sequence of GI and genotype 3 (GI.3), whereas the capsid sequence belongs to GII and genotype 4 (GII.4) (Nayak et al., 2008). To date, no animal norovirus have been detected in human stool, but some serological evidence hints to possible transmission from animals to humans (Villabruna et al., 2019). Some viruses that are categorized as human norovirus today might have originated from animal source. Targeted surveillance including samples of both humans and animals that are in close contact is required to record a trans-species transmission event.

11.3.13 HEPATITIS A VIRUS AND HEPATITIS E VIRUS

These viruses are mainly transmitted via the fecal-oral route. Zoonotic infections by hepatitis E virus (HEV) can occur via contact with infected animals or by consumption of contaminated foods of animal origin like meat, milk, and eggs. HEV was identified initially from Indian state of Jammu and Kashmir in the year 1978. The virus belongs to the family Hepeviridae. It infects many animals including wild and domestic pigs, camel, chicken, mongoose, rat, deer, and rabbits. Most acute hepatitis E infections in humans have been linked to consumption of contaminated pork or pig liver. Extra-hepatic diseases like neurological, hematological, cardiovascular, and renal syndromes can also be caused by the virus (Cossaboom et al., 2016; Khuroo et al., 2016). A systematic review of the global epidemiology of HEV estimates that nearly 939 million individuals have got infected ever which corresponds to 1 in 8 of the population and 15–110 million people are experiencing ongoing infection (Li et al., 2020). Globally, the seroprevalence rate was found to be 60% and 27% in domestic pigs and wild boars respectively which had ever encountered the disease whereas the HEV RNA positivity which indicates the active infection was 13% and 9.5% in domestic and wild swine, respectively. The HEV RNA positivity in commercial pork products was estimated to be 10% which puts forward the potential of its foodborne zoonotic transmission (Li et al., 2022). In developed countries, HEV infection is considered as a leading cause of viral hepatitis which are zoonotic. Out of the eight genotypes identified, genotypes 1, 2, 3 and 4 are frequently infecting humans. The transmission of disease in developing nations of Asia, Africa, and Latin America occurs mainly through contaminated water wherein genotypes 1 and 2 are the most common in occurrence. Genotypes 3 and 4 are frequent when the transmission is via fecal contamination or through consumption of contaminated or undercooked meat. While HEV1 and HEV2 shows endemic pattern of disease, HEV3 and HEV4 shows sporadic occurrence. HEV1 and HEV2 are found to cause disease in humans only while the other two produces disease both in humans as well as animals. A single case of zoonotic infection with genotype HEV7 was reported in UAE in a person who consumed meat and milk from an infected camel. Due to lack of FDA-approved testing methods, the exact disease occurrence in USA is undetermined (Damiris et al., 2022). Hepatitis A virus is typically transmitted through undercooked or raw seafood or other contaminated raw produce. WHO estimated 14 million HAV infection worldwide in the year 2010, but the occurrence of the disease is variable in different areas. India and Sub-Saharan Africa are grouped as highly endemic areas while Australia, USA, and Europe have low endemicity (Bachofen, 2018). Both these viruses are highly endemic in India and according to one report, sero-prevalence for HAV and HEV was found to be 92.68% and 17.05%, respectively, in individuals of India (Kotwal et al., 2014). Further, they reported that hand washing without soap, regular close contact with domestic animals, consumption of unpasteurized milk, and regular consumption of food outside home were risk factors for HAV ($p < 0.05$). While, for HEV, irregular hand washing, consumption of unpasteurized milk, and irregular consumption of freshly prepared food were risk factors ($p < 0.05$). Status of HEV in pig population of India is largely unknown as only few studies have been conducted till now. Although transmission through contact with infected animals have been proven, status of butchers and farmers still needs to be assessed in Indian context.

11.3.14 ROTAVIRUS

Rotavirus belongs to the family Reoviridae and it is the leading cause of severe diarrhea in neonates and infants in developed and developing countries including India. Humans get exposed to the virus by direct contact with the infected animals, through contaminated food and water and at times through blood transfusion (Wilhelm et al., 2015). The virus is highly variable in its genetics and thus classified into eight serogroups. Though reports of zoonotic transmission exists, human-to-human transmission is more frequent. A study in Southern Italy during 2015 recorded 16.87% prevalence of rotavirus in mussels (Fusco et al., 2017). A study for the detection of zoonotic viruses in pig fecal samples revealed 2.7% prevalence of rotavirus on assessing the presence of viral RNA (Lyoo et al., 2020). In year 2013, an estimated 47,100 rotavirus deaths occurred in India, 22% of all deaths due to rotavirus that occurred globally that year (Tate et al., 2016). In year 2016, Government of India launched vaccination program against rotavirus in phased manner, but the data of diarrheal cases after vaccination against rotavirus has not been analyzed yet. Though burden of rotavirus illness has been extensively mapped across the world, very little attention has been paid to identify the other potential modes of transmission of virus. The surveillance of circulating rotaviruses in the human population has revealed the presence of uncommon strains that are commonly found in domestic animals. Studies on unusual rotaviruses from livestock and companion animals are limited and the genomes of only a handful of these strains have been analyzed so far through which rare interspecies transmission and reassortment events have been proven. Surveillance for other groups of rotavirus besides Rotavirus A is also required.

11.3.15 FOODBORNE PARASITES

Apart from bacteria and viruses, the parasitic organisms especially the protozoa, nematodes, cestodes, and trematodes contribute widely to the foodborne infections. *Cryptosporidium*, *Giardia*, *Cyclospora*, and *Toxoplasma* are the major foodborne parasitic protozoa worldwide. Cryptosporidium and Giardia are widespread in aquatic environment, and easily spread via contaminated drinking water. 8.6 million cases of cryptosporidiosis and 3759 deaths were reported worldwide in 2010. It is the second major cause of mortality in children younger than 5 years due to diarrheal diseases. Anthroponotic and zoonotic transmission as well as through contaminated food and water are common routes of acquiring infection. Waterborne outbreaks of cryptosporidiosis had been reported in 1993 in USA wherein about four lakh individuals encountered the disease (Ryan et al., 2018). Recent researches has found that the disease can induce some long-term effects like lack of cognitive skills in malnourished children and economic losses in terms of livestock production. The parasite rapidly multiplies in the gut cells of the host damaging its intestinal epithelium. The host can shed a huge number of oocysts in feces leading to environmental contamination and spread. *Cryptosporidium parvum* is zoonotic while most of the other species are not. The parasite causes disease in many species of animals including cattle, pig, sheep, goat, equines, and many wild species. *Cryptosporidium* is the major cause of neonatal diarrhea in calves and at times in lambs and kids (Innes et al., 2020). Seasonal pattern of disease occurrence was reported as a hike in late spring and summer in some parts of Europe, and during spring in United Kingdom which may be associated with the lambing season and subsequent shedding of oocysts by young animals. The seasonality of disease occurrence has also been discussed by various authors in countries such as Ireland, New Zealand, Canada, and Netherlands. Epidemiological investigation of cryptosporidiosis in France during 2017–2019 revealed high occurrence of IIa and IId subtypes of *Cryptosporidium parvum* suggestive of frequent zoonotic transmission (Costa et al., 2020). Global prevalence of cryptosporidiosis was estimated to be 7.6% with maximum frequency in Mexico (69.6%), followed by Nigeria (34%) and Bangladesh (42.5%), according to a meta-analysis study by Dong et al. (2020). It has been reported that *Cryptosporidium* is responsible for 7% of children deaths due to diarrhea in India (Troeger et al., 2017). In general, parasitic protozoa do not multiply in foods, but they can survive in food stored under moist, cool, or damp conditions for months.

T. gondii is the potential foodborne pathogen which can be transmitted by consumption of raw milk or undercooked meat where the infective tissue cyst lodges. This parasite is responsible for causing abortion in pregnant women and may also cause chorioretinitis and encephalitis. *T. gondii* is incriminated as a fatal foodborne protozoan zoonotic disease in USA. Moreover, *T. gondii* is reported to be chronically infecting one-third of the global human population. A meta-analysis estimated a global prevalence of 33.8% latent infection in pregnant women wherein South America reported the highest prevalence of 56.2% (Almeria and Dubey, 2021). European Food Safety Authority reported that 60% of all toxoplasma infections are meat borne wherein pork, beef, and mutton are the major sources to contribute (Koutsoumanis, 2018). The prevalence of *T. gondii* in poultry and livestock during the period from 2000 to 2019, when estimated using random-effect model, was estimated to be 28.3%. The prevalence was highest in 2014 and lowest in 2013 in Asia (Hajimohammadi et al., 2022).

Taenia solium cysticercosis and trichinellosis are assumed to be 100% foodborne, but indirectly. Taeniasis/cysticercosis mainly affects the people in developing countries of Africa, Asia, and Latin America. It is second major cause of non-diarrheal foodborne deaths (WHO, 2015). Taeniasis and neurocysticercosis (NCC) are common in areas where animal husbandry practices are such that pigs and cattle come into contact with human feces. The metacestode stage of the larva is infective while ingestion of eggs also can cause infection with *T. solium* and *T. saginata* in humans. Cysticercosis is placed under the WHO–FAO list of neglected zoonotic diseases. The disease in endemic in developing nations of Asia, Latin America, and Africa while the long-absent cases are re-emerging in Western Europe as a result of immigration and trade. The infection can result in mild or non-specific gastrointestinal symptoms to fatal neurological manifestations (Symeonidou et al., 2018). The disease is prevalent in all states of India, although the prevalence varies between the states (NHP, 2019). The National Institute of Mental Health and Neuro Sciences (NIMHANS), Bangalore, reported NCC in 2% of unselected series of epilepsy patients. In a study from New Delhi, NCC accounted for 2.5% of all intracranial space occupying lesions. An unusual feature of the disease was seen that more than 95% of Indian patients with NCC are vegetarians. The transmission of *T. solium* eggs through contaminated water and salads might be one of the reasons for this finding. Consumption of infected pork and wild animal meat results in trichinellosis. Trichinellosis in animals is a notifiable disease and EC regulation 1375/2015 proposed mandatory active surveillance during meat inspection (Van der Giessen et al., 2021). Southwestern area of China is endemic to human Trichinellosis where eight outbreaks of the disease were reported between the period of 2009 and

2020 with 0.4% case fatality rate (Zhang et al., 2022). Seropositivity of anti-trichinella IgG antibodies in North Eastern Iran was calculated to be 2.6%. Consumption of game meat by certain communities there could have exposed the people to trichinellosis (Koohsar et al., 2021). Number of reported cases of human Trichinellosis was 117 in the EU in 2020. The infection rate was nil in the fattening pigs or breeding pigs while 0.0001% positive cases were observed in pigs which were not kept under controlled housing conditions which confirms that altering farming pattern can be a significant factor to control the disease (EFSA-ECDC, 2020). The exact burden due to parasitic contaminants in food is still largely unknown in India.

11.4 GLOBAL STATUS OF FBDs

World Health Organization estimates that 1 in 10 people are affected with foodborne infections each year resulting in a death count of 420,000. The highest burden of the diseases are found in South-East Asian and African regions. The diarrheal diseases which are the most common form are particularly caused by consumption of undercooked meat and dairy products, contaminated by Norovirus, *Campylobacter*, *Salmonella*, and *E. coli*. While pathogenic *E. coli* and *Campylobacter* are more frequent in low and high income countries, non-typhoidal *Salmonella* is a concern over all regions of the globe. Foodborne entero-pathogenic microorganisms are found to produce 1.5 billion cases of diarrhea per year particularly in children below 3 years of age resulting in more than 3 million deaths (EFSA-ECDC, 2016).

The African region of WHO is with the highest rate of FBD with 91 million people being infected per year with death reports of 137,000 of which 32,000 people die of non-typhoidal Salmonella. *Taenia solium* is responsible for 10% of the total foodborne infections in this region. FBD outbreak reports in South Africa during the period between 2013 and 2017 was reviewed and estimated to be 327 in number where above 11,000 reported cases with 0.4% case fatality were present. The etiology for most of the outbreaks were foodborne bacterial pathogens (Shonhiwa et al., 2019).

The diseases like toxoplasmosis and taeniasis which are spread through raw or undercooked meat are important foodborne zoonosis as far as Central and South America are concerned. In these countries, 77 million people fall ill each year due to foodborne infections.

FBD burden in Eastern Mediterranean region is estimated to be 100 million cases per year of which 32 million are children under 5 years of age. Along with the diarrheal diseases which accounts to 70% of the total, typhoid, brucellosis, and hepatitis A are reasons for death of about 37,000 people each year.

Lowest burden of FBDs are estimated in WHO European region with 23 million people getting diseased each year of which 15 million cases accounts to Norovirus infection, 5 million Campylobacteriosis cases, and 1 million foodborne Toxoplasmosis. Around 400 deaths are recorded annually in European region due to Listeriosis caused by consumption of contaminated meat and milk products, RTE meals as well as raw vegetables. According to EFSA-ECDC annual report, the most reported cases of FBD in EU is Campylobacteriosis followed by Salmonellosis in 2020 and the trend was stable from 2016 to 2020. Yersiniosis, ETEC, and Listeriosis are among the other infections which were less reported comparatively.

South-East Asian region is next to African region as far as foodborne infections burden is considered. Above 150 million cases are reported each year annually and deaths toll to 175,000. In the year 2020, China witnessed 7073 outbreaks of foodborne illnesses wherein 143 reported deaths occurred among 37,454 affected individuals. About 41.7% accounted to the foodborne microbial pathogens while poisonous mushrooms caused 58% outbreaks (Li et al., 2021).

In Western Pacific Region, 125 million people fall ill annually due to contaminated food and about 50,000 deaths occur. In this region, aflatoxin is the major burden rather than diarrheal diseases. Annually, above 10,000 people are affected with hepatic cancer due to this (WHO, 2015).

11.5 INDIAN STATUS OF FBDs

The burden of FBD in India is largely unknown. Majority outbreaks of FBDs go unreported, unrecognized, or un-investigated and may only be reported after major health damage has occurred. Recent estimates state that at least one out of 12 people fall ill due to food-related diseases in India, the number of foodborne illnesses are projected to grow by as much as 70% resulting in one out of 9 people falling ill by 2030, largely driven by population and changing consumption patterns (Kristkova et al., 2017). It is important to note that the expected FBDs increase might be still higher given the likely underestimation of the current FBDs burden. The analysis of IDSP data from the year 2011–2015 revealed that FBOs together with acute diarrheal diseases constitute nearly half of all reported outbreaks. Moreover, for the

FBDs, passive surveillance systems represent only the tip of the iceberg because most patients have mild symptoms of short duration and thus do not visit doctor and many of those who seek medical care may not have the laboratory test performed to determine a specific etiology. A review of recorded FBD outbreaks in India from 1980 to 2016 revealed that *S. aureus*, *Vibrio* spp., *Salmonella* spp., *E. coli*, *Y. enteroclitica*, and Norwalk-like virus are some important microbial pathogens responsible for foodborne illnesses (CD Alert, 2017). Different foods implicated in FBOs in India are milk and milk products such as dahi, khoa, butter milk, sweets, kheer; meat, poultry, fish, fowl, sea food such as prawns; cooked and uncooked rice; samosa, tamarind; and cooked as well as uncooked vegetables.

The total number of cases and deaths due to acute diarrheal diseases in India were 1,41,66,574 cases and 1555 deaths in year 2016 and 12,927,212 cases and 1331 deaths in year 2017 (NHP, 2018). Community studies revealed that every child under 5 years of age had two or three episodes of diarrhea every year, and 400,000–500,000 children under 5 years died from diarrhea. Viral hepatitis was estimated about 12 cases per 100,000, although urban communities had as high as 100 per 100,000. As per estimates by Kristkova et al. (2017), if the current scenario of foodborne illnesses is not mitigated, the number of people infected may increase from 100 million people today to 170 million in 2030.

11.6 STRATEGIES AND PREVENTIVE APPROACHES

Farm to fork approach is required to ensure food safety which involves series of steps to be taken by multidisciplinary team involving veterinarians, microbiologists, environmentalists, industry people, and most importantly consumers. Since the foodborne zoonotic diseases and their causative agents differ between the regions, the preventive approach should be tailor-made to each region at the national level. The existing difference in the burden of FBD between low- and high-income regions clearly suggests that a major fraction of the current burden is still avoidable through implementation of general principles of food safety system. An enforcement of food safety standards based on risk assessment in particular region along with effective surveillance networks will aid in prevention of FBDs to a greater extent. The following strategies may be implemented for effective prevention and control of foodborne zoonotic diseases:

11.6.1 STRATEGY I—SURVEILLANCE AND RESPONSE

- Prioritization of important foodborne zoonotic diseases on the basis of data available and interactions with planners, policy makers, scientists, extension workers, NGOs, etc., from different professional and social organizations.
- Ensure that priority zoonotic diseases are agreed among all the relevant sectors.
- Identification of network laboratories and working group for each pathogen across the country.
- Use of validated diagnostics and common protocols for pathogen detection and identification.
- Formulation of statistical plan for collection of samples from the entire value chain during surveillance.
- Information sharing and interactions between the core groups, while considering reliability of the data based on its sources.
- Strengthen public health systems to support surveillance and to implement prevention and control programs.
- Improve ability to communicate electronically with state and local health departments.

11.6.2 STRATEGY II—MICROBIOLOGICAL RISK ASSESSMENT AND RISK MANAGEMENT

- Risk assessment studies to be carried out for important foodborne zoonotic pathogens.
- Identification and implementation of different risk management options throughout the food chain from farm to fork for different food production systems.
- Establishing microbial standards based on the MRA studies in Indian context.
- Consider the food supply chain as a whole and look at how contamination might occur. Then, identify the individual control points where contamination can be minimized or eliminated. The recent adoption of WGS-based monitoring in several countries provides new prospects for defining control objectives more precisely, reducing food contamination, and reducing the enormous burden of disease.

11.6.3 STRATEGY III—APPLIED RESEARCH

- Reliable, rapid, simple, and affordable diagnostic tests need to be developed.
- Rapid screening test may be devised for detection of food intoxication, especially to test foods which are prepared and served in mass congregations.
- Alternatives to drugs and antimicrobials need to be explored.
- Genome analysis of foodborne pathogens to record interspecies transmission and to gain insights in the field of pathogen behavior.
- Research on prevention and control throughout food chain.

11.6.4 STRATEGY IV—COORDINATED APPROACH AND AWARENESS CAMPAIGNS

- Establish regional labs where experts from human health, animal health and agricultural scientists work together to contain FBDs with an integrated approach.
- Coordinated mechanisms to strengthen multisectoral collaboration.
- Awareness campaigns on zoonoses through interactions and trainings for professionals, high risk groups, common masses, and people engaged in livestock industry.
- Need to instill true sense of sanitation and hygiene among various stakeholders through motivation, education, and reward/recognition using mass media.
- Kitchen hygiene is an area where basic food safety education is to be provided so as to avoid risky food practices.
- Advocacy on the role of local welfare bodies, municipal corporation, and NGOs for prompt management and disposal of biohazard waste and garbage.

11.7 CONCLUSION

FBDs cause huge morbidity and considerable mortality and pose significant obstacle to socioeconomic development worldwide. However, information regarding magnitude of burden due to specific pathogens is sketchy in our country. On the other hand, precise information on the burden of foodborne zoonotic diseases is essentially required for taking suitable decision by policy-makers and to allocate appropriate resources for food safety control and intervention efforts. The prevention and control measures should be initiated by veterinarian's from primary production level through good animal husbandry practices (GAHP) and biosecurity measures. Further, during production and processing level good manufacturing practices (GMP) and good hygienic practices (GHP) and implementation of HACCP plan should be followed for reducing FBDs. There is an urgent need for prudent use of antibiotics in both human and veterinary medicine in order to mitigate the rising antimicrobial resistance scenario. One health approach not only improves the understanding of epidemiological nature of diseases, but also aid in effective implementation of control measures. It is increasingly necessary to consider holistically all aspects of FBD in the One Health Framework.

11.7.1 RECOMMENDATIONS/ACTIONABLE POINTS

- Formulate a "Network program on FBDs" with the mandate of conducting surveillance studies on foodborne zoonotic diseases followed by risk assessment studies for most prevalent/important pathogens.
- Formation of Rapid Response Teams comprising multisubject specialists for investigation of FBOs.
- Inclusion of veterinary public health specialist in all IDSP labs.
- Creation of a digital database like on the lines of "FoodNet" of CDC USA.
- Integrating food safety into nutrition and food security policies and programs.
- Incorporate zoonoses including foodborne infections and emerging infectious diseases in the National Health Policy.
- Mandatory training of butchers on hygienic slaughter practices and making it compulsory that only those personnel who have completed this training can do dressing.
- Training of street food vendors in basics of hygiene and food safety and issue of certificates so as to improve consumer confidence.
- A content on foodborne zoonotic disease and food safety may be made part of course curriculum of school and for undergraduate courses of home science and hospitality management.

KEYWORDS

- **foodborne zoonoses**
- **control strategies**
- **public health**
- **socioeconomic development**
- **healthy food**

REFERENCES

Ali, M.; Gupta, S. S.; Arora, N.; Khasnobis, P.; Venkatesh, S.; Sur, D.; Nair, G. B.; Sack, D. A.; Ganguly, N. K.; Identification of Burden Hotspots and Risk Factors for Cholera in India: An Observational Study. *PLoS One* **2017,** *12* (8), e0183100.

Arai, S.; Yamaya, S.; Ohtsuka, K.; Konishi, N.; Obata, H.; Ooka, T.; ... Hara-Kudo, Y. Detection of Escherichia Albertii in Retail Oysters. *J. Food Prot.* **2022,** *85* (1), 173–179.

Assefa, A.; Bihon, A. A Systematic Review and Meta-Analysis of Prevalence of *Escherichia coli* in Foods of Animal Origin in Ethiopia. *Heliyon* **2018,** *4* (8), e00716.

Bachofen, C. Selected Viruses Detected on and in Our Food. *Curr. Clin. Microbiol. Rep.* **2018,** *5* (2), 143–153.

Bancerz-Kisiel, A.; Szweda, W. Yersiniosis—A Zoonotic Foodborne Disease of Relevance to Public Health. *Annal. Agric. Environ. Med.* **2015,** *22* (3).

Bapat, P. R.; Dodkey, R. S.; Shekhawat, S. D.; Husain, A. A.; Nayak, A. R.; Kawle, A. P.; and Kashyap, R. S. Prevalence of Zoonotic Tuberculosis and Associated Risk Factors in Central Indian Populations. *J. Epidemiol. Glob. Health* **2017,** *7* (4), 277–283.

Barbuddhe, S. B.; Malik, S. V. S.; Kumar, J. A.; Kalorey, D. R.; Chakraborty, T. Epidemiology and Risk Management of Listeriosis in India. *Int. Jo. Food Microb.* **2012,** *154* (3), 113–118.

Batra, P.; Mathur, P.; Misra, M. C. *Aeromonas* spp.: An Emerging Nosocomial Pathogen. *J. Lab. Phys.* **2016,** *8* (1), 1.

Benevenia, R.; Arnaboldi, S.; Dalzini, E.; Todeschi, S.; Bornati, L.; Saetti, F.; ... Losio, M. N. Foodborne Botulism Survey in Northern Italy from 2013 to 2020: Emerging Risk or Stable Situation?. *Food Control.* **2022,** *132*, 108520.

Bennett, S. D.; Walsh, K. A.; Gould, L. H. Foodborne Disease Outbreaks Caused by *Bacillus cereus*, *Clostridium perfringens*, and *Staphylococcus aureus*—United States, 1998–2008. *Clin. Infect. Dis.* **2013**, *57*, 425–433.

Brahmabhatt, M. N. Prevalence of *Aeromonas* spp. in Chicken Samples Collected from Retail Shops of Anand (Gujarat). *J. Vet. Public Health,* **2011,** *9* (2), 115–117.

Caggiano, G.; Dambrosio, A.; Ioanna, F.; Balbino, S.; Barbuti, G. de Giglio, O.; Diella, G.; Lovero, G.; Rutigliano, S.; Scarafile, G.; Baldassarre, A.; Vimercati. L.; Musti, M.; Montagna, M. T. Prevalence and Characterization of Methicillin-resistant *Staphylococcus* aureus Isolates in Food Industry Workers. *Ann Ig* **2016,** *28,* 8–14.

Peck, M. W.; Smith, T. J.; Anniballi, F.; Austin, J. W.; Bano, L.; Bradshaw, M.; ... Stringer, S. C. Historical Perspectives and Guidelines for Botulinum Neurotoxin Subtype Nomenclature. *Toxins* **2017,** *9* (1), 38.

CD Alert *Foodborne Diseases and Food Safety in India. Monthly Newsletter of the National Centre for Disease Control Directorate General of Health Services*; Government of India, 2017.

CFSPH. Reptile-Associated Salmonellosis [Online] 2013. http://www.cfsph.iastate.edu/Factsheets/pdfs/reptile_associated_salmonellosis.pdf

Chaubey, K. K.; Singh, S. V.; Gupta, S.; Singh, M.; Sohal, J. S.; Kumar, N.; ...; Dhama, K. Mycobacterium Avium Subspecies Paratuberculosis—An Important Foodborne Pathogen of High Public Health Significance with Special Reference to India: An Update. *Vet. Q.* **2017,** *37* (1), 282–299.

Chen, X.; Zhu, Q.; Yu, F.; Zhang, W.; Wang, R.; Ye, X.; Jin, L.; Liu, Y.; Li, S.; Chen, Y. Serology, Virulence and Molecular Characteristics of Vibrio Parahaemolyticus Isolated from Seafood in Zhejiang province. *PLoS One* **2018,** *13,* e0204892. [Google Scholar] [crossref]

Chhabra, P.; Dhongade, R. K.; Kalrao, V. R.; Bavdekar, A. R.; Chitambar, S. D. Epidemiological, Clinical, and Molecular Features of Norovirus Infections in Western INDIA. *Journal of medical virology*, **2009,** *81* (5), 922–932.

Cossaboom, C. M.; Heffron, C. L.; Cao, D.; Yugo, D. M.; Houk-Miles, A. E.; Lindsay, D. S.; Zajac, A. M.; Bertke, A. S.; Elvinger, F.; Meng, X.-J. Risk Factors and Sources of Foodborne Hepatitis E Virus Infection in the United States. *J. Med. Virol.* **2016,** *88* (9), 1641–1645.

Costa, W. L. R.; Ferreira, J. D. S.; Carvalho, J. S.; Cerqueira, E. S.; Oliveira, L. C.; de Almeida, R. C. Methicillin-resistant *Staphylococcus aureus* in Raw Meats and Prepared Foods in Public Hospitals in Salvador, Bahia, Brazil. *J. Food Sci.* **2015,** *80* (1), M147–M150. https://doi.org/10.1111/1750-3841.12723

Costa, D.; Razakandrainibe, R.; Valot, S.; Vannier, M.; Sautour, M.; Basmaciyan, L.; ... Favennec, L. Epidemiology of Cryptosporidiosis in France from 2017 to 2019. *Microorganisms* **2020,** *8* (9), 1358.

Crago, B.; Ferrato, C.; Drews, S. J.; Svenson, L. W.; Tyrrell, G.; Louie, M. Prevalence of *Staphylococcus aureus* and Methicillin-resistant *S . Aureus* (MRSA) in Food Samples Associated with Foodborne Illness in Alberta, Canada from 2007 to 2010. *Food Microbiol.* **2012,** *32,* 202–205. https://doi.org/10.1016/j.fm.2012.04.012

Dallagassa, C. B.; Surek, M.; Vizzotto, B. S.; Prediger, K. C.; Moriel, B.; Wolf, S.; ... Fadel-Picheth, C. M. Characteristics of an Aeromonas Trota Strain Isolated from Cerebrospinal Fluid. *Microb. Pathog.* **2018,** *116,* 109–112.

Damiris, K.; Meybodi, M. A.; Niazi, M.; Pyrsopoulos, N. Hepatitis E in Immunocompromised Individuals. *World J. Hepatol.* **2022,** *14* (3), 482.

Das, A.; Vinayasree, V.; Santhosh, C. R.; Hari, S. S. Surveillance of *Aeromonas* sobria and Aeromonas Hydrophila from Commercial Food Stuffs and Environmental Sources. *J. Exp. Sci.* 2012.

De Macedo Couto, R.; Santana, G. O.; Ranzani, O. T.; Waldman, E. A. One Health and Surveillance of Zoonotic Tuberculosis in Selected Low-income, Middle-income and High-Income Countries: A Systematic Review. *PLoS Neglect. Trop. Dis.* **2022,** *16* (6), e0010428.

Deng, X.; den Bakker, H. C.; Hendriksen, R. S. Genomic Epidemiology: Whole-Genome-sequencing-powered Surveillance and Outbreak Investigation of Foodborne Bacterial Pathogens. *Annu. Rev. Food Sci. Technol.* **2016,** *7* (1), 353–374.

Dewey-Mattia, D.; Manikonda, K.; Hall, A. J.; Wise, M. E.; Crowe, S. J. Surveillance for Foodborne Disease Outbreaks—United States, 2009–2015. *MMWR Surveill Summ.* **2018,** *67* (10):, 1–11.

Dewey-Mattia, D.; Manikonda, K.; Hall, A. J.; Wise, M. E.; Crowe, S. J. Surveillance for Foodborne Disease Outbreaks—United States, 2009–2015. *MMWR Surveillance Summaries* **2018,** *67* (10), 1.

Didugu, H.; Krishnaiah, N.; Rao, M. T.; Ramanipushpa, R. N. Antibiogram of Aeromonas Species Isolated From Livestock Products. *Int. J. Sci. Environ. Technol.* **2016,** *5* (2), 475–478.

Doimari, S.; Singh, V.; Kumari, R.; Kumar, M. S.; Singh, D. K. In Vitro Antimicrobial Susceptibility of *Brucella* Species Isolated from Human and Animals in India. *J. Antibiot. Res.* **2018,** *3* (1), 101.

Dong, S.; Yang, Y.; Wang, Y.; Yang, D.; Yang, Y.; Shi, Y.; ... Zhou, Y. Prevalence of Cryptosporidium Infection in the Global Population: A Systematic Review and Meta-Analysis. *Acta Parasitol.* **2020,** *65* (4), 882–889.

EFSA, (European Food Safety Authority) Assessment of the Public Health Significance of Methicillin Resistant Staphylococcus Aureus (MRSA) in Animals and Foods. Scientific Opinion of the Panel on Biological Hazards. *EFSA J.* **2009,** *993,* 1–73.

EFSA. European Food Safety Authority and European Centre for Disease Prevention and Control. *EFSA J.* **2015,** *13*, 4329.

El Qouqa, I. A.; El Jarou, M. A.; Samaha, A. S. A.; Al Afifi, A. S.; Al Jarousha, A. M. K. Yersinia Enterocolitica Infection among Children Aged Less than 12 Years: A Case–control Study. *Int. J. Infect. Dis.* **2011,** *15* (1), e48–e53.

European Food Safety Authority (EFSA). The European Union One Health 2018 Zoonoses Report. *EFSA J.* **2019,** *17* (12), e05926. Https://doi.org/10.2903/j.efsa.2019.5926

European Food Safety Authority (EFSA). European Centre for Disease Prevention and Control (ECDC) The European Union Summary Report on Trends and Sources of Zoonoses; Zoonotic Agents and Food-Borne Outbreaks in 2017.

European Food Safety Authority, and European Centre for Disease Prevention and Control. The European Union one Health 2020 Zoonoses Report. *EFSA J.* **2021,** *19* (12), e06971.

Farahmand, M.; Moghoofei, M.; Dorost, A.; Shoja, Z.; Ghorbani, S.; Kiani, S. J.; ... Tavakoli, A. Global Prevalence and Genotype Distribution of Norovirus Infection in Children with Gastroenteritis: A Meta-analysis on 6 Years of Research from 2015 to 2020. *Rev. Med. Virol.* **2022,** *32* (1), e2237.

Felicio dasilva, et al. Risk Ranking of Pathogens in Ready-to-Eat Unprocessed Foods of Non-animal Origin (FoNAO) in the EU: Initial Evaluation Using Outbreak Data (2007–2011). *Int. J. Food Microbiol.* **2015,** *195,* 9–19. doi:10.1016/j.ijfoodmicro.2014.11.005.

Figueras, M. J.; Beaz-Hidalgo, R.; Hossain, M. J.; Liles, M. R. Taxonomic Affiliation of New Genomes Should be Verified Using Average Nucleotide Identity and Multilocus Phylogenetic Analysis. *Genome Announc.* **2014,** *2* (6), e00927–e00914.

Food and Agriculture Organization of the United Nations (FAO) and World Health Organization (WHO). *Risk Assessment of Vibrio Parahaemolyticus in Seafood: Interpretative Summary and Technical Report.* Microbiological Risk Assessment Series No. 16. Rome, Italy. http://www.fao.org/3/a-i2225e.pdf

Fredriksson-Ahomaa, M.; London, L.; Skrzypczak, T.; Kantala, T.; Laamanen, I.; Biström, M.; Gadd, T. Foodborne Zoonoses Common in Hunted Wild Boars. *Ecohealth,* **2020,** *17* (4), 512–522.

Fusco, G.; Di Bartolo, I.; Cioffi, B.; Ianiro, G.; Palermo, P.; Monini, M.; Amoroso, M. G. Prevalence of Foodborne Viruses in Mussels in Southern Italy. *Food Environ. Virol.* **2017,** *9* (2), 187–194.

Gray, J. A.; Chandry, P. S.; Kaur, M.; Kocharunchitt, C.; Bowman, J. P.; Fox, E. M. Characterisation of Listeria Monocytogenes Food-associated Isolates to Assess Environmental Fitness and Virulence Potential. *Int. J. Food Microbiol.* **2021,** *350,* 109247. http://dx.doi.org/10.1016/j.ijfoodmicro.2021.109247. PMID:34023680

Grutsch, A. A.; Nimmer, P. S.; Pittsley, R. H.; mckillip, J. L. *Bacillus* spp. As Pathogens in the Dairy Industry. In *Foodborne Diseases*; Academic Press, 2018; pp 193–211.

Hachemi, A.; Zenia, S.; Denia, M.F.; Guessoum, M.; Hachemi, M.M.; Ait-Oudhia, K. Epidemiological Study of Sausage in Algeria: Prevalence, Quality Assessment, and Antibiotic Resistance of *Staphylococcus aureus* Isolates and the Risk Factors Associated with Consumer Habits Affecting Foodborne Poisoning, *Vet. World* **2019,** *12* (8), 1240–1250.

Hajimohammadi, B.; Ahmadian, S.; Firoozi, Z.; Askari, M.; Mohammadi, M.; Eslami, G.; ... Boozhmehrani, M. J. A Meta-Analysis of the Prevalence of Toxoplasmosis in Livestock and Poultry Worldwide. *Ecohealth* **2022,** 1–20.

Ingram, M.; John, J. S.; Applewhaite, T.; Gaskin, P.; Springer, K.; and Indar, L. Population-based Estimates of Acute Gastrointestinal and Foodborne Illness in Barbados: A Retrospective Cross-sectional Study. *J. Health Popul. Nutr.* **2013,** *31* (4 Suppl 1), S81.

Innes, E. A.; Chalmers, R. M.; Wells, B.; Pawlowic, M. C. A One Health Approach to Tackle Cryptosporidiosis. *Trends Parasitol.* **2020,** *36* (3), 290–303.

Jain. D.; Sinha, S.; Prasad, N. K.; Pandey, M. C. *Campylobacter* Species and Drug Resistance in a North Indian Rural Community. *Trans. R. Soc. Trop. Med. Hyg.* **2005,** *99,* 207–214.

Jajere, S. M. A Review of *Salmonella enterica* with Particular Focus on the Pathogenicity and Virulence Factors, Host Specificity and Antimicrobial Resistance Including Multidrug Resistance. *Vet. World* **2019,** *12* (4), 504.

Jeffery, I. A.; Karim, S. Botulism. 2017.

Jessberger, N.; Dietrich, R.; Granum, P. E.; Märtlbauer, E. The *Bacillus cereus* Food Infection as Multifactorial Process. *Toxins* **2020**, *12*, 701.

John, N.; Vidyalakshmi, V. B.; Hatha, A. M. Effect of ph and Salinity on the Production of Extracellular Virulence Factors by Aeromonas from Food Sources. *J. Food Sci.* **2019,** *84* (8), 2250–2255.

Jordan, K.; Mcauliffe, O. Listeria Monocytogenes in Foods. In *Biological Emerging Risks in Foods (Advances in Food and Nutrition Research*), Vol. 86; Rodriguez-Lazaro, D., Ed.; Cambridge: Elsevier, 2018; pp 181–213. http://dx.doi.org/10.1016/bs.afnr.2018.02.006

Kaur, S.; Malik, S. V. S.; Vaidya, V. M.; Barbuddhe, S. B. *Listeria monocytogenes* in Spontaneous Abortions in Humans and Its Detection by Multiplex PCR. *J. Appl. Microbiol.* **2007,** *103* (5), 1889–1896.

Khuroo, M. S.; Khuroo, M. S.; Khuroo, N. S. Hepatitis E: Discovery, Global Impact, Control and Cure. *World J. Gastroenterol.* **2016,** *22* (31), 7030.

Knific, T.; Ocepek, M.; Kirbiš, A.; Krt, B.; Prezelj, J.; Gethmann, J. M. Quantitative Risk Assessment of Exposure to Mycobacterium avium subsp. Paratuberculosis (MAP) via Different Types of Milk for the Slovenian Consumer. *Foods* **2022,** *11* (10), 1472.

Koohsar, F.; Naddaf, S. R.; Rokni, M. B.; Mirjalali, H.; Mohebali, M.; Shafiei, R.; Mowlavi, G. Serological Detection of Trichinellosis among Suspected wild Boar Meat Consumers in North and Northeast of Iran. *Iran. J. Parasitol.* **2021**, *16* (2), 253.

Kotwal, A.; Singh, H.; Verma, A. K.; Gupta, R. M.; Jain, S.; Sinha, S.; Joshi, R. K.; et al. A Study of Hepatitis A and E Virus Seropositivity Profile Amongst Young Healthy Adults in India. *Med. J. Armed Forces India* **2014,** *70* (3), 225–229.

Koutsoumanis, K. Public Health Risks Associated with Food-borne Parasites. *EFSA J.* 2018.

Kristkova, Z. S.; Grace, D.; Kuiper, M. *The Economics of Food Safety in India: A Rapid Assessment*; Wageningen University and Research, 2017.

Kumar M. S. Molecular Studies on Antibiotic Resistance and Virulence Genes of *Campylobacter* Isolates from Diverse Origin. Ph.D. Thesis, submitted to Deemed University, Indian Veterinary Research Institute, Izatnagar, 2016.

Kumar, A.; Agarwal, R. K.; Bhilegaonkar, K. N.; Shome, B. R.; Bachhil, V. N. Occurrence of *Campylobacter jejuni* in Vegetables. *Int. J. Food Microbiol.* **2001,** *67* (1–2), 153–155.

Kumar, M.; Suman, T. P.; Ramees, H.; Dhanze, S.; Anjay, G.; Dubal, Z. B.; Sivakumar, M.; Kumar, A. Occurrence of Thermophilic Campylobacters in Foods of Animal Origin, Animal Faeces and Human Stool. *J. Vet. Pub. Health* **2015,** *13* (2), 105–109.

Lalitha, K. V.; Surendran, P. K. Occurrence of *Clostridium botulinum* in Fresh and Cured Fish in Retail Trade in Cochin (India). *Int. J. Food Microbiol.* **2002,** *72* (1–2), 169–174.

Lee, H.; Yoon, Y. Etiological Agents Implicated in Foodborne Illness World Wide. *Food Sci. Anim. Resour.* **2021,** *41* (1), 1.

Lee, N.; Chan, M. C.; Wong, B.; Choi, K. W.; Sin, W.; Lui, G.; ... Leung, W. K. Fecal Viral Concentration and Diarrhea in Norovirus Gastroenteritis. *Emerg. Infect. Dis.* **2007,** *13* (9), 1399.

Lenzi, A.; Marvasi, M.; Baldi, A. Agronomic Practices to Limit Pre- and Post-harvest Contamination and Proliferation of Human Pathogenic Enterobacteriaceae in Vegetable Produce. *Food Control* **2021,** *119,* 107486.

Leung, V. H.; Phan, Q.; Costa, C. E.; Nishimura, C.; Pung, K.; Horn, L.; Sosa, L. Notes from the Field: Clostridium Perfringens Outbreak at a Catered Lunch—Connecticut, September 2016. *MMWR Morb. Morb. Mortal. Wkly. Rep.* **2017,** *66,* 940–941.

Li, C.; Zeng, H.; Ding, X.; Chen, Y.; Liu, X.; Zhou, L.; Wang, X.; Cheng, Y.; Hu, S.; Cao, Z.; Liu, R.; Yin, C. Perinatal Listeriosis Patients Treated at a Maternity Hospital in Beijing, China, from 2013–2018. *BMC Infect. Dis.* **2020,** *20* (1), 601. http://dx.doi. Org/10.1186/s12879–020-05327-6. PMID:32799811.

Li, H.; Guo, Y.; Tian, T.; Guo, W.; Liu, C.; Liang, X.; ... Fu, P. Epidemiological Analysis of Foodborne Botulism Outbreaks—China, 2004–2020. *China CDC Week.* **2022,** 1–5.

Li, H.; Li, W.; Dai, Y.; Jiang, Y.; Liang, J.; Wang, S.; ... Guo, Y. Characteristics of Settings and Etiologic Agents of Foodborne Disease Outbreaks—China, 2020. *China CDC Week.* **2021,** *3* (42), 889.

Li, P.; Ji, Y.; Li, Y.; Ma, Z.; Pan, Q. Estimating the Global Prevalence of Hepatitis E Virus in Swine and Pork Products. *One Health* **2022,** *14,* 100362.

Li, P.; Liu, J.; Li, Y.; Su, J.; Ma, Z.; Bramer, W. M.; ... Pan, Q. The Global Epidemiology of Hepatitis E Virus Infection: A Systematic Review and Meta-analysis. *Liver Int.* **2020,** *40* (7), 1516–1528.

Liao, Y.; Hong, X.; Wu, A.; Jiang, Y.; Liang, Y.; Gao, J.; ... Kou, X. Global prevalence of Norovirus in Cases of Acute Gastroenteritis from 1997 to 2021: An Updated Systematic Review and Meta-analysis. *Microb. Pathog.* **2021,** *161,* 105259.

Liu, X.-Y.; Hu, Q.; Xu, F.; Ding, S.-Y.; Zhu, K. Characterization of *Bacillus cereus* in Dairy Products in China. *Toxins* **2020,** *12,* 454.

Lyoo, E. L.; Park, B. J.; Ahn, H. S.; Han, S. H.; Go, H. J.; Kim, D. H.; ... Choi, I. S. Detection and Genetic Analysis of Zoonotic Hepatitis E Virus, Rotavirus, and Sapovirus in Pigs. *Korean J. Vet. Res.* **2020,** *60* (2), 61–68.

Menon, V. K.; George, S.; Sarkar, R.; Giri, S.; Samuel, P.; Vivek, R.; ., Gray, J. J. Norovirus Gastroenteritis in a Birth Cohort in Southern India. *PLoS One*, **2016,** *11* (6), e0157007.

Muchaamba, F.; Barmettler, K.; Treier, A.; Houf, K.; Stephan, R. Microbiology and Epidemiology of Escherichia albertii—An Emerging Elusive Foodborne Pathogen. *Microorganisms* **2022,** *10* (5), 875.

Mughini-Gras, L.; Benincà, E.; mcdonald, S. A.; de Jong, A.; Chardon, J.; Evers, E.; Bonačić Marinović, A. A. *A Statistical Modelling Approach for Source Attribution Meta-analysis of Sporadic Infection with Foodborne Pathogens*; Zoonoses and Public Health, 2022.

Mulder, A. C.; van de Kassteele, J.; Heederik, D.; Pijnacker, R.; Mughini-Gras, L.; Franz, E. Spatial Effects of Livestock Farming on Human Infections with Shiga Toxin-producing *Escherichia coli* O157 in Small but Densely Populated Regions: The Case of the Netherlands. *Geohealth* **2020,** *4* (11), e2020gh000276.

Munck, N.; Smith, J.; Bates, J.; Glass, K.; Hald, T.; Kirk, M. D. Source Attribution of Salmonella in Macadamia Nuts to Animal and Environmental Reservoirs in Queensland, Australia. *Foodborne Pathog. Dis.* **2020,** *17* (5), 357–364.

National Health Profile. *13th Issue. Central Bureau of Health Intelligence Directorate General of Health Services Ministry of Health and Family Welfare*; Government of India WHO Collaborating Centre on Family of International Classifications, 2018.

National Health Profile. *14th Issue. Central Bureau of Health Intelligence Directorate General of Health Services Ministry of Health and Family Welfare*; Government of India WHO Collaborating Centre on Family of International Classifications, 2019.

Nayak, M. K.; Balasubramanian, G.; Sahoo, G. C.; Bhattacharya, R.; Vinje, J.; Kobayashi, N.; Sarkar, M. C.; Bhattacharya, M. K.; Krishnan, T. Detection of a Novel Intergenogroup Recombinant Norovirus from Kolkata, India. *Virology* **2008,** *377* (1), 117–123.

Odo, S. E.; Uchechukwu, C. F.; Ezemadu, U. R. Foodborne Diseases and Intoxication in Nigeria: Prevalence of Escherichia coli 0157: H7, Salmonella, Shigella and Staphylococcus aureus. *J. Adv. Microbiol.* **2021,** *20* (12), 84–94.

OIE. *Manual of Standards for Diagnostic Tests and Vaccines*, 7th ed.; OIE Press; France, 2016; p 4.

Olaimat, A. N.; Al-Holy, M. A.; Shahbaz, H. M.; Al-Nabulsi, A. A.; Abu Ghoush, M. H.; Osaili, T. M.; Ayyash, M. M.; Holley, R. A.; Emergence of Antibiotic Resistance in Listeria Monocytogenes Isolated from Food Products: A Comprehensive Review. *Compr. Rev. Food Sci. Food Saf.* **2018,** *17* (5), 1277–1292.

Olea-Popelka, F.; Muwonge, A.; Perera, A.; Dean, A. S.; Mumford, E.; Erlacher-Vindel, E.; ... Fujiwara, P. I. Zoonotic Tuberculosis in Human Beings Caused by Mycobacterium Bovis—A Call for Action. *Lancet Infect. Dis.* **2017,** *17* (1), e21–e25.

Ongor, H.; Cetinkaya, B.; Karahan, M.; Bulut, H. Evaluation of Immunomagnetic Separation-Polymerase Chain Reaction in Direct Detection of *Brucella abortus* and *Brucella melitensis* from Cheese Samples. *Foodborne Pathog. Dis.* **2006,** *3,* 245–250.

Oniciuc, E. A.; Ariza-Miguel, J.; Bolocan, A. S.; et al. Foods from Black Market at EU Border as a Neglected Route of Potential Methicillin-resistant *Staphylococcus aureus* Transmission. *Int. J. Food Microbiol.* **2015,** *209,* 34–38. https://doi.org/10.1016/j.ijfoodmicro.2014.11.015

Osuna-Chávez, R.; López-Robles, G.; Celaya-Michel, H.; Sosa-Castañeda, J.; Nieblas-López, M.; Ibarra-Zazueta, C. Seroprevalence of Mycobacterium Avium Subspecie

Paratuberculosis in Suspect Cattle Slaughtered at Slaughterhouse from Sonora, Mexico. *Abanico veterinario* **2021,** *11*.

Outreach Programme on Zoonotic Diseases (OPZD). *Annual Report* 2014–15, Submitted to ICAR New Delhi.

Outreach Programme on Zoonotic Diseases (OPZD). *Annual Report* 2015–16, Submitted to ICAR New Delhi.

Outreach Programme on Zoonotic Diseases (OPZD). *Annual Report* 2016–17, Submitted to ICAR New Delhi.

Outreach Programme on Zoonotic Diseases (OPZD). *Annual Report* 2017–18, Submitted to ICAR New Delhi.

Outreach Programme on Zoonotic Diseases (OPZD). *Annual Report* 2018–19, Submitted to ICAR New Delhi.

Ozana, V.; Hruska, K.; Sechi, L. A. Neglected Facts on Mycobacterium Avium Subspecies Paratuberculosis and Type 1 Diabetes. *Int. J. Mol. Sci.* **2022,** *23* (7), 3657.

Pal, M. Brucellosis: A Highly Infectious Foodborne Zoonotic Disease of Public Health Concern. *Madridge J. Food Tech.* **2018,** *S1* (1), 1–3.

Pal, M.; Beshatu, F.; Ritwick, D. Significance of Yersinia Entrocolitica as Emerging Zoonotic Food Borne Pathogen. *J. Natural Hist*, **2013,** *9*, 36–45.

Pattis, I.; Cressey, P. J.; Lopez, L. D.; Roos, R.; Horn, B.; Soboleva, T. *Annual Report Concerning Foodborne Disease in New Zealand 2017*; Ministry for Primary Industries: Manatū Ahu Matua, 2019.

Pires, S. M.; Fischer-Walker, C. L.; Lanata, C. F.; Devleesschauwer, B.; Hall, A. J.; Kirk, M. D.; et al. Aetiology-specific Estimates of the Global and Regional Incidence and Mortality of Diarrhoeal Diseases Commonly Transmitted Through Food. *PLoS One* **2015,***10* (12), e0142927.

Prasad, K. N.; Mathur, S. K.; Dhole, T. N.; Ayyagari, A. Antimicrobial Susceptibility and Plasmid Analysis of *Campylobacter jejuni* Isolated from Diarrheal Patients and Healthy Chickens in Northern India. *J. Diarrheal. Dis. Res.* **1994**, *12* (4), 270–273.

Qi, R.; Huang, Y. T.; Liu, J. W.; Sun, Y.; Sun, X. F.; Han, H. J.; ... Yu, X. J. Global Prevalence of Asymptomatic Norovirus Infection: A Meta-analysis. *Eclinicalmedicine* **2018,** *2,* 50–58.

Riahi, S. M.; Ahmadi, E.; Zeinali, T. Global Prevalence of Yersinia Enterocolitica in Cases of Gastroenteritis: A Systematic Review and Meta-analysis. *Int. J. Microbiol.* 2021.

Ronholm, J.; Nasheri, N.; Petronella, N.; Pagotto, F. Navigating Microbiological Food Safety in the Era of Whole-genome Sequencing. *Clin. Microbiol. Rev.* **2016,** *29* (4), 837–857.

Ryan, U.; Hijjawi, N.; Xiao, L. Foodborne Cryptosporidiosis. *Int. J. Parasitol.* **2018,** *48* (1), 1–12.

Scallan, E.; et al. Foodborne Illness Acquired in the United States—Major Pathogens. *Emerg. Infect. Dis.* **2011,** *17,* 7–15. doi: 10.3201/eid1701.p11101

Shakuntala, I.; Das, S.; Ghatak, S.; Milton, A. A. P.; Sanjukta, R.; Puro, K. U.; Aparajita, D.; Sukhadeo, B. B.; Sen, A. Prevalence, Characterization, and Genetic Diversity of *Listeria monocytogenes* Isolated from Foods of Animal Origin in North East India. *Food Biotechnol.* **2019,** *33* (3), 237–250.

Sharma, I.; Kumar, A.; Pramanik, A. K. Isolation and Identification of Mesophillic Aeromonas Bacteria from Meat and Fish Foods of North East India. *J. Pure Appl. Microbiol.* **2009,** *3* (2), 517–526.

Shonhiwa, A. M.; Ntshoe, G.; Essel, V.; Thomas, J.; Mccarthy, K. A Review of Foodborne Diseases Outbreaks Reported to the Outbreak Response Unit, National Institute for Communicable Diseases, South Africa, 2013–2017. *Int. J. Infect. Dis.* **2019,** *79,* 73.

Silva, N. C. C.; Guimarães, F. F.; Manzi, M. P.; Budri, P. E.; Gómez-Sanz, E.; Benito, D.; Langoni, H.; Rall, V. L. M.; Torres, C. Molecular Characterization and Clonal Diversity of Methicillin-Susceptible *Staphylococcus aureus* in Milk of Cows with Mastitis in Brazil. *J. Dairy Sci.* **2013,** *96* (11), 6856–6862.

Singh, B. B.; Dhand, N. K.; Gill, J. P. S. Economic Losses Occurring due to Brucellosis in Indian Livestock Populations. *Prev. Vet. Med.* **2015,** *119* (3–4), 211–215.

Singh, B. B.; Khatkar, M. S.; Aulakh, R. S.; Gill, J. P. S.; Dhand, N. K. Estimation of the Health and Economic Burden of Human Brucellosis in India. *Prev. Vet. Med.* **2018,** *154,* 148–155.6

Sivakumar, M. Phenotypic and Genotypic Characterization of Methicillin Resistant *Staphylococcus aureus* and ESBL Producing *Escherichia coli* Isolated from Street Vended Foods of Animal Origin and Associated Environment. M.V.Sc. Thesis, submitted to Deemed University, Indian Veterinary Research Institute, Izatnagar, 2016.

Smith, T. C. Livestock-associated Staphylococcus Aureus: The United States Experience. *PLoS Pathog.* **2015,** *11* (2), 1–8. https://doi.org/10.1371/journal.ppat.1004564

Srinivasan, S.; Easterling, L.; Rimal, B.; Niu, X. M.; Conlan, A. J.; Dudas, P.; Kapur, V. Prevalence of Bovine Tuberculosis in India: A Systematic Review and Meta-analysis. *Transbound. Emerg. Dis.* **2018,** *65* (6), 1627–1640.

Symeonidou, I.; Arsenopoulos, K.; Tzilves, D.; Soba, B.; Gabriël, S.; Papadopoulos, E. Human Taeniasis/Cysticercosis: A Potentially Emerging Parasitic Disease in Europe. *Ann. Gastroenterol.* **2018,** *31* (4), 406.

Tate, J. E.; Burton, A. H.; Boschi-Pinto, C.; Parashar, U. D.; World Health Organization–Coordinated Global Rotavirus Surveillance Network, Agocs, M.; Ranjan Wijesinghe, P. Global, Regional, and National Estimates of Rotavirus Mortality in Children<5 Years of Age, 2000–2013. *Clin. Infec. Dis.* **2016,** *62* (Suppl 2), S96–S105.

Thomas, J.; Govender, N.; mccarthy, K. M.; Erasmus, L. K.; Doyle, T. J.; Allam, M.; Ismail, A.; Ramalwa, N.; Sekwadi, P.; Ntshoe, G.; Shonhiwa, A.; Essel, V.; Tau, N.; Smouse, S.; Ngomane, H. M.; Disenyeng, B.; Page, N. A.; Govender, N. P.; Duse, A. G.; Stewart, R.; Thomas, T.; Mahoney, D.; Tourdjman, M.; Disson, O.; Thouvenot, P.; Maury, M. M.; Leclercq, A.; Lecuit, M.; Smith, A. M.; Blumberg, L. H. Outbreak of Listeriosis in South Africa Associated with Processed Meat. *N. Engl. J. Med.* **2020,** *382* (7), 632–643. http://dx.doi.org/10.1056/nejmoa1907462. PMID:32053299.

Troeger, C.; Forouzanfar, M.; Rao, P. C.; Khalil, I.; Brown, A.; Reiner Jr, R. C.; Alemayohu, M. A. Estimates of Global, Regional, and National Morbidity, Mortality, and Aetiologies of Diarrhoeal Diseases: A Systematic Analysis for the Global Burden of Disease Study 2015. *Lancet Infec. Dis.* **2017,** *17* (9), 909–948.

Van der Giessen, J.; Deksne, G.; Gómez-Morales, M. A.; Troell, K.; Gomes, J.; Sotiraki, S.; ... Robertson, L. J. Surveillance of Foodborne Parasitic Diseases in Europe in a One Health Approach. *Parasite Epidemiol. Cont.* **2021,** *13,* e00205.

Velebit, B.; Djordjevic, V.; Milojevic, L.; Babic, M.; Grkovic, N.; Jankovic, V.; Yushina, Y. In *The Common Foodborne Viruses: A Review*. IOP Conference Series: Earth and Environmental Science; IOP Publishing, September, 2019. Vol. 333, No. 1, p 012110.

Villabruna, N.; Koopmans, M. P.; de Graaf, M. Animals as Reservoir for human Norovirus. *Viruses* **2019,** *11* (5), 478.

World Health Organization. WHO Estimates of the Global Burden of Foodborne Diseases: Foodborne Disease Burden Epidemiology Reference Group 2007–2015. No. 9789241565165. World Health Organization, 2015.

Yang, K.; Wang, A.; Fu, M.; Wang, A.; Chen, K.; Jia, Q.; Huang, Z. Investigation of Incidents and Trends of Antimicrobial Resistance in Foodborne Pathogens in Eight Countries from Historical Sample Data. *Int. J. Environ. Res. Public Health* **2020,** *17* (2), 472.

Yucun, C.; Jin, H.; Shihong, Y.; Liyang, P.; Xiaohui, L.; Muyuan, M. Epidemiological Characteristics of Human Brucellosis in Dalian, Liaoning, 2014–2019. *Disease Surveillance* **2021,** *36* (1), 59–62. doi: 10.3784/jbjc.202008180279

Zhang, X. Z.; Wang, Z. Q.; Cui, J. Epidemiology of Trichinellosis in the People's Republic of China During 2009–2020. *Acta Tropica* **2022,** 106388.

CHAPTER 12

Omics Technologies: An Overview of Application of Proteomics in Meat Quality and Safety Assurance

KIRAN MOHAN[1], NAVEENA BASAPPA MAHESWARAPPA[2], S. WILFRED RUBAN[3], and RITUPARNA BANERJEE[2]

[1]*Department of Livestock Products Technology, Veterinary College, KVAFSU, Bidar, Karnataka, India*

[2]*ICAR-National Research Centre on Meat, Chengicherla, Hyderabad, Telangana, India*

[3]*Department of Livestock Products Technology, Veterinary College Bengaluru, KVAFSU, Bidar, Karnataka, India*

ABSTRACT

Over the last two decades, sophisticated "-OMICs" technologies like genomics, proteomics, peptidomics, transcriptomics, metabolomics, and lipidomics have been explored successfully to address food quality challenges. Additionally, these strong methodologies have been extensively employed to understand the biological mechanisms for phenotypic variance in technical and sensory quality attributes of muscle foods. Omics-based technologies allow researchers to investigate complex biological processes quantitatively and integratively using a systems biology approach. Food proteomics is a rapidly growing area that is gaining importance in product innovation, food safety, and food quality. Proteomics is the application of technology for the identification and quantification of a cell's, tissue's, or organism's total protein content. It complements other "omics" technologies such as genomics and transcriptomics in elucidating the identity of an

Engineering Principles for Food Processing Technology and Product Realization. Leonardo Sepúlveda Torre, Porteen Kannan, A. K. Haghi, PhD (Eds.)

organism's proteins and in comprehending their structure and function. The complexity of proteomics arises from the need to examine and classify all of a genome's protein signatures. This chapter highlights the role of proteomics in food research, the underlying assumptions, methodological considerations, and bioinformatic techniques used in the field. Further, it summarizes proteome data from the most important meat species used in food production and examines significant food concerns from a proteomic perspective. This chapter seeks to offer an overview of the various proteomics methods, their recent advancements, and their application in research and analysis in order to better highlight meat quality and safety problems.

12.1 INTRODUCTION

Proteomics may be characterized as a method of determining a cell's structure, sequence, and quantity of proteins, along with their interactions with other proteins and the degree of modification (Campbell, 2003). Many different variables are evaluated with respect to meat quality, including color or appearance, palatability (texture/tenderness/juiciness), nutritional factors such as composition (proteins/minerals/fatty acids), as well as meat safety in broad sense (Leheska et al., 2008; Verbeke et al., 2009). In contrast to genomes, the proteome is dynamic and changes in response to changes in the body's physiological state. Meat quality may be genetically influenced by many quantitative factors, such as the fiber characteristics or metabolic processes of postmortem or living muscle. Environmental factors and the procedure of meat processing influence meat quality just as much as the animal's genetic make-up.

Meat authentication is a procedure used to make sure the product is as described on the label. Biomarkers that can be used as indications of meat quality and authenticity are now being studied following recent advances in the fields of omics and proteomics. Food safety can be compromised by economic-motivated adulteration, which also has a detrimental impact on consumers' trust. Food safety and quality regulators are increasingly concerned about the authenticity of muscle foods because of their growing global reputation as a source of negative practices. Authentication of foods has been addressed by a variety of analytical techniques, including biochemical, immunological, mass spectrometry-based proteomic, and metabolomic approaches. There is no limit to the types of foods that may be studied using mass spectrometry-based methods, which are reliable, credible, and sensitive in their use. Protein fractionation techniques including OFFGEL and GELFrEE fractionation, two-dimensional gel electrophoresis, the use of various mass spectrometers in authentication, the use of lateral flow assay techniques as point-of-care tools, and detection of additional food frauds such geographical origin, additives, processing methods, etc. are explored in this paper.

12.2 TOOLS OF PROTEOMICS

It is difficult to perform proteomics because of the broad variety of proteins. An animal's genome, which has about 20,000 genes, has the ability to generate 1.8 million distinct types of protein (Jensen, 2004). An ideal proteomics approach can identify as many protein products as possible in a sensitive, reproducible, and quantitative way. Methods such as protein extraction, purification, and identification are popular in proteome research. Proteomics research makes use of a variety of techniques, including 2-DE gel electrophoresis, mass spectrometry, and bioinformatics. Proteomics is the study of proteins in a cell under specified conditions (Vaidyanathan and Goodacre, 2003). Purification of the crude sample is essential for all proteome investigations before any further analysis and characterization can be undertaken.

Electrophoresis of proteins using SDS-PAGE (sodium dodecyl sulphate polyacrylamide gel electrophoresis) has been considered as the best method for many biochemical studies. Expasy, a publicly accessible collection of proteomics data, reveals that 2-DE is the most frequently used technique in large-scale proteomics (http://www.expasy.ch). As of the mid-1970s, modern 2DE technology uses IEF with highly reproducible immobilized pH gradient (IPG) strips, a first-dimension isoelctric focusing technology. The first dimension (IEF) entails the separation of proteins according to their charge, followed by their molecular weight, which is performed using SDS-PAGE (2nd dimension). The DIGE (differential gel electrophoresis) approach, which was very recently created, allows for the direct measurement of abundance changes on a massive level without interference from gel-to-gel variation. Nanometric detection limits are possible with these dyes. Lametsch and Bendixen (2001) revealed that for each gel, 1000 individual protein spots may be separated by 2-DE using optimal extraction, separation, and staining protocols.

There is a rising need for speedy and accurate protein identification in light of the rapid advancement of numerous current and accomplished genome sequencing activities. Proteins in nanogram levels may be identified using mass spectrometry, which is a very effective method. It is possible for a basic cell lysate to have hundreds of thousands of peptides after a simple proteolytic digestion. Since the primary goal of many proteomic approaches is protein identification using mass spectrometry (MS), it is common practise to compare the MS results with protein sequence databases, generated by genomics study (Friedman, 2011).

Mass spectrometry technologies have advanced significantly and are increasingly being used to authenticate food. The mass spectrometry technique enables the precise measurement of the molecular mass of charged molecules. Analytes are ionized in the gas phase and separated by mass-to-charge ratio using mass analyzers and detectors, and the mass-to-charge ratio of each analyte is determined by the ionization source and mass analyzer. These instruments are either one-stage instruments (MSn) or many stages of instrumentation, depending on how many analyzers are used in tandem or in hybrid arrangements, such as Q3 or Q-TOF. Because of the simultaneous advancements in liquid chromatography (LC) and capillary electrophoresis for in-line separation, mass spectrometry analysis of proteins at the sequence/peptide level became feasible. It is now possible to identify peptides with femtomole (I0–15 mole) sensitivity in complicated biological materials using these recent technical developments in mass-spectrometry methods (Mallick and Kuster, 2010). Peptide-mass fingerprinting by MALDI-TOF MS is the most extensively used methodology for protein detection from 2-DE. High-throughput detection of femtomole amounts of proteins and peptides is partly owing to this technology's specificity. The amino acid sequence or peptides, which are generated after digestion with a protease, can be used to identify individual proteins in the data obtained by the mass spectrometry (Henzel et al., 1993). Electrospray-ionization triple quadrupole mass spectrometers are another frequent type of mass spectrometer. Bioinformatics approaches can then be used to explore databases which match these mass spectral fingerprints obtained from the experiment (Eng et al., 1994).

The theory behind spectroscopic meat analysis is that moisture, proteins, and lipids found in meat produce multiple spectra at different wavelengths (Wang et al., 2018). In recent years, spectral technologies have been investigated in food quality and safety because of their nondestructive nature. Other advantages of spectroscopic methods include ease of sample preparation and the absence of the demand for sample pretreatment (Wang et al., 2018). The

Raman scattering effect is used to analyze the composition and structure of a material in Raman spectroscopy (Hu et al., 2019). This noninvasive method does not need sample preparation or a high sample volume, and it may even be used on opaque materials (Hu et al., 2019; Lee et al., 2017). The type of the molecules and their concentration may be determined using Raman spectroscopy's fingerprints of vibrational frequency and intensity (Lee et al., 2017).

12.3 MEAT QUALITY AND SAFETY

Meat color is a critical quality characteristic that has a significant impact on customer purchasing decisions (Nair and Zhai, 2020). The color of raw meat is influenced by differences in reflection produced by protein denaturation and myoglobin (Aaslyng, 2002; Miller, 2002). Thus, proteomic techniques can assist in the comprehension of variations in meat color as well as the examination of the underlying determinants of muscle-specific meat color. There were 22 proteins or protein fragments that were found to be differently expressed in a sarcoplasmic proteins proteomic investigation of pig Semimembranosus muscle with high and low lightness values (L*) (Sayd et al., 2006). Overexpression of mitochondrial respiratory chain enzymes, hemoglobin, and chaperones was found in darker meat with greater levels of oxidative metabolism, while glycolysis enzymes were found in the lighter meat. Using pyruvate kinase, which was found to be more abundant in the Pectoralis major of broilers than in the Supracoracoideus, Zapata et al. (2012) hypothesized that the fast pH fall associated with "pale poultry muscle syndrome" might be detected using pyruvate kinase in the Pectoralis major of broilers.

Researchers have studied the postmortem variance in tenderness among species, breeds, ages, muscles, and processing settings using various proteomic methods. Pig meat tenderness may be linked to postmortem proteome changes, as initially reported by Lametsch et al. (2003). Meat tenderness is strongly linked to postmortem breakdown of actin and myosin heavy chain, according to these researchers. Many scientists believe that heat shock proteins (HSPs) might serve as biomarkers for meat tenderness (Laville et al., 2009; Polati et al., 2012). To maintain cellular proteins from stress-induced denaturation and probable loss of function, as well as to stabilize the muscle filament assembly throughout the stressful period, they may play a role in the tenderization of meat (Feder and Hoffman, 1999; Feasson et al., 2002; Perng et al., 1999). Several proteins, including uroplakin-1b, complement C1q subcomponent subunit B, myosin-IIIa, aspartate aminotransferase, glycogen phosphorylase, cytosolic carboxypeptidase 3, and phosphatidylinositide transfer protein isoform, were found by Kiran et al. (2016) that may be suitable biomarkers for the differentiating of young and spent Indian water buffalo meat.

Water holding capacity (WHC) is significant for a variety of reasons, including sensory appeal and economics, but it is also important because of the consequences of this capacity on meat quality. When contractile proteins are denatured and myofibrils shrink, the result is a loss of water and soluble components in the form of drip. This occurs as a result of the contraction of the muscles (Bertram et al., 2004; Offer et al., 1989). In order to better understand the underlying processes of this characteristic and explore reasonable protein biomarkers for WHC, a multitude of proteomics studies have been conducted. 2-DE has revealed a number of protein markers that may be linked to a decrease in the amount of drip loss in porcine muscle. Yu et al. (2009) reported that a decrease in heat shock protein expression was linked to a decrease in the amount of drip loss in muscle, indicating poor meat quality.

12.4 PROTEOMICS IN FOOD SCIENCE

New and exciting proteomic methods are being used to detect proteins in food matrixes and to examine protein–protein interactions in foods, along with protein–protein interactions study among food components. A sensitive evaluation of changes in protein structure occurring at specific amino acid locations may be made using this technique to identify components formed during processing processes.

For understanding the physiological and technical functions of amphiphilic proteins found in wheat kernels, a proteomic study of these proteins was carried out (Amiour et al., 2002). Proteomics has resulted in the development of markers for certain food processing methods as well as indicators for the quality of processed food products (Van Der Werf et al., 2001). Researchers were able to detect the distinct proteins generated due to heat stress in tomato using 2-DE (Iwahashi and Hosoda, 2000). The proteomic techniques can be used to explore biomarkers that are indicative of certain food spoilage or pathogenic microorganisms. Food allergies are becoming more and more of a concern for the food sector as a whole. Genes or allergenic proteins have been identified and effectively used in the study of allergic illnesses (Toda and Ono, 2002). The 2-DE and MALDI-MS are useful methods for the separation and investigation of complex protein mixtures, and they can also provide crucial information on digestion and bioavailability patterns of dietary proteins. Biological goods, such as vaccinations, blood products, and therapies, are made possible by biotechnology-derived therapeutic proteins. Recently, food quality analysis has seen the application of proteomics, which shows great promise (Kvasnicka, 2003).

12.5 APPLICATION OF PROTEOMICS IN MEAT QUALITY

An essential step toward a more complete knowledge of the complex biological systems that govern physiology and disease in animals can be achieved via the use of proteomics and other postgenomic methods. Animals are raised on a huge scale for the purpose of producing animal products for human consumption. Thus, substantial biological investigation is directed toward elucidating the biological characteristics that influence the quantity and quality of these products. However, much of the data collected from trials on farm animals such as swine and cattle is useful not just to farm animal sciences, but also to our knowledge of complicated molecular systems of health and illness in humans.

As in many other fields of life science, the application of proteomics and other postgenomic techniques in animal sciences represents an important step toward a more full understanding of the complicated biological processes that drive physiology and disease in living organisms. Farm animals are raised on a large scale by large-scale companies with the objective of generating animal products for human use. As a result, the majority of biological research is devoted to understanding the biological characteristics that determine the production and quality of these products. In addition to advancing our knowledge of human health and disease molecular systems, much of the data gathered from animal research, such as those on pigs and cattle, is also important in the field of farm animal science.

Meat quality is a wide phrase used in meat science to refer to a multitude of different factors depending on the objective. The necessity of a safe product is self-evident to the consumer. In addition, customers are looking for qualities that are linked to their physiological and hedonic demands in order to satisfy dietary requirements, but rather to the industry's competition. Several proteome reviews published in the 2000s focused on technological breakthroughs in proteomics, postgenomics technologies, and their limits in improving product quality (Bendixen, 2005; Mullen et al., 2006; Paredi et al., 2012). Researchers have employed proteomic research to link one or two qualitative characteristics to a collection of putative biomarkers in order to better understand the physiology and biology of animals, muscles, cells, and molecules. By analyzing the protein content of the Semitendinosus muscle of Charolais cows using 2DGE, Bouley et al. (2004) uncovered 129 different proteins. Among the proteins found were those involved in cell structure, cell defence, muscle metabolism, and the contractile machinery. In addition, the proteomic method discovered other proteins, including ATP synthase subunit alpha and alpha actinin 3, that may be involved in the modulation of muscle calpains (Brule et al., 2010). Postmortem proteomics research indicated that more than 100 unique protein loci had altered, most likely owing to proteolysis (Lametsch and Bendixen, 2001). Troponin T, actin, -crystallin, myokinase, creatine kinase, and mitochondrial ATPase were shown to be altered in pork aged 72 hours by Morzel et al. (2004). Changes in cryptic proteins and myozenin were also discovered. It was shown that the amount of stress-related proteins in the sarcoplasmic fraction reduced during ageing (Di Luca et al., 2013). It was discovered that Hsp70-1B, a minor heat shock protein, was a biomarker for decreased tenderness in beef among the 21 proteins tested in the study (Picard et al., 2014). Morzel et al. (2004) found that pigs transported the day before slaughter had a lower level of mitochondrial ATPase activity than pigs transported the day before slaughter. Protein profiling of chicken hearts was performed by Amid et al. (2012) to identify troponin I and alpha cardiac muscle actin 1 as putative electrically stunned markers. A mass spectrometry technique developed by Von Bargen et al. (2013) can identify even the tiniest amounts of pork and horse meat in halal beef samples. It was shown that the 2DE protein profiles of raw and processed meat products differed among pig, chicken, cow, turkey, duck, and goose species (Montowska and Pospiech, 2013).

Proteins that are critical to meat quality, such as pork color, drip loss, or shear force, can be identified via proteomic investigations, which look at a subset of the cellular protein composition at a particular time and often include several hundred proteins (Laville et al., 2009). Bouley et al. (2004) used 2-DE and MS to identify 75 gene products in bovine skeletal muscle and found that considerable cross-species identifications were achievable across bovine, human, mouse, and rat. Doherty et al. (2004) used 2-DE to separate soluble extracts from muscle homogenates from broiler and layer chickens, and then performed MALDI-TOF mass spectrometry on the targeted spots. A study by Hamelin et al. (2006) employed 2-DE and MALDI-MS to analyze the molecular basis of muscle fiber type-associated variability and found the differential expression of proteins involved in oxidative metabolism like citric acid cycle enzymes, proteins with functions linked to oxidative metabolism, oxidative stress, and higher protein turnover. Houbak et al. (2008) used 2D gel proteome analysis to investigate the breakdown of bovine muscle proteins by enzymes like proteasome and ubiquitous calpains. Hwang et al. (2005) used a proteomics technique to examine postmortem proteolysis in pig longissimus muscle and to investigate the association of proteolytic degradation to drip loss, WB-shear force, and lightness value during chilling ageing. It was suggested that developing muscle-specific processing strategies to improve beef color was necessary by Joseph et al. (2010), who studied the sarcoplasmic proteome to distinguish between muscles that were stable to liable in color. With DIGE and PMF Hollung et al. (2009) investigated the effect on adductor muscle size of age and breed.

The welfare of animals, as well as genetics, may have an impact on the quality of meat. After a time of reduced food intake, compensatory development in pigs has been linked to the tenderness of pork (Hollug et al., 2007; Kristensen et al., 2007). This property was studied by Lametsch et al. (2006) using proteomics, and they identified a correlation between this trait and the differential expression of protein. Stress and glycolytic proteins were shown to be decreased using 2D electrophoresis and MALDI TOF MS, which were employed in combination with a standard sample preparation. It has been shown by Liu et al. (2014) that a high-fat diet that alters the expression of stress proteins in muscle 2-DE was utilized in conjunction with MS to assess the quality of pig meat. A total of 27 proteins were found to have altered in response to ageing and changes in meat quality parameters. The majority of the proteins contained cytoskeletal and metabolic proteins and degradation products associated with those proteins. Research by Di Luca et al. (2013) using gel-based techniques (2D-DIGE and WB) indicated that fresh pork meat had a lower concentration of stress response proteins than dried pork meat.

12.6 APPLICATION OF PROTEOMICS IN MEAT SAFETY AND AUTHENTICITY

Authentication of meat has long been a top priority for producers, authorities, and consumers in the contemporary period. A clear and detailed description of the meat that people consume, especially in processed forms when visual distinction of components is only feasible, is required by health-conscious customers. The issues of food authenticity in meat and meat products are categorized by Montowska and Pospiech (2012) which are shown in Table 12.1.

TABLE 12.1 Major Categories of Food Authenticity Issues Found in Meat and Meat Products

Sl. No.	Category of food fraud	References
1	Substituting cheaper meat species for more expensive ones	Mane et al. (2012); Girish et al. (2015)
2	Substances of animal origin (Offal, mechanically recovered meat) that have not been declared	Velioglu et al. (2018); Fengou et al. (2021); Surowiec et al. (2011)
3	Contradictions in the product's labeling or specifications	Pinto et al. (2015)
4	Sale of previously frozen fresh beef	Alamprese et al. (2016)
5	Inadequate geographic origin labeling of items	Zhao et al. (2020)
6	Differentiation between meat from domesticated animals and meat from wild animals	Amaral et al. (2015)
7	Inadvertent inclusion of plant or dairy proteins	Salomé et al. (2021)
8	Product sales of nonecological items as ecological products	Stawitz et al. (2017)

Proteomics can significantly contribute in the knowledge of the processes of infection, antibiotic resistance, and biofilm development of these food-borne pathogens. An extensive body of literature exists addressing the discoveries that have been made possible by proteomics methods. A direct technique for the identification of enterohemorrhagic *Escherichia coli* O157:H7 was previously reported by Ochoa and Harrington (2005). In a recent study, Fagerquist and colleagues employed MALDI-TOF-TOF and tandem mass spectrometry (MS/MS) to identify distinct subtypes of Shiga toxin-producing *E. coli* (STEC) (Fagerquist et al., 2014). Recent studies employing 2D electrophoresis for the identification of allergens in various foods include studies on beer (Picariello et al., 2015), milk (Hettinga et al., 2015), and beef (Apostolovic et al., 2014).

Peptide markers indicating buffalo meat contamination in sheep meat combinations as low as 0.5% were identified using a proteomic technique, which was successful in both raw and cooked meat mixtures (Naveena et al., 2017; 2018). The researchers Fornal and Montowska (2019) used liquid chromatography and high-resolution Q-TOF mass analyzers to effectively detect unique peptide markers in highly processed composite food products from duck, goose, and chicken. Montowska and Pospiec (2013) discovered interspecies differences in 2-DE protein patterns in raw and processed meat from cow, pig, chicken, turkey, duck, and goose. They also revealed changes in 2-DE protein patterns in raw and processed meat from turkey, duck, and geese (2013). As potential markers, scientists suggested employing proteins that have been shown to be thermally stable and stable over long periods of time in the beef resource. Chicken contamination was detected by Sentandreu et al. (2010) utilizing OFFGEL isoelectric focusing, followed by MALDI–TOF MS and LC–ESI–MS/MS/MS analysis of species-specific peptide biomarkers in a combination of pig and chicken meat. A metabolomics and lipidomics technique was used by Trivedi et al. (2016) to identify pork adulteration in beef. Volatile molecules can be used as a biomarker for the detection of meat adulteration since various meat species have distinct flavors. Using an electronic nose and GC-MS, the research teams of Nurjulianaet al. (2011) and Haddie et al. (2015) were able to detect meat adulteration. Hu et al. (2017) used Fourier transform infrared spectroscopy to identify 10% of animal offal in ground beef with 99% accuracy (FT-IRS). Lee et al. (2018) recently used the RS approach to effectively identify beef tallow from pork lard, chicken fat from duck oil, and other animal fats as well. It was shown that 46 proteins were differently expressed in the muscle proteome of sheep exposed to stunning compared to non-stunning, showing alterations in the quantity of proteins involved in structural, catalytic, and stress-related processes (Kiran et al., 2019; 2021).

12.7 FUTURE TRENDS

Protein/peptide biomarkers can be used to verify the authenticity of food, although they are still a relatively new approach. However, they have the potential to be used in the future because of their high sensitivity, resilience, multiplexing capacity, and high-throughput and predictive power. A key limitation of DNA approaches is that DNA degradation occurs in highly processed samples (Buckey et al., 2009), peptide-based methods avoid this problem. Proteomic tool advancements have also led to their application in addressing safety and authenticity concerns such as the identification of meat species, detection of animal byproducts, non-meat components and tissues, traceability, identification of genetically modified ingredients, chemical residues, and other harmful substances. Understanding the impact on transportation, stunning, slaughter stress, halal verification, and difficulties linked to animal welfare are some of the possible areas where proteomic methods are being applied (Banerjee et al., 2022). Researchers should make an attempt in future research to include essential aspects that have been overlooked in the majority of previous studies. The tests will become more sensitive and specific when high-resolution MS instruments improve their resolution. Between initial discovery investigations and final validation trials, targeted MS/MS acquisition methods can help to fill the gap between the two. Innovative food proteome/matrix bioinformatic tools, as well as the existing ones, will be critical in advancing the fields of integrative biology and systems biology in the food industry. The use of high-throughput technologies, such as mass spectrometry and NMR-based metabolomics, mass spectrometry and NMR-based metabolomics, quantitative proteomics, and lipidomics, is a recent trend in farm animal research (Wang et al., 2012; Zhao et al., 2012). The scientific community should work together to share publicly available databases and data sets and to produce certified reference materials and standard operating procedures (Ortea et al., 2016).

12.8 CONCLUSION

The use of proteomic techniques is speeding up the understanding of the biological pathways and mechanisms that support meat quality. The meat industry has a long-term need and a significant technological difficulty in analyzing several targets on a single-test platform. As a potent tool for studying protein alterations generated by varied situations, proteomics is indispensable. Global protein analyses are supposed to provide a more direct knowledge of protein function and regulation than gene analyses. Genes and proteins that govern crop design and/or stress resistance in a wide variety of settings must be discovered in order to promote biological crop improvement in the midst of present food insecurity issues. The use of a wide array of procedures to protect consumers by screening for adulteration in the meat value supply chain is underway. In order to improve the precision and reproducibility of proteomic and metabolomic approaches, efforts should be made to standardize the extraction of analytes. Integrating genomes, transcriptomics, proteomics, and metabolomics will improve the accuracy and precision of authentication results, as would a more comprehensive approach. Proteomics investigations in meat science have been limited by a lack of access to high-throughput proteomic approaches that need expensive apparatus and highly qualified technical staff that are typically unaffordable among most meat science research centers. It is also necessary to expand collaboration between scientists, proteomics platforms, and the meat industry itself in order to solve these obstacles.

KEYWORDS

- **food production**
- **omics technology**
- **bioinformatics**
- **proteomics methods**
- **meat quality**
- **meat safety**

REFERENCES

Aaslyng, M. D. Quality Indicators for Raw Meat. *Meat Processing: Improving Quality*, 157–174.

Alamprese, C.; Amigo, J. M.; Casiraghi, E.; Engelsen, S. B. Identification and Quantification of Turkey Meat Adulteration in Fresh, Frozen-Thawed and Cooked Minced Beef by FT-NIR Spectroscopy and Chemometrics. *Meat Sci.* **2016,** *121*, 175–181.

Amaral, J. S.; Santos, C. G.; Melo, V. S.; Costa, J.; Oliveira, M. B. P.; Mafra, I. Identification of Duck, Partridge, Pheasant, Quail, Chicken and Turkey Meats by Species-Specific PCR Assays to Assess the Authenticity of Traditional Game Meat Alheira Sausages. *Food Control* **2015,** *47*, 190–195.

Amid, A.; Samah, N. A.; Yusof, F. Identification of Troponin I and Actin, Alpha Cardiac Muscle 1 as Potential Biomarkers for Hearts of Electrically Stimulated Chickens. *Proteome Sci.* **2012,** *10* (1), 1–8.

Amiour, N.; Merlino, M.; Leroy, P.; Branlard, G. Proteomic Analysis of Amphiphilic Proteins of Hexaploid Wheat Kernels. *Proteomics* **2002,** *2*(6), 632–641.

Apostolovic, D.; Tran, T. A. T.; Hamsten, C.; Starkhammar, M.; Cirkovic Velickovic, T.; van Hage, M. Immunoproteomics of Processed Beef Proteins Reveal Novel Galactose-α-1, 3-Galactose-Containing Allergens. *Allergy* **2014,** *69* (10), 1308–1315.

Banerjee, R.; Maheswarappa, N. B.; Mohan, K.; Biswas, S.; Batabyal, S. Proteomic Technologies and Their Application for Ensuring Meat Quality, Safety and Authenticity. *Curr. Proteomics* **2022,** *19* (2), 128–141.

Bendixen, E. The Use of Proteomics in Meat Science. *Meat Sci.* **2005,** *71* (1), 138–149.

Bertram, H. C.; Whittaker, A. K.; Shorthose, W. R.; Andersen, H. J.; Karlsson, A. H. Water Characteristics in Cooked Beef as Influenced by Ageing and High-Pressure Treatment—An NMR micro imaging study. *Meat Sci.* **2004,** *66* (2), 301–306.

Bouley, J.; Chambon, C.; Picard, B. Mapping of Bovine Skeletal Muscle Proteins Using Two-Dimensional Gel Electrophoresis and Mass Spectrometry. *Proteomics* **2004,** *4* (6), 1811–1824.

Brulé, C.; Dargelos, E.; Diallo, R.; Listrat, A.; Bechet, D.; Cottin, P.; Poussard, S. Proteomic Study of Calpain Interacting Proteins During Skeletal Muscle Aging. *Biochimie* **2010,** *92* (12), 1923–1933.

Buckley, M.; Collins, M.; Thomas-Oates, J.; Wilson, J. C. Species Identification by Analysis of Bone Collagen Using Matrix-Assisted Laser Desorption/Ionisation Time-of-Flight Mass Spectrometry. *Rapid Commun. Mass Spectrometry* **2009,** *23* (23), 3843–3854.

Campbell, P. A Cast of Thousands. *Nat Biotechnol.* **2003,** *21*, 213.

Di Luca, A.; Elia, G.; Mullen, A. M.; Hamill, R. M. Monitoring Post Mortem Changes in Porcine Muscle Through 2-D DIGE Proteome Analysis of Longissimus Muscle Exudate. *Proteome Sci.* **2013,** *11* (1), 1–14.

Di Pinto, A.; Bottaro, M.; Bonerba, E.; Bozzo, G.; Ceci, E.; Marchetti, P. et al. Occurrence of Mislabeling in Meat Products Using DNA-Based Assay. *J. Food Sci. Technol.* **2015,** *52* (4), 2479–2484.

Doherty, M. K.; McLean, L.; Beynon, R. J. Avian Proteomics: Advances, Challenges and New Technologies. *Cytogenetic Genome Res.* **2007,** *117* (1–4), 358–369.

Eng, J. K.; McCormack, A. L.; Yates, J. R. An Approach to Correlate Tandem Mass Spectral Data of Peptides with Amino Acid Sequences in a Protein Database. *J. Am. Soc. Mass Spectrometry* **1994,** *5* (11), 976–989.

Fagerquist, C. K.; Zaragoza, W. J.; Sultan, O.; Woo, N.; Quiñones, B.; Cooley, M. B.; Mandrell, R. E. Top-Down Proteomic Identification of Shiga Toxin 2 Subtypes from Shiga Toxin-Producing *Escherichia coli* by Matrix-Assisted Laser Desorption Ionization–Tandem Time of Flight Mass Spectrometry. *Appl. Environ. Microbiol.* **2014,** *80* (9), 2928–2940.

Feasson, L.; Stockholm, D.; Freyssenet, D.; Richard, I.; Duguez, S.; Beckmann, J. S.; Denis, C. Molecular Adaptations of Neuromuscular Disease-Associated Proteins in Response to Eccentric Exercise in Human Skeletal Muscle. *J. Physiol.* **2002,** *543* (1), 297–306.

Feder, M. E.; Hofmann, G. E. Heat-Shock Proteins, Molecular Chaperones, and the Stress Response: Evolutionary and Ecological Physiology. *Annu. Rev. Physiol.* **1999,** *61* (1), 243–282.

Fengou, L. C.; Lianou, A.; Tsakanikas, P.; Mohareb, F.; Nychas, G. J. E. Detection of Meat Adulteration Using Spectroscopy-Based Sensors. *Foods* **2021,** *10* (4), 861.

Fornal, E.; Montowska, M. Species-Specific Peptide-Based Liquid Chromatography–Mass Spectrometry Monitoring of Three Poultry Species in Processed Meat Products. *Food Chem.* **2019,** *283*, 489–498.

Friedman, D. B. An Introduction to Proteomics Technologies for the Genomics Scientist. *Genomics: Essential Methods* **2010,** 307–323.

Girish, P. S.; Anjaneyulu, A. S. R.; Viswas, K. N.; Shivakumar, B. M.; Anand, M.; Patel, M.; Sharma, B. Meat Species Identification by Polymerase Chain Reaction-Restriction Fragment Length Polymorphism (PCR-RFLP) of Mitochondrial 12S rRNA Gene. *Meat Sci.* **2005,** *70* (1), 107–112.

Haddi, Z.; El Barbri, N.; Tahri, K.; Bougrini, M.; El Bari, N.; Llobet, E.; Bouchikhi, B. Instrumental Assessment of Red Meat Origins and Their Storage Time Using Electronic Sensing Systems. *Anal. Methods* **2015,** *7*(12), 5193–5203.

Hamelin, M.; Sayd, T.; Chambon, C.; Bouix, J.; Bibé, B.; Milenkovic, D. et al. Proteomic Analysis of Ovine Muscle Hypertrophy. *J. Anim. Sci.* **2006,** *84* (12), 3266–3276.

Henzel, W. J.; Billeci, T. M.; Stults, J. T.; Wong, S. C.; Grimley, C.; Watanabe, C. Identifying Proteins from Two-Dimensional Gels by Molecular Mass Searching of Peptide Fragments in Protein Sequence Databases. *Proc. Natl. Acad. Sci.* **1993,** *90* (11), 5011–5015.

Hettinga, K. A.; Reina, F. M.; Boeren, S.; Zhang, L.; Koppelman, G. H.; Postma, D. Set al. Difference in the Breast Milk Proteome Between Allergic and Non-Allergic Mothers. *PloS One* **2015,** *10* (3), e0122234.

Hollung, K.; Grove, H.; Færgestad, E. M.; Sidhu, M. S.; Berg, P. Comparison of Muscle Proteome Profiles in Pure Breeds of Norwegian Landrace and Duroc at Three Different Ages. *Meat Sci.* **2009,** *81* (3), 487–492.

Houbak, M. B.; Ertbjerg, P.; Therkildsen, M. In Vitro Study to Evaluate the Degradation of Bovine Muscle Proteins Post-Mortem by Proteasome and μ-Calpain. *Meat Sci.* **2008,** *79* (1), 77–85.

Hu, R.; He, T.; Zhang, Z.; Yang, Y.; Liu, M. Safety Analysis of Edible Oil Products via Raman Spectroscopy. *Talanta* **2019,** *191*, 324–332.

Hu, Y.; Zou, L.; Huang, X.; Lu, X. Detection and Quantification of Offal Content in Ground Beef Meat Using Vibrational Spectroscopic-Based Chemometric Analysis. *Sci. Rep.* **2017,** *7* (1), 1–9.

Hwang, I. H.; Park, B. Y.; Kim, J. H.; Cho, S. H.; Lee, J. M. Assessment of Postmortem Proteolysis by Gel-Based Proteome Analysis and Its Relationship to Meat Quality Traits in Pig Longissimus. *Meat Sci.* **2005,** *69* (1), 79–91.

Iwahashi, Y.; Hosoda, H. Effect of Heat Stress on Tomato Fruit Protein Expression. *Electrophoresis* **2000,** *21* (9), 1766–1771.

Jensen, O. N. Modification-Specific Proteomics: Characterization of Post-Translational Modifications by Mass Spectrometry. *Curr. Opin. Chem. Biol.* **2004,** *8* (1), 33–41.

Joseph, P.; Suman, S. P.; Rentfrow, G.; Li, S.; Beach, C. M. Proteomics of Muscle-Specific Beef Color Stability. *J. Agric. Food Chem.* **2012,** *60* (12), 3196–3203.

Kiran, M.; Maheswarappa, N. B.; Banerjee, R.; Ch, V.; Rapole, S. Impact of Stunning Before Slaughter on Expression of Skeletal Muscles Proteome in Sheep. *Anim. Biotechnol.* **2021,** 1–8.

Kiran, M.; Naveena, B. M.; Reddy, K. S.; Shahikumar, M.; Reddy, V. R.; Kulkarni, V. V. et al. Understanding Tenderness Variability and Ageing Changes in Buffalo Meat: Biochemical, Ultrastructural and Proteome Characterization. *Animal* **2016,** *10* (6), 1007–1015.

Kiran, M.; Naveena, B. M.; Reddy, K. S.; Shashikumar, M.; Reddy, V. R.; Kulkarni, V. V. et al. Muscle-Specific Variation in Buffalo (B ubalus bubalis) Meat Texture: Biochemical, Ultrastructural and Proteome Characterization. *J. Text. Stud.* **2015,** *46* (4), 254–261.

Kiran, M.; Naveena, B. M.; Smrutirekha, M.; Reddy, P. B.; Rituparna, B.; Kumar, Y. P. et al. Traditional Halal Slaughter Without Stunning Versus Slaughter with Electrical Stunning of Sheep (Ovis aries). *Meat Sci.* **2019,** *148*, 127–136.

Kristensen, L.; Therkildsen, M.; Aaslyng, M. D.; Oksbjerg, N.; Ertbjerg, P. Compensatory Growth Improves Meat Tenderness in Gilts But Not in Barrows. *J. Anim. Sci.* **2004,** *82* (12), 3617–3624.

Kvasnička, F. Proteomics: General Strategies and Application to Nutritionally Relevant Proteins. *J. Chromatogr. B* **2003,** *787* (1), 77–89.

Lametsch, R.; Bendixen, E. Proteome Analysis Applied to Meat Science: Characterizing Post Mortem Changes in Porcine Muscle. *J. Agric. Food Chem.* **2001,** *49* (10), 4531–4537.

Lametsch, R.; Karlsson, A.; Rosenvold, K.; Andersen, H. J.; Roepstorff, P.; Bendixen, E. Postmortem Proteome Changes of Porcine Muscle Related to Tenderness. *J. Agric. Food Chem.* **2003,** *51* (24), 6992–6997.

Lametsch, R.; Kristensen, L.; Larsen, M. R.; Therkildsen, M.; Oksbjerg, N.; Ertbjerg, P. Changes in the Muscle Proteome After Compensatory Growth in Pigs. *J. Anim. Sci.* **2006,** *84* (4), 918–924.

Laville, E.; Sayd, T.; Morzel, M.; Blinet, S.; Chambon, C.; Lepetit, J. et al. Proteome Changes During Meat Aging in Tough and Tender Beef Suggest the Importance of Apoptosis and Protein Solubility for Beef Aging and Tenderization. *J. Agric. Food Chem.* **2009,** *57* (22), 10755–10764.

Lee, J. Y.; Park, J. H.; Mun, H.; Shim, W. B.; Lim, S. H.; Kim, M. G. Quantitative Analysis of Lard in Animal Fat Mixture Using Visible Raman Spectroscopy. *Food Chem.* **2018,** *254*, 109–114.

Lee, T. H.; Wani, W. A.; Koay, Y. S.; Kavita, S.; Tan, E. T. T.; Shreaz, S. Recent Advances in the Identification and Authentication Methods of Edible Bird's Nest. *Food Res. Int.* **2017,** *100*, 14–27.

Leheska, J. M.; Thompson, L. D.; Howe, J. C.; Hentges, E.; Boyce, J.; Brooks, J. C. et al. Effects of Conventional and Grass-Feeding Systems on the Nutrient Composition of Beef. *J. Anim. Sci.* **2008,** *86* (12), 3575–3585.

Liu, J.; He, J.; Yu, J.; Mao, X.; Zheng, P.; Huang, Z. et al. Birth Weight Alters the Response to Postnatal High-Fat Diet-Induced Changes in Meat Quality Traits and Skeletal Muscle Proteome of Pigs. *Br. J. Nutr.* **2014,** *111* (10), 1738–1747.

Mallick, P.; Kuster, B. Proteomics: A Pragmatic Perspective. *Nat. Biotechnol.* **2010,** *28* (7), 695–709.

Mane, B. G.; Mendiratta, S. K.; Tiwari, A. K.; Bhilegaokar, K. N. Detection of Adulteration of Meat and Meat Products with Buffalo Meat Employing Polymerase Chain Reaction Assay. *Food Anal. Methods* **2012,** *5* (2), 296–300.

Miller, R. K. Factors Affecting the Quality of Raw Meat. *Meat Processing* **2002,** 27–63.

Montowska, M.; Pospiech, E. Is Authentication of Regional and Traditional Food Made of Meat Possible? *Crit. Rev. Food Sci. Nutr.* **2012,** *52* (6), 475–487.

Montowska, M.; Pospiech, E. Species-Specific Expression of Various Proteins in Meat Tissue: Proteomic Analysis of Raw and Cooked Meat and Meat Products Made from Beef, Pork and Selected Poultry Species. *Food Chem.* **2013,** *136* (3–4), 1461–1469.

Morzel, M.; Chambon, C.; Hamelin, M.; Santé-Lhoutellier, V.; Sayd, T.; Monin, G. Proteome Changes During Pork Meat Ageing Following Use of Two Different Pre-Slaughter Handling Procedures. *Meat Sci.* **2004,** *67* (4), 689–696.

Mullen, A. M.; Stapleton, P. C.; Corcoran, D.; Hamill, R. M.; White, A. Understanding Meat Quality Through the Application of Genomic and Proteomic Approaches. *Meat Sci.* **2006,** *74* (1), 3–16.

Nair, M. N.; Zhai, C. Application of Proteomic Tools in Meat Quality Evaluation. In *Meat Quality Analysis*; Academic Press, 2020; pp 353–368.

Naveena, B. M.; Jagadeesh, D. S.; Babu, A. J.; Rao, T. M.; Kamuni, V.; Vaithiyanathan, S. et al. OFFGEL Electrophoresis and Tandem Mass Spectrometry Approach Compared with DNA-Based PCR Method for Authentication of Meat Species from Raw and Cooked Ground Meat Mixtures Containing Cattle Meat, Water Buffalo Meat and Sheep Meat. *Food Chem.* **2017,** *233*, 311–320.

Naveena, B. M.; Jagadeesh, D. S.; Kamuni, V.; Muthukumar, M.; Kulkarni, V. V.; Kiran, M.; Rapole, S. In-Gel and OFFGEL-Based Proteomic Approach for Authentication of Meat Species from Minced Meat and Meat Products. *J. Sci. Food Agric.* **2018,** *98* (3), 1188–1196.

Nurjuliana, M.; Man, Y. C.; Hashim, D. M.; Mohamed, A. K. S. Rapid Identification of Pork for Halal Authentication Using the Electronic Nose and Gas Chromatography Mass Spectrometer with Headspace Analyzer. *Meat Sci.* **2011,** *88* (4), 638–644.

Ochoa, M. L.; Harrington, P. B. Immunomagnetic Isolation of Enterohemorrhagic *Escherichia coli* O157: H7 from Ground Beef and Identification by Matrix-Assisted Laser Desorption/Ionization Time-of-Flight Mass Spectrometry and Database Searches. *Anal. Chem.* **2005,** *77* (16), 5258–5267.

Offer, G.; Knight, P.; Jeacocke, R.; Almond, R.; Cousins, T.; Elsey, J. et al. The Structural Basis of the Water-Holding, Appearance and Toughness of Meat and Meat Products. *Food Struct.* **1989,** *8* (1), 17.

Ortea, I.; Cañas, B.; Gallardo, J. M. Selected Tandem Mass Spectrometry Ion Monitoring for the Fast Identification of Seafood Species. *J. Chromatogr. A* **2011,** *1218* (28), 4445–4451.

Paredi, G.; Sentandreu, M. A.; Mozzarelli, A.; Fadda, S.; Hollung, K.; de Almeida, A. M. Muscle and Meat: New Horizons and Applications for Proteomics on a Farm to Fork Perspective. *J. Proteomics* **2013,** *88*, 58–82.

Perng, M. D.; Cairns, L.; Van Den IJssel, P.; Prescott, A.; Hutcheson, A. M.; Quinlan, R. A. Intermediate Filament Interactions Can be Altered by HSP27 and alphaB-Crystallin. *J. Cell Sci.* **1999,** *112* (13), 2099–2112.

Picard, B.; Gagaoua, M.; Micol, D.; Cassar-Malek, I.; Hocquette, J. F.; Terlouw, C. E. Inverse Relationships Between Biomarkers and Beef Tenderness According to Contractile and Metabolic Properties of the Muscle. *J. Agric. Food Chem.* **2014,** *62* (40), 9808–9818.

Picariello, G.; Mamone, G.; Nitride, C.; Ferranti, P. Proteomic Analysis of Beer. In *Proteomics in Food Science*; Academic Press, 2017; pp 383–403.

Polati, R.; Menini, M.; Robotti, E.; Millioni, R.; Marengo, E.; Novelli, E. et al. Proteomic Changes Involved in Tenderization of Bovine Longissimus Dorsi Muscle During Prolonged Ageing. *Food Chem.* **2012,** *135* (3), 2052–2069.

Salomé, M.; Huneau, J. F.; Le Baron, C.; Kesse-Guyot, E.; Fouillet, H.; Mariotti, F. Substituting Meat or Dairy Products with Plant-Based Substitutes Has Small and Heterogeneous Effects on Diet Quality and Nutrient Security: A Simulation Study in French Adults (INCA3). *J. Nutr.* **2021,** *151* (8), 2435–2445.

Sayd, T.; Morzel, M.; Chambon, C.; Franck, M.; Figwer, P.; Larzul, C. et al. Proteome Analysis of the Sarcoplasmic Fraction of Pig Semimembranosus Muscle: Implications on Meat Color Development. *J. Agric. Food Chem.* **2006,** *54* (7), 2732–2737.

Sentandreu, M. A.; Fraser, P. D.; Halket, J.; Patel, R.; Bramley, P. M. A proteomic-Based Approach for Detection of Chicken in Meat Mixes. *J. Proteome Res.* **2010,** *9* (7), 3374–3383.

Stawitz, C. C.; Siple, M. C.; Munsch, S. H.; Lee, Q.; Derby, S. R. Financial and Ecological Implications of Global Seafood Mislabeling. *Conserv. Lett.* **2017,** *10* (6), 681–689.

Surowiec, I.; Fraser, P. D.; Patel, R.; Halket, J.; Bramley, P. M. Metabolomic Approach for the Detection of Mechanically Recovered Meat in Food Products. *Food Chem.* **2011,** *125* (4), 1468–1475.

Toda, M.; Ono, S. J. Genomics and Proteomics of Allergic Disease. *Immunology* **2002,** *106* (1), 1–10.

Trivedi, D. K.; Hollywood, K. A.; Rattray, N. J.; Ward, H.; Trivedi, D. K.; Greenwood, J. et al. Meat, the Metabolites: An Integrated Metabolite Profiling and Lipidomics Approach for the Detection of the Adulteration of Beef with Pork. *Analyst* **2016,** *141* (7), 2155–2164.

Vaidyanathan, S.; Goodacre, R. Proteome and Metabolome Analyses for Food Authentication. In *Food Authenticity and Traceability*; Michele L. Wood Head Publishing, 2003.

Vanderwerf, S. M.; Cooper, M. J.; Stetsenko, I. V.; Lutsenko, S. Copper Specifically Regulates Intracellular Phosphorylation of the Wilson's Disease Protein, a Human Copper-Transporting ATPase. *J. Biol. Chem.* **2001,** *276* (39), 36289–36294.

Velioglu, H. M.; Sezer, B.; Bilge, G.; Baytur, S. E.; Boyaci, I. H. Identification of Offal Adulteration in Beef by Laser Induced Breakdown Spectroscopy (LIBS). *Meat Sci.* **2018,** *138*, 28–33.

Verbeke, W.; Pérez-Cueto, F. J.; de Barcellos, M. D.; Krystallis, A.; Grunert, K. G. European Citizen and Consumer Attitudes and Preferences Regarding Beef and Pork. *Meat Sci.* **2010,** *84* (2), 284–292.

von Bargen, C.; Dojahn, J.; Waidelich, D.; Humpf, H. U.; Brockmeyer, J. New Sensitive High-Performance Liquid Chromatography–Tandem Mass Spectrometry Method for the Detection of Horse and Pork in Halal Beef. *J. Agric. Food Chem.* **2013,** *61* (49), 11986–11994.

Wang, G. J.; Zhou, G. Y.; Ren, H. W.; Xu, Y.; Yang, Y.; Guo, L. H.; Liu, N. Peptide Biomarkers Identified by LC–MS in Processed Meats of Five Animal Species. *J. Food Compos. Anal.*, *73*, 47–54.

Yu, J.; Tang, S.; Bao, E.; Zhang, M.; Hao, Q.; Yue, Z. The Effect of Transportation on the Expression of Heat Shock Proteins and Meat Quality of *M.* Longissimus dorsi in Pigs. *Meat Sci.* **2009,** *83* (3), 474–478.

Zapata, I.; Reddish, J. M.; Miller, M. A.; Lilburn, M. S.; Wick, M. Comparative Proteomic Characterization of the Sarcoplasmic Proteins in the Pectoralis Major and Supracoracoideus Breast Muscles in 2 Chicken Genotypes. *Poultry Sci.* **2012,** *91* (7), 1654–1659.

Zhao, J.; Li, A.; Jin, X.; Pan, L. Technologies in Individual Animal Identification and Meat Products Traceability. *Biotechnol. Biotechnol. Equip.* **2020,** *34* (1), 48–57.

Zhao, Y.; Zhang, B.; Chen, G.; Chen, A.; Yang, S.; Ye, Z. Tracing the Geographic Origin of Beef in China on the Basis of the Combination of Stable Isotopes and Multielement Analysis. *J. Agric. Food Chem.* **2013,** *61* (29), 7055–7060.

CHAPTER 13

Food Waste Revalorization as a Functional Food or Ingredient

ESTRADA-GIL L. E.[1], N. D. CERDA-CEJUDO[1], CHÁVEZ-GONZÁLEZ M. L.[1], A. C. FLORES-GALLEGOS A. C.[1], M. GOVEA-SALAS[2], CONTRERAS-ESQUIVEL J. C.[1], C. N. AGUILAR-GONZÁLEZ[1], and J. A. ASCACIO-VALDÉS[1]

[1]Bioprocesses & Bioproducts Group, Food Research Department, School of Chemistry, Universidad Autónoma de Coahuila, Saltillo, Coahuila, México

[2]Laboratory of Nanobiociences, School of Chemistry, Universidad Autónoma de Coahuila, Ing. J. Cárdenas Valdéz S/N, República, Saltillo, México

ABSTRACT

One of the greatest problems humanity faces nowadays is the loss of food and the discarding of many different food sources that are not appealing to the consumer, with little knowledge that these so-called "wastes" are what holds one of the greatest sources of high-value bioactive compounds. Peels, leaves, stems, and seeds of many fruits and vegetables have already been studied and they possess phenolic compounds, and dietetic fibers among many other biologically active compounds that may prove to benefit the consumer. In this research the different methods to apply agricultural by-products from different fruits and vegetables are shown, varying from their use in bakery products to yogurt ingredients to fortify and enhance the already healthy attributes some foods have.

Engineering Principles for Food Processing Technology and Product Realization. Leonardo Sepúlveda Torre, Porteen Kannan, A. K. Haghi, PhD (Eds.)

13.1 INTRODUCTION

One of the world's greatest problems in modern times is not the production of food, as we can produce enough food to nourish every single person on the planet, but instead, the biggest problem humanity faces at this time is the untimely loss of food before it can be consumed as it is estimated by the World Food Programme that at least one-third of the food produced globally is lost in retail and the fridges of many people around the world (WFP, 2020).

Data from the Food and Agriculture Organization (FAO) estimates that around 14% of the food of the world is lost in the process of the shipping of the products from harvest to retail which represents a loss of $400 billion annually and an estimated amount of 17% of all the food purchased and kept in households and retail goes to waste (FAO, 2021).

COVID-19 also changed the amount of waste that is produced worldwide, as people stayed locked down in their households many food products like fruits, vegetables, poultry, and beef were lost as the government lockdown caused huge food losses in food market retail as well as in agricultural production.

Not only food waste is an economical problem, but it is also an ethical, environmental, and social problem that will just keep on increasing as time goes by and food prices keep going up. International organizations, NGOs, governments, and many branches of the United Nations have declared that this is an issue of utmost importance as it affects all levels of society. As an ethical and social problem, it must be addressed that many people around the globe do not possess access to a source of food or are unable to buy food products for the constant uprising in prices (Edjabou et al., 2016; Ponis et al., 2017; Roodhuyzen et al., 2017). As an environmental problem, the high amounts of food wastes such as peels, seeds, pulps, and rotten or spoiled foods produce high amounts of greenhouse gases and not only that, as an example, many residues from the coffee industry are left behind in roads just to decompose, this waste does not only contaminate the air with gases but also the soil due to high amounts of antinutrients present in the pulp and peel of the coffee, just to state an example, contributing to climate change and future problems in the use of soil (Arya et al., 2022).

The valorization approach for food waste can differ from each different kind of waste as not all of them are suitable for human consumption as they are produced, some may contain carbohydrates, proteins, fats, and other nutraceutical compounds that can be reused to obtain energy, biofuels, enzymes, antioxidant extracts, novel biodegradable materials, and other commercial products (Duenas and Garciá-Estévez, 2020).

Food waste is by no means, inevitable as it is a vicious circle, where the production and consumption of food will always produce waste, but how we approach this waste is part of the answer. Nowadays, companies, producers, and retailers possess new policies for zero waste, for example, the wine industry donates the bagasse of the grapes that were used in the production of wine to extract and obtain high-value compounds which can be used to elaborate different products like fungicides (da Silva et al., 2022), feed for shrimps (Rosas et al., 2022), and even cookies with antioxidant activity (Fontana et al., 2022).

Functional ingredients represent one of the best alternatives in the treatment of food waste, as many of the wastes such as peels, seeds, bagasse, and leaves of fruits, vegetables, cereals, and legumes have many bioactive compound molecules which may confer the consumer great properties to maintain health and prevent chronic diseases such as diabetes type 2, cardiovascular disease, mental diseases, and some kinds of cancer. The usage of these compounds has an extensive background as there are studies that show the possibility of the usage of certain agricultural food wastes (AFW) to enrich pasta with phenolic compounds (Melini et al., 2020). Another usage for agricultural waste is the elaboration of kombucha as stated in a study by Liu et al., 2022. Another option for the usage of agricultural waste is the elaboration of a functional bakery, where bananas, mango, pineapple, pomegranate, and other peels were used in the elaboration of bakery products (Ajila et al., 2008; Eshak, 2016; Ismail et al., 2014; Noor Aziah et al., 2011; Wu and Shiau, 2015).

In this work, the usage of food waste will be explored in novel ways to reuse and reduce the waste produced by the agricultural industry as new functional foods, ingredients, and extracts to obtain high-added value biological compounds will be reviewed.

13.2 FRUIT AND VEGETABLE WASTE REVALORIZATION

Fruits and vegetables are highly perishable food that has a high index of being thrown to waste or considered as a loss in certain cases, but it must be considered that loss and waste are not the same as loss can be considered when the food is no longer edible, and waste is when the food does not get eaten once in a household or retail market. Food loss is a problem that is bigger in developing countries as they do not possess the technology, infrastructure, or production practices that ensure the safety of food production.

Food waste is most likely to happen in developed countries as the consumer has higher aesthetical standards where a fruit or vegetable that is not eye-appealing to the customer will not get purchased and sooner than later it will rot. In East and Southeast Asia as well as in sub-Saharan Africa some studies show around 50% of the fruit and vegetables in those regions are lost due to poor storage and production conditions; this calls for measures not only in the valorization of this waste but also on the improvement of conditions on which food is produced and stored (FAO, 2017, 2020).

The use of fruit and vegetable wastes as a functional ingredient has been studied for a long time; this makes it possible to find different studies that take advantage of these wastes, and here are some of the most recent examples of how these wastes have been used.

13.3 FRUIT WASTE AS FUNCTIONAL INGREDIENTS

13.3.1 PROSOPIS NIGRA

Prosopis nigra also known as “Algarrobo negro” is a legume tree from South America that can produce a pod that is very sweet, yellow and with purple spots and it is used to produce flours and a fermented alcoholic beverage (María J. Pérez et al., 2018).

This tree is very much appreciated in the regions of Argentina where it is grown because practically all of the trees can be very well exploited, the bark of this tree is very useful as it can be used as a great astringent and dye, its wood is very useful as it has many uses for example construction, charcoal and firewood, and the leaves are great fertilizers (María J. Pérez et al., 2014).

A study by María J. Pérez et al. (2020) utilized the waste obtained from the flour manufacturing process, which was processed to obtain powders with a fine particle size where F1, F2, and F3 were obtained, each with a different size. All powders had similar properties as they all possessed great fiber and fructose content, which are some desirable characteristics a powder destined to be consumed can have. As macronutrients correspond the three samples contained a similar amount of glucose around 65–67 g/100 g of powder, 11.5–12.84 g/ 100 g of powder of proteins, and a low-fat content with barely 0.87–1.44 g g/100 g of powder, presence of Mg, K, Na, Cu, Fe, and Zn can be detected in this potential functional ingredient which makes it a great option for revalorization and usage in food production.

HPLC/ESI/MS analysis of the powder detected the presence of various polyphenolic compounds such as quercetin, kaempferol, coumaric acid, ferulic acid, and vicenin II among others, the same compounds which have been studied and proven to possess antioxidant, anti-inflammatory, neuroprotective, vascular-protective, anticarcinogenic, antiplatelet, antihypertensive and hypoglycemic activities just to mention some of them that can grant a great positive effect on the consumer if this functional ingredient is consumed (Bento-Silva et al., 2018; Cattaneo et al., 2019; Xiao et al., 2016).

13.3.2 APPLE WASTE

One of the most common kinds of waste in the food industry is the peel of the fruit, as it is often considered to not be useful anymore, although some are a great font of fiber and edible. In some fruits the peels constitute as much as 40 to 50% of the total weight of the product and in others it is just around 20%, and this varies from fruit to fruit. Taking this information into account fruit processing industries that only use the pulp of the fruit to elaborate their products, generate a great deal of peel waste that just goes into the trash to generate more pollution and create a bigger environmental problem for future generations. Apple is by far one of the world's most popular fruits as it is the fourth most widely produced crop fruit in the world with a production of 87 million tons (Wani et al., 2022).

A study by Lazari et al. (2019) demonstrated that it was possible to incorporate peel and pulp from apples into a low-calorie ice cream product, the peel was added in two different ways, cooked and natural, and three formulations were made, each different in the amount and kind of ingredients used but they all contained cream, apple peel and pulp, skimmed milk, condensed milk, and emulsifier. Among the three formulations, the best one obtained an amount of 80.8 kcal, 11.3 g of carbohydrates, 1.36 g of protein, and 2.92 g of lipids for every 100 g of ice cream. In this same study also, sensory evaluations were performed obtaining great results for acceptations with high values of 92% for sweetness, 92.2% of acceptation on flavor, and 84.7 acceptations for texture.

Apple pomace is by far one of the most popular functional ingredients obtained from agricultural waste, as it is composed of the core, peel, seed, pulp, and kernel of the fruit and it represents almost 20–35% of the fruit's weight when it is fresh (Rabetafika et al., 2014). It is estimated that around 4 million tons of apple pomace are produced worldwide by the juice industry, making it a great problem to dispose of this waste effectively (J. M. Costa et al., 2022).

A study by Curutchet et al. (2021) demonstrates how useful apple pomace is, in the elaboration of cakes as a fiber source, as 40% of it is constituted of dietary fiber which in long term can help in the treatment of obesity, atherosclerosis, coronary heart diseases, intestine cancer, diabetes and can be used as a potential prebiotic by stimulating the growth of probiotic bacteria (Kruczek et al., 2017). For every 100 g, the fiber-enriched cake had 275 kcal, 43 g of carbohydrate, 4.8 g of protein, and 9.5 g of lipids of which 1.1 g are from saturated fat and 6.3 g of fiber, compared to the regular cake which had 332 kcal, 52 g of carbohydrate, 11 g of protein, 8.7 g of lipids out of which 3.2 are from saturated fat and 1 g of fiber. The results benefit each individually as a regular cake possesses more protein than its counterpart, but its saturated fat content is higher than the fiber-enriched cake, also it is important to remark the fact that the fiber content cannot be compared as this cake made with apple pomace possesses greater fiber content with 6.3 g. A sensorial evaluation was also performed in this test and the results show that both cakes are very well accepted by consumers.

13.3.3 MANGO WASTE

Mango (*Mangifera* indica) is one of the most popular fruits harvested in tropical countries around the world, with a production of 2.3 million tons per year in 2021 (FAO, 2022b), mango is most commonly used in the elaboration of liquors, juices, and for household consumption, but only the pulp of the fruit is used in these types of consumption, leaving behind both the peel and the seed which represent a great deal of the weight of the fruits with around 50–60% of the total weight (Tirado-Kulieva et al., 2022).

The usage of mango peel flour has been reported as an ingredient in the elaboration of yogurt, as it possesses a great amount of fiber which can be considered prebiotic and polyphenols which may confer the product with antioxidant activity among others that may be of interest to the consumer (Pérez-Chabela et al., 2022). Yogurt is a fermented food that is very much consumed worldwide as it is a great source of probiotics; it is a fermented food with the presence of lactic acid bacteria such as the *Lactobacillus* family. The introduction of mango peel as a fiber source is capable of making yogurt formulations a type of symbiotic food, meaning that it is a combination of probiotics and prebiotics that give the consumer/host great health benefits by improving the survival and colonization of selected bacteria in the gastrointestinal tract (Fazilah et al., 2018).

It is well established that mango peel is a great source of dietary fiber, but it is also a great antioxidant, as extracts with high polyphenolic content can be obtained from it. Mango peel nanoparticles have been used in many different studies regarding quality enhancement and shelf-life improvement, for example in the elaboration of beef sausages where mango peel nanoparticles proved to be a great way to maintain microbial quality for this product when used in a concentration of 2% for 9 days (Boff et al., 2022).

Another study by Aamir et al. (2022) determined that mango peel could be additioned to multigrain cookies and fortify them with their nutrients. Cookies made with mango peel powder were analyzed and it was determined that they contained 6.66% moisture, 2.17% ashes, protein in 8.17% of, and fat at 1.67%. Sensory evaluation was performed on the cookies they proved to possess high sensory scores and showed better results for physicochemical and antioxidant analysis.

13.3.4 GRAPE WASTE

Grape (*Vitis vinifera sativa*) is one of the world's most cultivated fruits all around the world and it is of high relevance, not only for its consumption in wine but it is also consumed fresh and dried and the fruit's pomace possesses great nutritional value as well as biological compounds with pharmaceutical activities among which we can find, anti-inflammatory, antioxidant, antihypertensive, among many others. The pomace is constituted of skin, stems, seeds, and pulp which represents around 20% of the total weight of the fruit (M. Costa et al., 2022), all of this is obtained in the natural process of winemaking, as it is obtained by the crushing and pressing of the fruits, considering that the world's production of wine in 2021 was 260 million hectoliters, this represents a huge amount of waste produced every year.

With pomace there is a potential to fabricate flours, which can potentially be used in the elaboration of cookies, bread, cereal bars, pasta, vitamins, and juices (Oliveira et al., 2016) when pomace is used in the elaboration of these products they are conferred with high polyphenolic content as anthocyanins, flavonols, catechins, and proanthocyanins are present in this agricultural by-product. These compounds are the ones that confer the pomace of their important biological activities.

In a study by Marchiani et al. (2016) grape pomace is used as a functional ingredient in the elaboration of yogurt to improve and add dietary fiber and some antioxidant compounds, same that can help the probiotic bacteria in the yogurt to colonize the gastrointestinal tract. This yogurt was prepared using UHT whole milk with 36 g/kg of fat, 31 g/kg of protein, and 48 g/kg of carbohydrates, milk powder at 3% was added and the milk was inoculated with a culture of YO-MIX 401 (Santamaria, Burago di Molgora, Italy) which contained a mixture of *Streptococcus thermophilus* and *Lactobacillus dellbrueckii*. Milk was fermented at 42°C until a pH of 4.8 was obtained and then the grape pomace was added to the mix; there were 3 types of pomace used from different varieties such as Chardonnay, Moscato, and Pinot noir.

The addition of the pomace to the yogurt resulted in the finding of various phenolic compounds such as gallic acid, protocatechuic acid, PB1, THA, catechin, vanillic acid, epicatechin, rutin, and quercetin, same which are not present in normal yogurt. Although the presence of grape pomace changed its organoleptic characteristics, they were not very appealing to the sensory test subjects, as pinot noir additioned yogurt was not suitable for tests due to its strong smell and flavor, Chardonnay and Moscato varieties were able to be evaluated in sensory tests, but the results were not all that favorable. The addition of pomace to yogurt can be done but organoleptic characteristics were not of the liking of the evaluators, properties like sweetness, acidity, smell, and taste were not very favorable, a new strategy for the elaboration of yogurt additioned with grape pomace should be considered.

Pasta is one of the most consumed products worldwide as it is present in many places, from restaurants to households and retail markets and it can be acquired by almost everyone, so it has a great reach in the food industry, so the addition of enriched flours that may help in the enhancement of the nutritional traits of the product is of utmost importance. Grape pomace as it can be made into fine flour is ideal to do as the addition of this ingredient modifies the cooking properties, starch digestibility, and glycemic index of the pasta. Tolve et al. (2020) analyzed the elaboration of pasta by replacing durum wheat semolina with grape pomace; the pomace used was from the Corvina variety grape. The pasta obtained was of reddish color which only increased with the concentration of grape pomace used, this was due to the natural pigments of the fruit, the addition of pomace also added phenolic compounds to the pasta with the best treatment obtaining 2.57 mg of gallic acid equivalents (GAE)/g while uncooked and 1.81 mg GAE/g after cooking the pasta, this loss is normal as the boiling of water can degrade biological compounds such as phenolic compounds, the pasta also presented antioxidant activities in FRAP and ABTS assays. Fiber content changed in comparison to normal pasta, as grape pomace additioned pasta had an amount of 8.2 g of fiber/100 g of pasta compared to 3 g of fiber for the normal semolina pasta.

The results found in this study show the importance of grape pomace as the addition of it to pasta can increase its nutritional properties.

13.4 VEGETABLE WASTE AS FUNCTIONAL INGREDIENTS

13.4.1 CARROT WASTE

Carrots are a very popular vegetable as it is produced in many regions of the world; they originated in the region now known as Afghanistan and some parts of northern Iran and Pakistan (Simon, 2020). The global production of this vegetable was around 40 million tons per year in 2020 and China is the main producer with almost 50% of the total production, followed by Uzbekistan, Russia, the United States (USA), Ukraine, and Poland (FAO, 2022a). There are many kinds of carrot as there are many varieties around the world, the most dominating one is the orange kind as it can be seen everywhere but, white types of carrots still are being consumed in western and Eastern Europe, mainly as feed for livestock, red carrots can be found in Japan and yellow and purple in the Middle East.

Carrot pomace is an industrial by-product obtained by the elaboration of juices, as only 60–70% of juice yield is obtained from a carrot, although carrot pomace is a high nutritional value by-product as it has large amounts of carotenoids, vitamins, dietary fiber, and minerals (Umair et al., 2021). The utilization of carrot pomace can be done as a functional ingredient, one of the ways it can be used is to enhance the nutritional properties of baked products such as cakes. Kamiloglu et al. (2017) used the pomace of black carrot to improve the nutritional characteristics of a cake and analysis showed that the cake had the presence of the five major anthocyanins, neochlorogenic acid, chlorogenic acid, and caffeic acid were detected in the enriched cakes conferring this product great antioxidant activities.

Smoothies are a popular way to elaborate beverages that contain many kinds of fruits and vegetables, some are made with just water but others are made with a great variety of ingredients such as milk, ice cream, sugar, yogurt, and other milk derivates, the composition of some smoothies makes it viable to add some ingredients that may negatively alter the taste as they get covered up by the ingredients previously mentioned, this is the case for black carrot a study by Kidoń and Uwineza (2022) aimed to use black carrot with a mix of banana and pumpkin in 3 concentrations of sugar, high, low, and no sugar. The smoothies produced with these ingredients proved to be a great source of polyphenols as the total polyphenolic content ranged from

39.2 to 55.8 mg/100 g with a strong correlation with antioxidant activity. The mix of pumpkin and carrot had better carotenoid content while the mixture of banana and carrot showed better antioxidant activity, while the sensory evaluation was a success as the judges gave excellent scores of the smoothies in categories such as smell, taste, consistency, and overall acceptance.

13.4.2 CASSAVA WASTE

Cassava is a very starchy root and its leaves can be eaten as a source of protein, vitamins, and minerals; it originated in South America and has had a great journey all around the world as it is present also in Asia and Africa, ever since the production of cassava has been increasing mainly because it is a relatively easy crop to work with and it is one of the main foods for more than 800 million people worldwide. Cassava is now considered to have the potential to fight food insecurity, poverty, and hunger in many African countries (Otun et al., 2022).

Cassava peel is a material that has been studied for its high content of fibers, minerals, and sugars and has been used in animal feed, in a study by Rodrigues et al. (2021) cassava peel was used in the elaboration of flour which was packed with high polyphenolic content, some of the main one's found were gallic acid at 270.62 mg/100 g, catechin 14.87 mg/100 g, chlorogenic acid 0.79 mg/100 g, caffeic acid 0.19 mg/100 g, and rutin 0.96 mg/100 g. Also, antioxidant activity was evaluated in ABTS and DPPH assays with good results with 26.31 μM Trolox/g and 24.13 μM Trolox/g respectively.

13.4.3 BROCCOLI WASTE

Brassica oleracea is a biennial plant that belongs to the Brassicaceae family, it is eaten as a vegetable all over the world, and it is a popular vegetable that has been consumed ever since the time of the Greeks 2500 years ago. China is the largest producer of broccoli followed by India, the USA, Spain, Mexico, and Italy, not so long ago, broccoli consumption spiked due to the now popular awareness of healthy lifestyle and diet, in which broccoli apports an ingredient with high content of polyphenols, vitamins, mineral, fibers, and very low caloric content (Nagraj et al., 2020).

Recently, the scientific community has taken into consideration that not only the head and inflorescence of the vegetable contain the nutritional content mentioned, but also the leaves and stalks which represent a 70%

of the broccoli plant. This causes a great environmental problem that can be fixed by taking advantage of this waste; a study by Costa-Pérez et al. (2022) set out to identify which biological compounds were present in the stalk of the vegetable. To mention some of the polyphenolic content found the stalk had the presence of 5-caffeoylquinic acid, feruloyl-caffeoyl derivative, di-sinapoyl-gentiobioside II, 1,2,3'-Tri-sinapoyl-gentiobioside among others, which make the stalk an option to make a functional ingredient.

Núñez-Gómez et al. (2022) conducted a study to obtain different dietary fiber fractions from broccoli stalks to revalorize the vegetable waste and potentially obtain a new functional ingredient. The fiber fractions obtained from broccoli yielded 67% total fiber fraction and 70% for insoluble fiber fraction which makes this a great material to obtain fiber from; these fractions also contained a great amount of nutritional value as protein had 3.8–4 g/100 g, 19.3–24.4 g/100 g of total carbohydrates, 60.8–68.9 g/100 g of fiber, 54.0–54.3 g/100 g of total insoluble fiber, and 7.5–10.8 g/100 g of total ash. The implementation of a functional ingredient of these results found can be very beneficial to consumers as the number of nutrients found in this by-product of the food industry is astounding, and the amount of fibers present makes it an ideal candidate to elaborate bread with it.

The elaboration of bread with broccoli by-products has been studied by Lafarga et al. (2019) which demonstrated that the addition of powdered broccoli stalks and leaves into the formulation of bread at a concentration of 2% does not affect the acceptance of the product by the consumers, but there are notable changes such as color and texture. The addition of these compounds to the bread confers the product of increased nutritional characteristics and against all odds the antioxidant and phenolic compounds found in the stalks, resist the baking process making this a great option for consumption.

13.5 FUTURE PERSPECTIVES

New functional ingredients are found every day in the least of the expected places, the revalorization of the wastes that the food industry generates is now more important than ever as new options to lower the carbon footprint and lower the emission of greenhouse gases arise. As we have seen in this study it is possible to find gold, where you least expect it as new functional ingredients, which can be obtained from agricultural by-products which are often considered as low-value waste, but studies indicate that most of the biological importance factors reside in this so-called waste. Many researchers have found that in the peels of fruits we can find polyphenolic content which depending on the compound is the range of biological activities which we may find and may vary from antioxidant, antiproliferative, anti-inflammatory, antibiotic, prebiotic, etc.

Much research ends up just finding what biological properties are present in the studied subject and they never move up to the next stage, finding a practical use for the newfound information, more research is needed in functional food development as there is so much potential in the new ingredients that every day are being discovered out of food wastes. Environmentally this could create a long-term solution to one of humanity's greatest problems, food loss, and food discarding, as the valorization of every single aspect of an ingredient, from the peel, seed, pulp, stem, stalk, leaves, everything is an important aspect of a whole matrix. Ethically this could be a new beginning for new ways to open new food resources to entire populations who live in conditions where food is more like a luxury instead of a human right. Economically food industries would benefit so much from the usage of all the food source as waste management and food loss represents a big investment every year.

13.6 CONCLUSIONS

The usage of waste as a functional ingredient or to obtain new functional foods is one of the best ways technology can be applied to solve a problem, millions of tons of food are wasted every year and every year it turns worse, the valorization of the industry's by-products is one of the best strategies that can be performed to reduce the amount of waste that is generated, as we have seen it is possible to turn peels of fruit into new ingredients such as flour to prepare additioned and fortified products such as bakery products, yogurts, ice cream. Research is needed to find new possible ingredients out of materials usually considered waste to fight back food loss and while doing so, also fight contamination and global pollution.

ACKNOWLEDGMENTS

Autor Estrada-Gil L.E. thanks the Mexican Council for Science and Technology (CONACYT) for the financial support for the development of the project (Ph.D.) in Food Science and Technology offered by the Autonomous University of Coahuila.

KEYWORDS

- **food loss**
- **waste**
- **bioactive compounds**
- **phenolic compounds**
- **agricultural by-products**
- **yogurt ingredients**
- **healthy foods**

REFERENCES

Aamir, M.; Saleem, A.; Rakha, A.; Nadeem, M.; Ateeq, H.; Saeed, F.; Asghar, A.; Shah, Y. A.; Afzaal, M.; Javed, A.; Tufail, T.; Qamar, A. Functional Utilization of Mango Waste for Improving the Nutritional and Sensorial Properties of Multigrain Cookies. *J. Food Process. Preserv.* **2022**. https://doi.org/10.1111/jfpp.17173

Ajila, C. M.; Leelavathi, K.; Prasada Rao, U. J. S. Improvement of Dietary Fiber Content and Antioxidant Properties in Soft Dough Biscuits with the Incorporation of Mango Peel Powder. *J. Cereal Sci.* **2008,** *48* (2), 319–326. https://doi.org/10.1016/j.jcs.2007.10.001

Arya, S. S.; Venkatram, R.; More, P. R.; Vijayan, P. The Wastes of Coffee Bean Processing for Utilization in Food: A Review. *J. Food Sci. Technol.* **2022,** *59* (2), 429–444. https://doi.org/10.1007/s13197-021-05032-5

Bento-Silva, A.; Vaz Patto, M. C.; do Rosário Bronze, M. Relevance, Structure and Analysis of Ferulic Acid in Maize Cell Walls. *Food Chem.* **2018,** *246*, 360–378. https://doi.org/10.1016/j.foodchem.2017.11.012

Boff, J. M.; Strasburg, V. J.; Ferrari, G. T.; de Oliveira Schmidt, H.; Manfroi, V.; de Oliveira, V. R. Chemical, Technological, and Sensory Quality of Pasta and Bakery Products Made with the Addition of Grape Pomace Flour. *Foods* **2022,** *11* (23), 3812. https://doi.org/10.3390/foods11233812

Cattaneo, F.; Roco, J.; Alarcón, G.; Isla, M. I.; Jeréz, S. Prosopis Alba Seed Flour Improves Vascular Function in a Rabbit Model of High Fat Diet-Induced Metabolic Syndrome. *Heliyon* **2019,** *5* (8), e01967. https://doi.org/10.1016/j.heliyon.2019.e01967

Costa-Pérez, A.; Moreno, D. A.; Periago, P. M.; García-Viguera, C.; Domínguez-Perles, R. A New Food Ingredient Rich in Bioaccessible (poly)Phenols (and Glucosinolates) Obtained from Stabilized Broccoli Stalks. *Foods* **2022,** *11* (12), 1–17. https://doi.org/10.3390/foods11121734

Costa, J. M.; Ampese, L. C.; Ziero, H. D. D.; Sganzerla, W. G.; Forster-Carneiro, T. Apple Pomace Biorefinery: Integrated Approaches for the Production of Bioenergy, Biochemicals, and Value-Added Products—An Updated Review. *J. Environ. Chem. Eng.* **2022,** *10* (5), 108358. https://doi.org/10.1016/j.jece.2022.108358

Costa, M.; Alfaia, C. M.; Lopes, P. A.; Pestana, J. M.; Prates, J. A. M. Grape By-Products as Feedstuff for Pig and Poultry Production. *Animals* **2022,** *12* (17), 1–18. https://doi.org/10.3390/ani12172239

Curutchet, A.; Trias, J.; Tárrega, A.; Arcia, P. Consumer Response to Cake with Apple Pomace as a Sustainable Source of Fibre. *Foods* **2021,** *10* (3), 1–12. https://doi.org/10.3390/foods10030499

da Silva, C. M. S.; da Boit Martinello, K.; Lütke, S. F.; Godinho, M.; Perondi, D.; Silva, L. F. O.; Dotto, G. L. Pyrolysis of Grape Bagasse to Produce Char for Cu (II) Adsorption: A Circular Economy Perspective. *Biomass Convers. Biorefinery* **2022,** *Ii*. https://doi.org/10.1007/s13399-022-02792-8

Duenas, M.; Garciá-Estévez, I. Agricultural and Food Waste: Analysis, Characterization and Extraction of Bioactive Compounds and Their Possible Utilization. *Foods* **2020,** *9* (6), 4–6. https://doi.org/10.3390/foods9060817

Edjabou, M. E.; Petersen, C.; Scheutz, C.; Astrup, T. F. Food Waste from Danish Households: Generation and Composition. *Waste Manage.* **2016,** *52*, 256–268. https://doi.org/10.1016/j.wasman.2016.03.032

Eshak, N. S. Sensory Evaluation and Nutritional Value of Balady Flat Bread Supplemented with Banana Peels as a Natural Source of Dietary Fiber. *Ann. Agric. Sci.* **2016,** *61* (2), 229–235. https://doi.org/10.1016/j.aoas.2016.07.002

FAO. Policy Measures for Managing Quality and Reducing Post-Harvest Losses in Fresh Produce Supply Chains in South Asian Countries; 2017.

FAO. Fruit and Vegetables—Your Dietary Essentials. In *Fruit and Vegetables—Your Dietary Essentials*; FAO, 2020. https://doi.org/10.4060/cb2395en

FAO. *Food Loss and Waste*; Food and Agriculture Organization, 2021. https://www.fao.org/nutrition/capacity-development/food-loss-and-waste/en/

FAO. *FAOSTAT*; 2022a. https://www.fao.org/faostat/en/#data/QCL

FAO. Major Tropical Fruits: Market Review 2021. Statistical Compendium Rome FAO, 01, 18; 2022b.

Fazilah, N. F.; Ariff, A. B.; Khayat, M. E.; Rios-Solis, L.; Halim, M. Influence of Probiotics, Prebiotics, Synbiotics and Bioactive Phytochemicals on the Formulation of Functional Yogurt. *J. Funct. Foods* **2018,** *48*, 387–399. https://doi.org/10.1016/j.jff.2018.07.039

Fontana, M.; Murowaniecki Otero, D.; Pereira, A. M.; Santos, R. B.; Gularte, M. A. Grape Pomace Flour for Incorporation into Cookies: Evaluation of Nutritional, Sensory and Technological Characteristics. *J. Culinary Sci. Technol.* **2022,** 1–20. https://doi.org/10.1080/15428052.2022.2086956

Ismail, T.; Akhtar, S.; Riaz, M.; Ismail, A. Effect of Pomegranate Peel Supplementation on Nutritional, Organoleptic and Stability Properties of Cookies. *Int. J. Food Sci. Nutr.* **2014,** *65* (6), 661–666. https://doi.org/10.3109/09637486.2014.908170

Kamiloglu, S.; Ozkan, G.; Isik, H.; Horoz, O.; Van Camp, J.; Capanoglu, E. Black Carrot Pomace as a Source of Polyphenols for Enhancing the Nutritional Value of Cake: An In Vitro Digestion Study with a Standardized Static Model. *LWT* **2017,** *77*, 475–481. https://doi.org/10.1016/j.lwt.2016.12.002

Kidoń, M.; Uwineza, P. A. New Smoothie Products Based on Pumpkin, Banana, and Purple Carrot as a Source of Bioactive Compounds. *Molecules* **2022,** *27* (10). https://doi.org/10.3390/molecules27103049

Kruczek, M.; Gumul, D.; Kačániová, M.; Ivanišhová, E.; Mareček, J.; Gambuś, H. Industrial Apple Pomace By-Products as a Potential Source of Pro-Health Compounds in Functional

Food. *J. Microbiol. Biotechnol. Food Sci.* **2017,** *7* (1), 22–26. https://doi.org/10.15414/jmbfs.2017.7.1.22-26

Lafarga, T.; Gallagher, E.; Bademunt, A.; Viñas, I.; Bobo, G.; Villaró, S.; Aguiló-Aguayo, I. Bioaccessibility, Physicochemical, Sensorial, and Nutritional Characteristics of Bread Containing Broccoli Co-Products. *J. Food Process. Preserv.* **2019,** *43* (2), 1–11. https://doi.org/10.1111/jfpp.13861

Lazari, M.; Aguiar-Oliveira, E.; de Oliveira, D. S.; Kamimura, E. S.; Maldonado, R. R. Production of Low-Calorie Ice Cream Utilizing Apple Peel and Pulp. *J. Culinary Sci. Technol.* **2019,** *17* (6), 481–490. https://doi.org/10.1080/15428052.2018.1489323

Liu, Y.; Zheng, Y.; Yang, T.; Mac Regenstein, J.; Zhou, P. Functional Properties and Sensory Characteristics of Kombucha Analogs Prepared with Alternative Materials. *Trends Food Sci. Technol.* **2022,** *129*, 608–616. https://doi.org/10.1016/j.tifs.2022.11.001

Marchiani, R.; Bertolino, M.; Belviso, S.; Giordano, M.; Ghirardello, D.; Torri, L.; Piochi, M.; Zeppa, G. Yogurt Enrichment with Grape Pomace: Effect of Grape Cultivar on Physicochemical, Microbiological and Sensory Properties. *J. Food Qual.* **2016,** *39* (2), 77–89. https://doi.org/10.1111/jfq.12181

Melini, V.; Melini, F.; Acquistucci, R. Phenolic Compounds and Bioaccessibility Thereof in Functional Pasta. *Antioxidants* **2020,** *9* (4), 1–30. https://doi.org/10.3390/antiox9040343

Nagraj, G. S.; Chouksey, A.; Jaiswal, S.; Jaiswal, A. K. Broccoli. In: *Nutritional Composition and Antioxidant Properties of Fruits and Vegetables*; 2020; pp 5–17. https://doi.org/10.1016/B978-0-12-812780-3.00001-5

Noor Aziah, A. A.; Lee Min, W.; Bhat, R. Nutritional and Sensory Quality Evaluation of Sponge Cake Prepared by Incorporation of High Dietary Fiber Containing Mango (Mangifera indica var. Chokanan) Pulp and Peel Flours. *Int. J. Food Sci. Nutr.* **2011,** *62* (6), 559–567. https://doi.org/10.3109/09637486.2011.562883

Núñez-Gómez, V.; González-Barrio, R.; Baenas, N.; Moreno, D. A.; Periago, M. J. Dietary-Fibre-Rich Fractions Isolated from Broccoli Stalks as a Potential Functional Ingredient with Phenolic Compounds and Glucosinolates. *Int. J. Mol. Sci.* **2022,** *23* (21), 13309. https://doi.org/10.3390/ijms232113309

Oliveira, R. M.; Oliveira, F. M.; Hernandes, J. V.; Jacques, A. C. Composição Centesimal De Farinha De Uva Elaborada Com Bagaço Da Indústria Vitivinícola. In: *Mio Cid e D. Sebastião*, Vol. 1; Peter Lang, 2016; pp 2–7. https://doi.org/10.3726/978-3-653-06456-8/4

Otun, S.; Escrich, A.; Achilonu, I.; Rauwane, M.; Alexis, J.; Lerma-Escalera, Morones-Ramírez, J. R.; Rios-Solis, L. The Future of Cassava in the Era of Biotechnology in Southern Africa. *Crit. Rev. Biotechnol.* **2022,** *0* (0), 1–19. https://doi.org/10.1080/07388551.2022.2048791

Pérez-Chabela, M. de L.; Cebollón-Juárez, A.; Bosquez-Molina, E.; Totosaus, A. Mango Peel Flour and Potato Peel Flour as Bioactive Ingredients in the Formulation of Functional Yogurt. *Food Sci. Technol. (Brazil)* **2022,** *42*. https://doi.org/10.1590/fst.38220

Pérez, M. J.; Zampini, I. C.; Alberto, M. R.; Isla, M. I. Prosopis nigra Mesocarp Fine Flour, A Source of Phytochemicals with Potential Effect on Enzymes Linked to Metabolic Syndrome, Oxidative Stress, and Inflammatory Process. *J. Food Sci.* **2018,** *83* (5), 1454–1462. https://doi.org/10.1111/1750-3841.14113

Pérez, M. J.; Cuello, A. S.; Zampini, I. C.; Ordoñez, R. M.; Alberto, M. R.; Quispe, C.; Schmeda-Hirschmann, G.; Isla, M. I. Polyphenolic Compounds and Anthocyanin Content of *Prosopis nigra* and *Prosopis alba* Pods Flour and Their Antioxidant and Anti-Inflammatory Capacities. *Food Res. Int.* **2014,** *64*, 762–771. https://doi.org/10.1016/j.foodres.2014.08.013

Pérez, M. J.; Rodriguez, I. F.; Zampini, I. C.; Cattaneo, F.; Mercado, M. I.; Ponessa, G.; Isla, M. I. *Prosopis nigra* Fruits Waste Characterization, a Potential Source of Functional Ingredients for Food Formulations. *LWT* **2020,** *132* (June). https://doi.org/10.1016/j.lwt.2020.109828

Ponis, S. T.; Papanikolaou, P.-A.; Katimertzoglou, P.; Ntalla, A. C.; Xenos, K. I. Household Food Waste in Greece: A Questionnaire Survey. *J. Clean. Product.* **2017,** *149*, 1268–1277. https://doi.org/10.1016/j.jclepro.2017.02.165

Rabetafika, H. N.; Bchir, B.; Blecker, C.; Richel, A. Fractionation of Apple By-Products as Source of New Ingredients: Current Situation and Perspectives. *Trends Food Sci. Technol.* **2014,** *40* (1), 99–114. https://doi.org/10.1016/j.tifs.2014.08.004

Rodrigues, C.; Silva, V.; Loyola, A.; Silva, M.; Augusti, R.; Melo, J.; Carlos, L.; Fante, C. Characterization and Identification of Bioactive Compounds in Agro-Food Waste Flours. *Química Nova* **2021,** *45* (4), 403–409. https://doi.org/10.21577/0100-4042.20170853

Roodhuyzen, D. M. A.; Luning, P. A.; Fogliano, V.; Steenbekkers, L. P. A. Putting Together the Puzzle of Consumer Food Waste: Towards an Integral Perspective. *Trends Food Sci. Technol.* **2017,** *68*, 37–50. https://doi.org/10.1016/j.tifs.2017.07.009

Rosas, V. T.; Mureb, R. A.; Monserrat, J. M.; Wasielesky, W.; Tesser, M. B. Inclusion of Grape Bagasse (Vitis sp.) in the Diet of White Shrimp (Litopenaeus vannamei) and Its Effects on Growth and Antioxidant System. *Aquac. Res.* **2022,** *53* (13), 4805–4813. https://doi.org/10.1111/are.15972

Simon, P. W. *Carrot Facts*; USDA, 2020. https://www.ars.usda.gov/midwest-area/madison-wi/vegetable-crops-research/docs/simon-carrot-facts/

Tirado-Kulieva, V. A.; Gutiérrez-Valverde, K. S.; Villegas-Yarlequé, M.; Camacho-Orbegoso, E. W.; Villegas-Aguilar, G. F. Research Trends on Mango By-Products: A Literature Review with Bibliometric Analysis. *J. Food Measure. Character.* **2022,** *16* (4), 2760–2771. https://doi.org/10.1007/s11694-022-01400-7

Tolve, R.; Pasini, G.; Vignale, F.; Favati, F.; Simonato, B. Effect of Grape Pomace Addition on the Technological, Sensory, and Nutritional Properties of Durum Wheat Pasta. *Foods* **2020,** *9* (3), 354. https://doi.org/10.3390/foods9030354

Umair, M.; Jabbar, S.; Nasiru, M. M.; Lu, Z.; Zhang, J.; Abid, M.; Murtaza, M. A.; Kieliszek, M.; Zhao, L. Ultrasound-Assisted Extraction of Carotenoids from Carrot Pomace and Their Optimization through Response Surface Methodology. *Molecules* **2021,** *26* (22), 6763. https://doi.org/10.3390/molecules26226763

Wani, S. A.; Naqash, F.; Wani, F. J. Current World Trade and Market Trends. In: *Apples*; Wani, S. A., Naqash, F., Wani, F. J., Eds., 1st ed.; Taylor & Francis, 2022; p 346. https://doi.org/https://doi.org/10.1201/9781003239925

WFP. *5 Facts About Food Waste and Hunger*, 2020. https://www.wfp.org/stories/5-facts-about-food-waste-and-hunger

Wu, M.-Y.; Shiau, S.-Y. Effect of the Amount and Particle Size of Pineapple Peel Fiber on Dough Rheology and Steamed Bread Quality. *J. Food Process. Preserv.* **2015,** *39* (6), 549–558. https://doi.org/10.1111/jfpp.12260

Xiao, J.; Capanoglu, E.; Jassbi, A. R.; Miron, A. Advance on the Flavonoid C -glycosides and Health Benefits. *Crit. Rev. Food Sci. Nutr.* **2016,** *56* (sup 1), S29–S45. https://doi.org/10.1080/10408398.2015.1067595

Index

H

I

J

L

Q

R

S

T